涡轮机械与推进系统出版项目
航空发动机技术出版工程

U0276496

完全流程手册：

从流程建模到管理的知识体系
卷一（中）

The Complete Business Process Handbook
Body of Knowledge from Process Modeling to BPM
Volume I

〔法〕M. V. 罗辛（Mark von Rosing）

〔德〕A. W. 舍尔（August-Wilhelm Scheer）　著

〔瑞士〕H. V. 谢尔（Henrik von Scheel）

张玉金　王占学　等 译

科 学 出 版 社

北 京

图字：01-2021-1780号

内 容 简 介

本书从流程概念的发展与演变出发，汇集大量来自政府机构、标准组织、企业、大学、研究机构以及行业专家的杰出贡献，将涉及流程管理的相关知识概念与理论加以系统整合，构建了流程管理领域全面的知识体系。

本书采取理论与实践分析相结合的方法，通过对流程管理知识、模板、专家建议以及最佳实践的介绍，从什么是流程管理、流程管理的思维方式、工作方式、建模方式、实施与治理方式、培训与指导方式六部分进行编著，将每一个复杂主题拆解成便于理解的知识点，提供了有关实现流程管理的流程、框架、方法的所有内容，为商用航空发动机研制企业构建基于流程管理的自主研发体系提供了一份全面的实用指南。

本书可为企业中高级管理人员、流程管理业务人员、管理咨询行业从业人员、大学管理类专业相关人员提供全面的学习参考，也可作为高校管理类教师和研究生的参考书。

图书在版编目(CIP)数据

完全流程手册：从流程建模到管理的知识体系. 卷一
= The Complete Business Process Handbook: Body
of Knowledge from Process Modeling to BPM Volume Ⅰ:
汉、英/(法)马克·冯·罗辛 (Mark von Rosing) 等
著；张玉金等译. — 北京：科学出版社，2021.12
(航空发动机技术出版工程)
国家出版基金项目　涡轮机械与推进系统出版项目
ISBN 978-7-03-068886-6

Ⅰ. ①完… Ⅱ. ①马… ②张… Ⅲ. ①航空发动机-系统工程-流程-手册-汉、英 Ⅳ. ①V23-65

中国版本图书馆CIP数据核字(2021)第101405号

责任编辑：徐杨峰／责任校对：谭宏宇
责任印制：黄晓鸣／封面设计：殷　靓

科 学 出 版 社 出版
北京东黄城根北街16号
邮政编码：100717
http://www.sciencep.com
南京展望文化发展有限公司排版
广东虎彩云印刷有限公司印刷
科学出版社发行　各地新华书店经销
*
2021年12月第　一　版　　开本：B5(720×1000)
2025年 2 月第五次印刷　总印张：91 3/4
总字数：1687 000
定价：**600.00元**
(如有印装质量问题，我社负责调换)

The Complete Business Process Handbook: Body of Knowledge from Process Modeling to BPM Volume I

Mark von Rosing, August-Wilhelm Scheer, Henrik von Scheel

ISBN: 9780127999593

《完全流程手册：从流程建模到管理的知识体系 卷一》（张玉金 王占学 等译）

ISBN: 9787030688866

注意

本书涉及领域的知识和实践标准在不断变化。新的研究和经验拓展我们的理解，因此须对研究方法、专业实践或医疗方法作出调整。从业者和研究人员必须始终依靠自身经验和知识来评估和使用本书中提到的所有信息、方法、化合物或本书中描述的实验。在使用这些信息或方法时，他们应注意自身和他人的安全，包括注意他们负有专业责任的当事人的安全。在法律允许的最大范围内，爱思唯尔、译文的原文作者、原文编辑及原文内容提供者均不对因产品责任、疏忽或其他人身或财产伤害及/或损失承担责任，亦不对由于使用或操作文中提到的方法、产品、说明或思想而导致的人身或财产伤害及/或损失承担责任。

涡轮机械与推进系统出版项目
顾问委员会

主任委员

张彦仲

委 员

（以姓名笔画为序）

航空发动机技术出版工程

专家委员会

航空发动机技术出版工程
编写委员会

主任委员
尹泽勇

副主任委员
李应红　刘廷毅

委　员
（以姓名笔画为序）

丁水汀	王太明	王占学	王健平	尤延铖
尹泽勇	帅　永	宁　勇	朱俊强	向传国
刘　建	刘廷毅	杜朝辉	李应红	李建榕
杨　晖	杨鲁峰	吴文生	吴施志	吴联合
吴锦武	何国强	宋迎东	张　健	张玉金
张利明	陈保东	陈雪峰	叔　伟	周　明
郑　耀	夏峥嵘	徐超群	郭　昕	凌文辉
陶　智	崔海涛	曾海军	戴圣龙	

秘书组
组　长　朱大明
成　员　晏武英　沙绍智

涡轮机械与推进系统出版项目

序

涡轮机械与推进系统涉及航空发动机、航天推进系统、燃气轮机等高端装备。其中每一种装备技术的突破都令国人激动、振奋，但是由于技术上的鸿沟，使得国人一直为之魂牵梦绕。对于所有从事该领域的工作者，如何跨越技术鸿沟，这是历史赋予的使命和挑战。

动力系统作为航空、航天、舰船和能源工业的"心脏"，是一个国家科技、工业和国防实力的重要标志。我国也从最初的跟随仿制，向着独立设计制造发展。其中有些技术已与国外先进水平相当，但由于受到基础研究和条件等种种限制，在某些领域与世界先进水平仍有一定的差距。在此背景下，出版一套反映国际先进水平、体现国内最新研究成果的丛书，既切合国家发展战略，又有益于我国涡轮机械与推进系统基础研究和学术水平的提升。"涡轮机械与推进系统出版项目"主要涉及航空发动机、航天推进系统、燃气轮机以及相应的基础研究。图书种类分为专著、译著、教材和工具书等，内容包括领域内专家目前所应用的理论方法和取得的技术成果，也包括来自一线设计人员的实践成果。

"涡轮机械与推进系统出版项目"分为四个方向：航空发动机技术、航天推进技术、燃气轮机技术和基础研究。出版项目分别由科学出版社和浙江大学出版社出版。

出版项目凝结了国内外该领域科研与教学人员的智慧和成果，具有较强的系统性、实用性、前沿性，既可作为实际工作的指导用书，也可作为相关专业人员的参考用书。希望出版项目能够促进该领域的人才培养和技术发展，特别是为航空发动机及燃气轮机的研究提供借鉴。

张彦仲

2019 年 3 月

航空发动机技术出版工程

序

  航空发动机被誉称为工业皇冠之明珠,实乃科技强国之重器。

  几十年来,我国航空发动机技术、产品及产业经历了从无到有、从小到大的艰难发展历程,取得了显著成绩。在世界新一轮科技革命、产业变革同我国转变发展方式的历史交汇期,国家决策进一步大力加强航空发动机事业发展,产学研用各界无不为之振奋。

  迄今,科学出版社于2019年、2024年两次申请国家出版基金,安排了"航空发动机技术出版工程",确为明智之举。

  本出版工程旨在总结、推广近期及之前工作中工程、科研、教学的优秀成果,侧重于满足航空发动机工程技术人员的需求,尤其是从学生到工程师过渡阶段的需求,借此也为扩大我国航空发动机卓越工程师队伍略尽绵力。本出版工程包括设计、试验、基础与综合、前沿技术、制造、运营及服务保障六个系列,2019年启动的前三个系列近五十册任务已完成;后三个系列近三十册任务则于2024年启动。对于本出版工程,各级领导十分关注,专家委员会不时指导,编委会成员尽心尽力,出版社诸君敬业把关,各位作者更是日无暇晷、研教著述。同道中人共同努力,方使本出版工程得以顺利开展、如期完成。

  希望本出版工程对我国航空发动机自主创新发展有所裨益。受能力及时间所限,当有疏误,恭请斧正。

2024 年 10 月修订

译 者 序

航空发动机的研制是一项复杂的系统工程,长久以来,我国航空发动机受跟踪研发模式的影响,尚未建立起完整、统一的自主研发体系,存在数据不共享、标准不统一、管理"两张皮"、运行效率低等情况,明显阻碍型号项目研制顺利推进。要实现国产商用航空发动机的自主研制,必须遵循航空发动机发展的客观规律,建立面向航空发动机产品全生命周期完整统一的流程体系,为独立自主研制出先进可靠的航空发动机夯实基础。

中国航发商用航空发动机有限责任公司自2009年成立以来,就明确提出"聚焦客户、流程主导、追求卓越、持续改进"的管理政策,以流程管理为主导,建设价值驱动型流程体系,以流程统领所有业务活动,以流程绩效评价业务结果,通过持续改进,不断推动业务高质量发展。商发公司矢志不渝地推进流程型企业的建设,围绕流程、组织和文化持续打造国际一流的企业运营系统:一是企业家面向系统求价值,管理者面向流程做改进,操作者面向作业做完善;二是企业作为价值创造系统,创新与变革是永远不变的主题,企业要始终动态地适应环境而改进,必须面对环境变化作战略取舍,面对客户需求优化产品价值链,面对资源需求优化供应链与资源保障;三是企业运营系统的核心要素:流程、人与文化,流程是主航道,必须持续地"清淤拓土筑基",人与文化要素所赋予的知识资本、信息资本与组织资本必须嵌入流程鲜活灵动地动作。

经过十余年单通道窄体干线客机发动机验证机研制走完全过程和全面推进产品研制的实践积累,商发公司在管理运营与产品研制等方面进行了全面探索与实践,初步形成了以产品研发体系为核心,涵盖公司20个业务领域的流程体系。同时,在不断地面向用户质量目标、聚焦价值创造、开展流程再造与数字化转型等创新变革活动中,逐步探索建立了面向商用航空发动机公司的体系运行管理知识体系,并尝试通过对供应商的管理体系延伸,进一步打造安全、可靠、稳定的供应链体系,建立产业链良好的演进生态,最终实现国产商用航空发动机的产品交付,达成客户满意。

本书总结吸纳了世界优秀企业实践,提供了一套流程优化剪裁模板、一套持续改进优化的方法论、一套文化重塑与习惯再造的利器。商发公司在流程体系建设实践中,借鉴了其中的原理与方法,特别是流程全生命周期以及实施与治理等部分,得到了很多启发和帮助。该书针对流程管理进行了全面的系统论述,并构建了较为完整的知识体系;该书作为流程管理的实用指南,详细揭示了流程管理中我们思考和使用流程的方式,深入论述了业务流程的本质,以及从流程建模到治理的完整知识。为了使广大的学习者和实践者能够准确掌握本书所包罗的理论与方法精髓,本译著将以中、英文逐页对照的方式出版发行。

企业管理就是以流程为核心、价值为目标、自我驱动的生命之旅,企业管理运作是一门技术,又是一门艺术,同时也是一门实践的系统工程。中国企业尤其是从事复杂系统工程与高端制造业的企业更应学习如何站在巨人的肩膀上不断成就自我,打造既有东方智慧又有西方商业理念优秀基因的一流企业。翻译和实践本书就是在汲取西方优秀的企业管理最佳实践,为我国航空制造企业提供参考借鉴。在此,要特别感谢支持本书翻译与校订工作,以及在过程中提供资料案例和提出宝贵建议的相关专家、同事,他们是黄博、陈楠、黄干明、陈天彧、项飞、黄飞、汤先萍、张滟滋、杨博文、吴帆、何宛文、陈婧怡、郑冰雷。希望这本译著能够进一步促进流程管理方法在我国航空制造企业乃至中国企业的管理变革中的实践,让东方智慧与西方商业理念的深度结合转化为企业治理效能,打造更多世界一流的中国企业,为国家打造央企"市营"新范式提供有益的借鉴和参考!

张玉金

2021 年 5 月

Foreword

This book has been put together to help you explore Business Process concepts and to understand what BPM really is all about.

We wrote this book for YOU—the individual. You may be a business executive, manager, practitioner, subject matter expert, student, or researcher. Or may be an ambitious career individual who wants to know more about business process concepts and/or BPM, what it is all about and how to apply it.

This, *The Complete Business Process Handbook*, provides a comprehensive body of knowledge written as a practical guide for you—by the authorities that have shaped the way we think and work with processes today. You hold the first of three books in the series in your hand.

- The first volume endows the reader with a deep insight into the nature of business process concepts and how to work with them. From BPM Ontology, semantics, and BPM Portfolio management, to the BPM Life Cycle, it provides a unique foundation within this body of knowledge.
- The second volume bridges theory and application of BPM in an advanced modeling context by addressing the subject of extended BPM.
- The third volume explores a comprehensive collection of real-world BPM lessons learned, best practices, and leading practices examples from award-winning industry leaders and innovators.

We wish you well on your Business Process journey and that is why we also have invested years putting this Handbook series together. To share the knowledge, templates, concepts, best and leading practices and to ensure high quality and standards, we have worked and coordinated with standard development organizations like International Organization for Standardization (ISO), Object Management Group (OMG), Institute of Electrical and Electronics Engineers (IEEE), North Atlantic Treaty Organization (NATO), Council for Scientific and Industrial Research (CSIR), MITRE—a Federally Funded Research and Development Center, European Committee for Standardization (CEN), The Security Forum, World Wide Web Consortium (W3C), and LEADing Practice.

We have also identified and worked with leading organizations and with their process experts/architects, and have described their practices. Among them are Lego, Maersk Shipping, Carlsberg, FLSmidth, the US Government, AirFrance, KLM, German Government, SaxoBank, Novozymes, the Canadian Government, US Department of Defense, Danish Defense, Johnson & Johnson, Dutch Railway, Australian Government, and many more. At last but not least the Global University Alliance consisting of over 400 universities, lecturers, and researchers have analyzed and examined what works, again and again (best practice), and what are the unique practices applied by these leading organizations (leading practices). They then identified common and repeatable patterns, which provide the basis for the BPM Ontology, BPM Semantics, the BPM standards, and the process templates found in this book.

原 书 序

本书已经整合在一起，可以帮助您探索业务流程概念，并了解业务流程管理（BPM）的真正含义。

我们为您编写的这本书。您可能是：业务主管、经理、工作者、某一领域专家、学生或研究员，也或许是一个雄心勃勃的职场人士，想要更多地了解业务流程概念和BPM是什么以及如何应用它。

这本《完全流程手册》，它为您提供一个全面的知识体系，作为一本实用指南，由那些塑造了我们今天思考和使用流程的方式的权威人士编写。您手里拿着的这本是这个系列的三卷书中的第一卷。

- 第一卷让读者深入了解业务流程概念的本质以及如何使用它们。从BPM本体论、语义、BPM项目组合管理，到BPM生命周期，它在这个知识体系中提供一个独特的基础。
- 第二卷通过处理扩展BPM主题，在高级建模背景中架起BPM理论和应用之间的桥梁。
- 第三卷探讨屡获殊荣的行业领导者和创新者的全面实际BPM经验教训、最佳实践和领导实践示例。

我们祝愿您在业务流程之旅中取得成功，这就是我们花费多年时间将本系列手册整合在一起的原因，分享知识、模板、概念、最佳和领导实践。为了确保高质量和高标准，我们与国际标准化组织（ISO）、对象管理组织（OMG）、电气和电子工程师协会（IEEE）、北大西洋公约组织（NATO）、科学与工业研究理事会（CSIR）、联邦资助的研究与发展中心（Federally Funded Research and Development Center，FFRDC）MITER、欧洲标准化委员会（CEN）、安全论坛（The Security Forum）、万维网联盟（W3C）和领导实践（LEADing Practice）等组织进行了协调。

我们还选择与领先的组织及其流程专家/架构师合作，并描述他们的实践。其中包括：乐高、马士基航运、嘉士伯、艾法史密斯（FLSmidth）、美国政府、法国航空、荷兰皇家航空、德国政府、盛宝银行、诺维信、加拿大政府、美国国防部（United States Department of Defense，DOD）、丹麦国防部、强生、荷兰铁路、澳大利亚政府等。最后但同样重要的是，由400多所大学的讲师和研究人员组成的全球大学联

We have worked years on this book, and as you just read, with contributions of standard bodies, governments, defense organizations, enterprises, universities, research institutes and individual thought leaders. We put these chapters and their subjects carefully together and hope you enjoy reading it—as much as we did writing, reviewing and putting it together.

Name	Organization
Mark von Rosing	Global University Alliance
August-Wilhelm Scheer	Scheer Group GmBH
Henrik von Scheel	LEADing Practices, Google Board
Adam D.M. Svendsen	Institute of Future Studies
Alex Kokkonen	Johnson & Johnson
Andrew M. Ross	Westpac
Anette Falk Bøgebjerg	LEGO Group
Anni Olsen	Carlsberg Group
Antony Dicks	NedBank
Asif Gill	Global University Alliance
Bas Bach	NS Rail
Bob J. Storms	LEADing Practices
Callie Smit	Reserve Bank
Cay Clemmensen	LEADing Practices
Christopher K. Swierczynski	Electrolux
Clemens Utschig-Utschig	Boehringer Ingelheim Pharma
Dan Moorcroft	QMR
Daniel T. Jones	Lean UK
David Coloma	Universitat Politècnica de Catalunya, Spain
Deb Boykin	Pfizer Pharmaceuticals
Dickson Hunja Muhita	LEADing Practices
Duarte Gonçalves	CSIR—Council for Scientific and Industrial Research
Ekambareswaran Balasubramanian	General Motors
Fabrizio Maria Maggi	University of Estonia
Fan Zhao	Florida Gulf Coast University
Fatima Senghore	NASA
Fatma Dandashi	MITRE
Freek Stoffel	LEADing Practices
Fred Cummins	OMG

盟（Global University Alliance）分析和检查哪些有效，以及这些领先组织独特的实践应用（领导实践）。然后，他们确定了常见和可重复的模式，这些模式为BPM本体、BPM语义、BPM标准和本书中的流程模板提供了基础。

正如您刚才所读，我们已经为编写这本书工作了多年，您将在本书中找到标准机构、政府、国防组织、企业、大学、研究机构和个人思想领袖的贡献。我们将这些章节及其主题精心放在一起，希望您就像我们写作、复习和整理一样喜欢阅读它。

姓名	组织
Mark von Rosing	Global University Alliance
August-Wilhelm Scheer	Scheer Group GmBH
Henrik von Scheel	LEADing Practices, Google Board
Adam D.M. Svendsen	Institute of Future Studies
Alex Kokkonen	Johnson & Johnson
Andrew M. Ross	Westpac
Anette Falk Bøgebjerg	LEGO Group
Anni Olsen	Carlsberg Group
Antony Dicks	NedBank
Asif Gill	Global University Alliance
Bas Bach	NS Rail
Bob J. Storms	LEADing Practices
Callie Smit	Reserve Bank
Cay Clemmensen	LEADing Practices
Christopher K. Swierczynski	Electrolux
Clemens Utschig-Utschig	Boehringer Ingelheim Pharma
Dan Moorcroft	QMR
Daniel T. Jones	Lean UK
David Coloma	Universitat Politècnica de Catalunya, Spain
Deb Boykin	Pfizer Pharmaceuticals
Dickson Hunja Muhita	LEADing Practices
Duarte Gonçalves	CSIR—Council for Scientific and Industrial Research
Ekambareswaran Balasubramanian	General Motors
Fabrizio Maria Maggi	University of Estonia
Fan Zhao	Florida Gulf Coast University
Fatima Senghore	NASA
Fatma Dandashi	MITRE
Freek Stoffel	LEADing Practices
Fred Cummins	OMG

Name	Organization
Gabriel von Scheel	LEADing Practices
Gabriella von Rosing	LEADing Practices
Gary Doucet	Government of Canada
Gert Meiling	Tommy Hilfiger
Gert O. Jansson	LEADing Practices
Hans Scheruhn	University of Harz, Gemany
Hendrik Bohn	Nedbank
Henk de Man	OMG, VeeBee
Henk Kuil	KLM, Air France
Henrik Naundrup Vester	iGrafx
Jacob Gammelgaard	FLSchmidt
James P. Womack	Cambridge University-Massachusetts Institute of Technology (MIT)
Jeanne W. Ross	Cambridge University-Massachusetts Institute of Technology (MIT)
Jeff Greer	Cardinal Health
Jens Theodor Nielsen	Danish Defense
John A. Zachman	Zachman International
John Bertram	Government of Canada
John Golden	iGrafx
John M. Rogers	Government of Australia
Jonnro Erasmus	CSIR—Council for Scientific and Industrial Research
Joshua Michael von Scheel	LEADing Practices
Joshua Waters	LEADing Practices
Justin Tomlinson	LEADing Practices
Karin Gräslund	RheinMain University-Wiesbaden Business School
Katia Bartels	Office Depot
Keith Swenson	Fujitsu
Kenneth Dean Teske	US Government
Kevin Govender	Transnet Rail
Klaus Vitt	German Federal Employment Agency
Krzysztof Skurzak	NATO ACT
LeAnne Spurrell	QMR
Lloyd Dugan	BPM.com
Lotte Tange	Carlsberg Group

姓名	组织
Gabriel von Scheel	LEADing Practices
Gabriella von Rosing	LEADing Practices
Gary Doucet	Government of Canada
Gert Meiling	Tommy Hilfiger
Gert O. Jansson	LEADing Practices
Hans Scheruhn	University of Harz, Gemany
Hendrik Bohn	Nedbank
Henk de Man	OMG, VeeBee
Henk Kuil	KLM, Air France
Henrik Naundrup Vester	iGrafx
Jacob Gammelgaard	FLSchmidt
James P. Womack	Cambridge University-Massachusetts Institute of Technology (MIT)
Jeanne W. Ross	Cambridge University-Massachusetts Institute of Technology (MIT)
Jeff Greer	Cardinal Health
Jens Theodor Nielsen	Danish Defense
John A. Zachman	Zachman International
John Bertram	Government of Canada
John Golden	iGrafx
John M. Rogers	Government of Australia
Jonnro Erasmus	CSIR—Council for Scientific and Industrial Research
Joshua Michael von Scheel	LEADing Practices
Joshua Waters	LEADing Practices
Justin Tomlinson	LEADing Practices
Karin Gräslund	RheinMain University-Wiesbaden Business School
Katia Bartels	Office Depot
Keith Swenson	Fujitsu
Kenneth Dean Teske	US Government
Kevin Govender	Transnet Rail
Klaus Vitt	German Federal Employment Agency
Krzysztof Skurzak	NATO ACT
LeAnne Spurrell	QMR
Lloyd Dugan	BPM.com
Lotte Tange	Carlsberg Group

Name	Organization
Mads Clausager	Maersk Group
Mai Phuong	Northrop Grumman Electronic Systems
Maria Hove	LEADing Practices
Maria Rybrink	TeliaSonera
Marianne Fonseca	LEADing Practices
Mark Stanford	iGrafx
Marlon Dumas	University of Tartu
Mathias Kirchmer	BPM-d
Maxim Arzumanyan	St. Petersburg University
Michael Tisdel	US Government, DoD
Michel van den Hoven	Philips
Mikael Munck	SaxoBank
Mike A. Marin	IBM Corporation
Mona von Rosing	LEADing Practices
Nathaniel Palmer	BPM.com, Workflow Management Coalition (WfMC)
Neil Kemp	LEADing Practices
Nils Faltin	Scheer Group GmBH
Partha Chakravartti	AstraZeneca
Patricia Kemp	LEADing Practices
Peter Franz	BPM-d
Philippe Lebacq	Toyota
Régis Dumond	French Ministry of Defense, NATO, ISO
Rich Hilliard	IEEE, ISO
Richard L. Fallon	Sheffield Hallam University
Richard N. Conzo	Verizon
Rod Peacock	European Patent Office
Rogan Morrison	LEAD Enterprise Architect Professional
Ronald N. Batdorf	US Government, DoD, Joint Staff
Sarel J. Snyman	SAP Solution Design
Scott Davis	Government of Canada
Simon Polovina	Sheffield Hallam University
Stephen White	IBM Corporation
Steve Durbin	Information Security Forum
Steve Willoughby	iGrafx

姓名	组织
Mads Clausager	Maersk Group
Mai Phuong	Northrop Grumman Electronic Systems
Maria Hove	LEADing Practices
Maria Rybrink	TeliaSonera
Marianne Fonseca	LEADing Practices
Mark Stanford	iGrafx
Marlon Dumas	University of Tartu
Mathias Kirchmer	BPM-d
Maxim Arzumanyan	St. Petersburg University
Michael Tisdel	US Government, DoD
Michel van den Hoven	Philips
Mikael Munck	SaxoBank
Mike A. Marin	IBM Corporation
Mona von Rosing	LEADing Practices
Nathaniel Palmer	BPM.com, Workflow Management Coalition (WfMC)
Neil Kemp	LEADing Practices
Nils Faltin	Scheer Group GmBH
Partha Chakravartti	AstraZeneca
Patricia Kemp	LEADing Practices
Peter Franz	BPM-d
Philippe Lebacq	Toyota
Régis Dumond	French Ministry of Defense, NATO, ISO
Rich Hilliard	IEEE, ISO
Richard L. Fallon	Sheffield Hallam University
Richard N. Conzo	Verizon
Rod Peacock	European Patent Office
Rogan Morrison	LEAD Enterprise Architect Professional
Ronald N. Batdorf	US Government, DoD, Joint Staff
Sarel J. Snyman	SAP Solution Design
Scott Davis	Government of Canada
Simon Polovina	Sheffield Hallam University
Stephen White	IBM Corporation
Steve Durbin	Information Security Forum
Steve Willoughby	iGrafx

Name	Organization
Sven Vollbehr	SKF
Thomas Boosz	German Government
Thomas Christian Olsen	NovoZymes
Tim Hoebeek	SAP
Tom Preston	Booz Allen Hamilton
Ulrik Foldager	LEADing Practices
Victor Abele	Government of Canada
Vincent Snels	Nationale Nederlanden
Volker Rebhan	German Federal Employment Agency
Wim Laurier	Université Saint-Louism Bruxelles
Ýr Gunnarsdottir	Shell
Yury Orlov	Smart Architects
Zakaria Maamar	Zayed University, United Arab Emirates

姓名	组织
Sven Vollbehr	SKF
Thomas Boosz	German Government
Thomas Christian Olsen	NovoZymes
Tim Hoebeek	SAP
Tom Preston	Booz Allen Hamilton
Ulrik Foldager	LEADing Practices
Victor Abele	Government of Canada
Vincent Snels	Nationale Nederlanden
Volker Rebhan	German Federal Employment Agency
Wim Laurier	Université Saint-Louism Bruxelles
Ýr Gunnarsdottir	Shell
Yury Orlov	Smart Architects
Zakaria Maamar	Zayed University, United Arab Emirates

Abbreviation Meaning

A2A	Application to application
AAIM	Agility adoption and improvement model
ACM	Adaptive case management
ADDI	Architect design deploy improve
API	Application programming interface
APQC	American productivity and quality center
B2B	Business to business
BAM	Business activity monitoring
BCM	Business continuity management
BEP	Break even point
BI	Business intelligence
BITE	Business innovation and transformation enablement
BOM	Business object management
BPA	Business process analysis
BPaaS	Business process as a service
BPCC	Business process competency center
BPD	Business process diagram
BPE	Business process engineering
BPEL	Business process execution language
BPEL4WS	Business process execution language for web services
BPG	Business process guidance
BPI	Business process improvement
BPM	Business process management
BPM CM	Business process management change management
BPM CoE	Business process management center of excellence
BPM LC	Business process management life cycle
BPM PM	Business process management portfolio management
BPMaaS	BPM as a service
BPMI	Business process management institute
BPMN	Business process model and notation
BPMS	Business process management system
BPO	Business process outsourcing
BPPM	Business process portfolio management
BPR	Business process reengineering
BRE	Business rule engine
BRM	Business rules management
CDM	Common data model
CE-BPM	Cloud-enabled BPM
CEAP	Cloud-enabled application platform
CEN	European committee for standardization
CEP	Complex event processing
CM	Configuration management
CMS	Content management system
COBIT	Control objectives for information and related technology

缩 略 词

A2A	应用到应用
AAIM	敏捷应用和改进模型
ACM	适应性案例管理
ADDI	架构师设计部署改进
API	应用程序编程接口
APQC	美国生产力和质量中心
B2B	业务到业务
BAM	业务活动监控
BCM	业务连续性管理
BEP	盈亏平衡点
BI	商务智能
BITE	业务创新和转型支持
BOM	业务对象管理
BPA	业务流程分析
BPaaS	业务流程即服务
BPCC	业务流程能力中心
BPD	业务流程图
BPE	业务流程工程
BPEL	业务流程执行语言
BPEL4WS	Web服务的业务流程执行语言
BPG	业务流程指导
BPI	业务流程改进
BPM	业务流程管理
BPM CM	业务流程管理变更管理
BPM CoE	业务流程管理卓越中心
BPM LC	业务流程管理生命周期
BPM PM	业务流程管理组合管理
BPMaaS	BPM即服务
BPMI	业务流程管理机构
BPMN	业务流程建模标记法
BPMS	业务流程管理系统

CPO	Chief process officer
CRM	Customer relationship management
CSF	Critical success factor
CSIR	Council for Scientific and Industrial Research
CxO	Chief x officer
DB	Database
DBMS	Database management system
DMS	Document management system
DNEAF	Domain neutral enterprise architecture framework
DSDM	Dynamic systems development method
EAI	Enterprise application integration
EITE	Enterprise innovation & transformation enablement
EMR	Enterprise-wide metadata repositories
EPC	Event-driven process chain
EPSS	Electronic performance support system
ERM	Entity relationship modeling
ERP	Enterprise resource planning
ESB	Enterprise service bus
FEAF	Federal enterprise architecture framework
FI	Financial
iBPM	Intelligent business process management
IDE	Integrated development environment
IE	Information engineering
IEEE	Institute of electrical and electronics engineers
ISO	International Organization for Standardization
ITIL	Information technology infrastructure library
KPI	Key performance indicator
L&D	Learning and development
LEADP	Layered enterprise architecture development and/or LEADing big in Practice
MDM	Master data management
NATO	North Atlantic Treaty Organisation
NIST	National Institute of Standards and Technology
OCM	Organizational change management
OLAP	Online analytic processing
OLTP	Online transaction processing
OMG	Object management group
PDC	Process data collection
PIM	Process instance management
PM	Portfolio management
PM	Project management
PMBOK	Project management body of knowledge
PMO	Project management offices
POA	Process oriented architecture
PPI	Process performance indicator
PPM	Project portfolios management
PPPM	Portfolio, program and project management
PRINCE	PRojects IN Controlled Environments
QM	Quality management

BPO	业务流程外包
BPPM	业务流程组合管理
BPR	业务流程再造
BRE	业务规则引擎
BRM	业务规则管理
CDM	通用数据模型
CE-BPM	支持云端的 BPM
CEAP	支持云的应用平台
CEN	欧洲标准化委员会
CEP	复杂事件处理
CM	配置管理
CMS	内容管理系统
COBIT	信息及相关技术控制目标
CPO	首席流程官
CRM	客户关系管理
CSF	关键成功因素
CSIR	科学和工业研究委员会
CxO	首席 x 官员
DB	数据库
DBMS	数据库管理系统
DMS	文件管理系统
DNEAF	领域中立的企业架构框架
DSDM	动态系统开发方法
EAI	企业应用程序集成
EITE	企业创新和转型支持
EMR	企业范围的元数据存储库
EPC	事件驱动的流程链
EPSS	电子绩效支持系统
ERM	实体关系建模
ERP	企业资源规划
ESB	企业服务总线
FEAF	联邦企业架构框架
FI	金融
iBPM	智能业务流程管理
IDE	集成开发环境
IE	信息工程
IEEE	电气和电子工程师协会
ISO	国际标准化组织
ITIL	信息技术基础架构库
KPI	关键绩效指标
L&D	学习和发展
LEADP	分层企业架构开发和/或领导实践
MDM	主数据管理

ROI	Return on investment
SBO	Strategic business objective
SCM	Supply chain management
SCOR	Supply chain operations reference model
SD	Sales and distribution
SNA	Social network analysis
SOA	Service oriented architecture
SPI	Service performance indicator
SRM	Supply relationship management
SW	Software
TCO	Total cost of ownership
TOGAF	The open group architecture framework
TQM	Total quality management
UI	User interface
ULM	Unified modeling language
USGAP	United States general accounting principles
VDML	Value delivery modeling language
VNA	Value network analysis
W3C	World Wide Web consortium
xBPMN	eXtended business process model and notation
XLM	Extensible markup language
XMI	Metadata interchange
XSD	XML schema definition

NATO	北大西洋公约组织
NIST	美国国家标准与技术研究院
OCM	组织变革管理
OLAP	联机分析处理
OLTP	联机事物处理
OMG	对象管理组织
PDC	过程数据收集
PIM	流程实例管理
PM	项目组合管理
PM*	项目管理
PMBOK	项目管理知识体系
PMO	项目管理办公室
POA	面向流程的体系结构
PPI	流程绩效指标
PPM	项目组合、项目集、项目管理
PPPM	投资组合、计划和项目管理
PRINCE	受控环境中的项目
QM	质量管理
ROI	投资回报率
SBO	战略业务目标
SCM	供应链管理
SCOR	供应链运作参考模型
SD	销售和分销
SNA	社会网络分析
SOA	面向服务的架构
SPI	服务绩效指标
SRM	供应关系管理
SW	软件
TCO	总拥有成本
TOGAF	开放组体系结构构框架
TQM	全面质量管理
UI	用户界面
UML	统一建模语言
USGAP	美国通用会计准则
VDML	价值交付建模语言
VNA	价值网络分析
W3C	万维网联盟
xBPMN	扩展业务流程模型和符号
XML	可扩展标记语言
XMI	元数据交换
XSD	XML 模式定义

---

* 正文中出现 PM 缩写时请对照原文。

Introduction to the Book

Prof. Mark von Rosing, Henrik von Scheel, Prof. August-Wilhelm Scheer

It is not a new phenomenon that the markets are changing; however, the business environment in which firms operate lies outside of themselves and their control. So, while it is their external environment, which is always changing, most changes on the outside affect the need for innovation and transformation on the inside of the organization. The ability to change the business and to manage their processes is symbiotic, which is, among others, one of the reasons for such a high Business Process Management (BPM) adoption rate in the market. It is, however, important to note that unlike some analysts might claim, the size of the market and its adoption is in no way an indicator of maturity. As a matter of fact, the maturity of many of the BPM concepts can have a low maturity, even though the adoption is widespread. So while the high demand for BPM as a management method and a software solution, and the maturing BPM capabilities develop and unfold, the challenge quickly develops to provide concise and widely accepted BPM definitions, taxonomies, standardized, and integrated process templates, as well as overall frameworks, methods, and approaches.

Written as the practical guide for you—by the authorities that have shaped the way we think and work with process today. This handbook series stands out as a masterpiece, representing the most comprehensive body of knowledge published on business process. The first volume endows the reader with a deep insight into the nature of business process, and a complete body of knowledge from process modeling to BPM, thereby covering what executives, managers, practitioners, students, and researchers need to know about:

- Future BPM trends that will affect business
- A clear and precise definition of what BPM is
- Historical evolution of process concepts
- Exploring a BPM Ontology
- In-depth look at the Process Semantics
- Comprehensive Frameworks, Methods, and Approaches
- Process analysis, process design, process deployment, process monitoring, and Continuous Improvement
- Practical usable process templates
- How to link Strategy to Operation with value-driven BPM
- How to build BPM competencies and establish a Center of Excellence
- Discover how to apply Social Media and BPM
- Sustainable-Oriented process Modeling
- Evidence-based BPM
- Learn how Value and Performance Measurement and Management is executed
- Explore how to enable Process Owners

本 书 介 绍

Mark von Rosing, Henrik von Scheel, August-Wilhelm Scheer

市场的变化并不是一个新的现象,但是企业的经营环境是不受自身控制的。因此,虽然外部环境一直在变化,但大多数外部变化都会影响组织内部的创新和转型需求。改变业务和管理流程的能力是共生的,这也是BPM在市场上被广泛应用的原因之一。不过,值得注意的是,与一些分析人员发表的意见不同,市场的规模和应用程度绝不是BPM成熟的主要标志。事实上,尽管许多BPM概念被广泛应用,但是成熟度依然很低,因此,对BPM作为管理方法和软件解决方案的高需求以及日益成熟的BPM能力,两者不断发展和展现的同时,在提供简洁并广泛接受的BPM定义、分类法、标准化集成流程模板以及总体框架、方法和途径等方面的挑战也在快速发展。

作为实用指南,本书作者塑造了BPM中我们思考和使用流程的方式。本手册系列作为杰作脱颖而出,代表了当前业务流程文档中最全面的知识体系。第一卷使读者深入了解业务流程的本质,以及从流程建模到BPM的完整知识体系,涵盖高管、经理、工作者、学生和研究人员需要了解的内容:

- 影响未来业务的BPM趋势;
- 一个对BPM清晰而精确的定义;
- 流程概念的历史演变;
- 探索BPM本体;
- 深入研究流程语义;
- 综合框架、方法和途径;
- 流程分析、流程设计、流程部署、流程监控以及持续改进(continuous improvement,CI);
- 实用的流程模板;
- 如何将战略与价值驱动的BPM运营联系起来;
- 如何建立BPM能力和卓越中心(center of excellence,CoE);
- 了解如何应用社交媒体和BPM;

- BPM Roles and Knowledge Workers
- Discover how to develop information models within the process models
- Uncovering Process Life cycle
- BPM Maturity
- BPM Portfolio Management and BPM Alignment
- BPM Change Management and BPM Governance
- Learning a structured way of Thinking, Working, Modeling, and Implementing processes.

This book is organized into various chapters that have been thoughtfully put together to communicate many times a complex topic into a replicable and manageable structure—that you as a reader can apply. Furthermore, the book is structured into six parts with the intention to guide you in turning business processes into real assets.

In Part I, we introduce a comprehensive "history of process concepts" from Sun Tzu's, to Taylorism, to Business Process Reengineering to Lean and BPMN, providing the reader with an in-depth understanding of the evolution of process thinking, approaches, and methods: a fundamental insight to what has shaped and what is shaping process thinking.

In Part II, we introduce the "Way of Thinking" around Business Process with focus on the value of Ontology, and a comprehensive BPM Ontology—the essential starting point that creates the guiding principles.

In Part III, we establish a "Way of Working" with Business Processes—the critical discipline of translating both strategic planning and effective execution. Exploring the current and future process trends that you need to be aware of with a detailed practical guide on how to apply them in areas such as BPM Life cycle, BPM Roles, process templates, evidence-based BPM, and many more.

In Part IV, we provide the essential guidance to help you in a "Way of Modeling" in traditional Process Modeling concepts to BPMN and Value-Oriented Process Modeling, how to work with and model Business Processes variations, as well as how to interlink information models and process models.

In Part V, we focus on the "Way of Implementation" and "Way of Governance"— the approach the practitioner follows in order to apply and steer what exists, spanning issues ranging from BPM change management, agile BPM, business process outsourcing, and holistic governance to project, program, and portfolio alignment.

In Part VI, we focus on the "Way of Training and Coaching"—to provide insight into ideal process expert, process engineer and process architecture training, from online to class-based learning and coaching.

While this book certainly can be read cover to cover, depending on where you are in your Business Process journey, you may wish to choose a different path. If you are new to Business Process concepts, you might start at the beginning, with Part I. If you are beginning a BPM project, or it has already begun its journey, or you are looking for inspiration, we recommend using the book as a reference tool to access it by the topic of interest.

But no matter how you plan on building your knowledge, the book has been designed and architected to be a guide and a handbook able to create the right way of thinking, working, modeling implementation, and governance.

- 面向可持续发展的流程建模；
- 基于证据的BPM；
- 了解如何执行战略、绩效测量和管理；
- 探索如何启用流程责任人；
- BPM角色和知识工作者；
- 了解如何在流程模型中开发信息模型；
- 揭示流程生命周期；
- BPM成熟度；
- BPM组合管理和BPM协调机制；
- BPM变革管理和BPM治理；
- 学习思维、工作、建模和实施流程的结构化方法。

本书由不同的章节构成，这些章节经过深思熟虑被组合在一起，将一个复杂的主题分解成一个可复制和可管理的结构，便于读者应用。此外，本书分为六个部分，旨在指导您将业务流程转化为公司的宝贵资产。

第一部分，我们介绍从孙子到泰勒主义（Taylorism）、业务流程再造（BPR）、精益和业务流程建模标记法（BPMN）的流程概念的演变历史，有助于读者形成对进化流程思维、途径和方法的深入理解，建立对流程思维及形成流程的基本认识。

第二部分，我们介绍围绕业务流程的思维方法，重点介绍本体论的价值，以及一个全面的BPM本体论，这是创建指导原则的基本出发点。

第三部分，我们建立业务流程的工作方法，这是战略规划和有效执行的关键。我们协助您探索您需要了解的当前和未来流程趋势，并提供详细的实践指导，包括：如何将其应用于诸如BPM生命周期、BPM角色、流程模板、基于证据的BPM等领域。

第四部分，我们为您提供基本的指导，帮助您以传统流程建模概念的建模方法为基础，实现BPMN和价值导向的流程建模，包括：如何应对与建模业务流程有关的变化，以及如何将信息模型和流程模型相互链接等。

第五部分，我们将重点放在实施方法和治理方法——从业者遵循的方法方面，尤其是如何面对存在的问题，包括从BPM变革管理、敏捷BPM、业务流程外包（BPO）和整体治理到项目、计划和组合调整等方面的各种问题。

第六部分，我们将重点放在"培训和指导的方法"上，从在线、课堂学习到经验指导，我们将深入学习流程专家、流程工程专家和流程架构专家的培训内容与知识体系。

尽管这本书可以从头到尾详细地进行阅读，但您也可根据您在业务流程中所处的环节选择不同的章节学习。如果您对业务流程概念不熟悉，我们建议您可以从第一部分开始。如果您正在计划或者已经开始了一个BPM项目，我们建议您使用这本书作为参考工具，研究其中您感兴趣的主题。

但是，不管您打算如何构建您的知识体系，这本书在设计和架构方面都是一本能够创建正确的思考、工作、建模和治理方式的指南和手册。

目　录

第三部分

完全流程手册：

从流程建模到管理的知识体系
卷一（中）

The Complete Business Process Handbook

Body of Knowledge from Process Modeling to BPM
Volume I

The BPM Way of Working

Henrik von Scheel, Mark von Rosing, August-Wilhelm Scheer

INTRODUCTION

In Part III, we establish a way of working with business processes—the critical discipline of translating strategic planning into effective process execution. We explore both the current and future business process trends that we advise you to be fully aware of in order to be best prepared for many of the coming business process changes on the global scale. We also provide a perspective regarding what is hype and what is real to allow you to make critical decisions about what to possibly adopt, transform, and innovate within your organization.

The way of working around business processes is structured to provide you with a practical guide for how to organize, classify, align, arrange, and quantify business process management (BPM) concepts and to select and use process objects and/or process templates/artifacts in the systemized and categorized way that they need to be applied and used within your process initiatives. This is described around important topics such as:

- What are the current and future business process trends?
- Building BPM competencies: the BPM Center of Excellence
- How to use process templates
- The various BPM roles
- Working with the BPM lifecycle
- Uncovering a detailed guide for how to work with process analysis, process design, process deployment, process monitoring, process maintenance, and continuous process improvement
- Determining the potential of working with BPM maturity models
- Discovering how BPM alignment management enables identification of duplicated business processes, tasks, roles, measures, and reports; unleashes reusability and unifies initiatives
- Realizing the impact and opportunities in intelligent business process management (iBPMN): From automation to orchestration—the realignment of BPM around service-oriented architecture (SOA)
- How evidence-based BPM seeks to instill a data-driven approach
- Understanding how social media and BPM fit together

The Complete Business Process Handbook. http://dx.doi.org/10.1016/B978-0-12-799959-3.00010-0

第三部分

3.1　BPM 的工作方法

Henrik von Scheel, Mark von Rosing, August-Wilhelm Scheer

介绍

在这一部分中,我们提出一种业务流程的工作方法,将战略规划转化为有效执行的关键。我们探索当前和未来的业务流程趋势,建议您充分了解这些趋势,以便为未来全球范围内的众多业务流程的变化做最全面的准备。同时还提供一个关于什么是虚假、什么是真实的观点,以使您能够对组织中可能采用、转换和创新的内容做出关键决策。

围绕业务流程运行的工作方式为您提供组织、分类、对齐、安排和量化 BPM 概念的实用指南,以使选择和使用流程对象、流程模板制品,并在流程计划中系统化和分门别类地应用,这些都是围绕以下重要主题描述的。

- 当前和未来的业务流程趋势是什么?
- 构建 BPM 能力：BPM CoE。
- 如何使用流程模板。
- 各种 BPM 角色。
- BPM 生命周期的运行。
- 提供一份关于如何进行流程分析、设计、部署、监控、维护和持续改进的详细指南。
- 评估使用 BPM 成熟度模型的可能性。
- 了解 BPM 一致性管理如何识别重复的业务流程、任务、角色、措施和报告,避免重复性并统一规划。
- 认识智能 BPM(iBPM)的影响和机遇：从自动化到统筹编排——面向 SOA 的 BPM 重组。
- 基于证据的 BPM 如何逐渐推广到数据驱动的方法中。
- 了解社交媒体和 BPM 是怎样结合在一起的。

Business Process Trends

Mark von Rosing, August-Wilhelm Scheer, Henrik von Scheel, Adam D.M. Svendsen, Alex Kokkonen, Andrew M. Ross, Anette Falk Bøgebjerg, Anni Olsen, Antony Dicks, Asif Qumer Gill, Bas Bach, Bob J. Storms, Callie Smit, Cay Clemmensen, Christopher K. Swierczynski, Clemens Utschig-Utschig, Dan Moorcroft, Daniel T. Jones, David Coloma, Deb Boykin, Dickson Hunja Muhita, Duarte Gonçalves, Fabrizio Maria Maggi, Fan Zhao, Fatima Senghore, Fatma Dandashi, Fred Cummins, Freek Stoffel, Gabriel von Scheel, Gabriella von Rosing, Gary Doucet, Gert Meiling, Gert O. Jansson, Hans Scheruhn, Hendrik Bohn, Henk de Man, Henk Kuil, Henrik Naundrup Vester, Jacob Gammelgaard, James P. Womack, Jeanne W. Ross, Jeff Greer, Jens Theodor Nielsen, John A. Zachman, John Bertram, John Golden, John M. Rogers, Jonnro Erasmus, Joshua von Scheel, Joshua Waters, Justin Tomlinson, Karin Gräslund, Katia Bartels, Keith D. Swenson, Kenneth Dean Teske, Kevin Govender, Klaus Vitt, Krzysztof Skurzak, LeAnne Spurrell, Lloyd Dugan, Lotte Tange, Mads Clausager, Maria Hove, Maria Rybrink, Marianne Fonseca, Mark Stanford, Marlon Dumas, Mathias Kirchmer, Maxim Arzumanyan, Michael D. Tisdel, Michel van den Hoven, Mikael Munck, Mike A. Marin, Mona von Rosing, Nathaniel Palmer, Neil Kemp, Nils Faltin, Partha Chakravartti, Patricia Kemp, Peter Franz, Philippe Lebacq, Rich Hilliard, Richard L. Fallon, Richard N. Conzo, Rod Peacock, Ronald N. Batdorf, Sarel J. Snyman, Scott Davis, Simon M. Polovina, Stephen White, Steve Durbin, Steve Willoughby, Sven Vollbehr, Thomas Boosz, Thomas Christian Olsen, Tim Hoebeek, Tom Preston, Ulrik Foldager, Victor Abele, Vincent Snels, Volker Rebhan, Wim Laurier, Yr Gunnarsdottir, Yury Orlov, Zakaria Maamar, Ekambareswaran Balasubramanian, Mai Phuong, Régis Dumond

INTRODUCTION

Business process and business process management (BPM) concepts have matured over the years and new technology, concepts, standards and solutions appear. In this chapter we will therefore focus on the current and future process trends. We will elaborate on the importance of trends, the maturity of the subject, giving a perspective on what emerging trends, industry trends, mega trends are, what is hyped at the moment, and what has reached a market adoption where it has started to become the de facto standard in terms of mega trends that has achieved a dominant position by public acceptance.

THE IMPORTANCE OF TRENDS

A trend is defined as a general direction in which something is developing or changing.[1] Trends involve looking at the statistical analysis of historical data over a selected time frame and charting the progression. If the data suggest consistent

The Complete Business Process Handbook. http://dx.doi.org/10.1016/B978-0-12-799959-3.00011-2

3.2　业务流程趋势

Mark von Rosing, August-Wilhelm Scheer, Henrik von Scheel, Adam D.M. Svendsen, Alex Kokkonen, Andrew M. Ross, Anette Falk Bøgebjerg, Anni Olsen, Antony Dicks, Asif Qumer Gill, Bas Bach, Bob J. Storms, Callie Smit, Cay Clemmensen, Christopher K. Swierczynski, Clemens Utschig-Utschig, Dan Moorcroft, Daniel T. Jones, David Coloma, Deb Boykin, Dickson Hunja Muhita, Duarte Gonçalves, Fabrizio Maria Maggi, Fan Zhao, Fatima Senghore, Fatma Dandashi, Fred Cummins, Freek Stoffel, Gabriel von Scheel, Gabriella von Rosing, Gary Doucet, Gert Meiling, Gert O. Jansson, Hans Scheruhn, Hendrik Bohn, Henk de Man, Henk Kuil, Henrik Naundrup Vester, Jacob Gammelgaard, James P. Womack, Jeanne W. Ross, Jeff Greer, Jens Theodor Nielsen, John A. Zachman, John Bertram, John Golden, John M. Rogers, Jonnro Erasmus, Joshua von Scheel, Joshua Waters, Justin Tomlinson, Karin Gräslund, Katia Bartels, Keith D. Swenson, Kenneth Dean Teske, Kevin Govender, Klaus Vitt, Krzysztof Skurzak, LeAnne Spurrell, Lloyd Dugan, Lotte Tange, Mads Clausager, Maria Hove, Maria Rybrink, Marianne Fonseca, Mark Stanford, Marlon Dumas, Mathias Kirchmer, Maxim Arzumanyan, Michael D. Tisdel, Michel van den Hoven, Mikael Munck, Mike A. Marin, Mona von Rosing, Nathaniel Palmer, Neil Kemp, Nils Faltin, Partha Chakravartti, Patricia Kemp, Peter Franz, Philippe Lebacq, Rich Hilliard, Richard L. Fallon, Richard N. Conzo, Rod Peacock, Ronald N. Batdorf, Sarel J. Snyman, Scott Davis, Simon M. Polovina, Stephen White, Steve Durbin, Steve Willoughby, Sven Vollbehr, Thomas Boosz, Thomas Christian Olsen, Tim Hoebeek, Tom Preston, Ulrik Foldager, Victor Abele, Vincent Snels, Volker Rebhan, Wim Laurier, Yr Gunnarsdottir, Yury Orlov, Zakaria Maamar, Ekambareswaran Balasubramanian, Mai Phuong, Régis Dumond

3.2.1　介绍

业务流程和BPM概念经过多年的发展已经日趋成熟,新的技术、概念、标准和解决方案不断涌现。因此,在本节中,我们将重点关注当前和未来的流程趋势,并将详细阐述趋势的重要性、主体的成熟度,同时展望什么是新兴趋势、行业趋势、大趋势,找出当前活跃的以及被市场所采用的标准,按照大趋势而言,这些标准已经被公众所接受并占据主导地位,成为事实上的标准。

3.2.2　趋势的重要性

趋势被定义为事物发展或变化的一般方向[1]。趋势的变现方式是在选定的时间范围内对历史数据进行统计分析,并用图形化展示。如果数据显示一致的增加、

increases, decreases, or even constancy or flatness, a trend exists. Businesses of all sizes use these kinds of data to help predict the future or shape strategic decisions.

So why are trends important? Because trends help you prepare for the future! From a business perspective, there are three main types of trend: emerging, industry, and mega trends. If organizations ignore any of them, the business drivers or trends may eventually evolve to become a direct threat to their existing business model. If embraced, they hold the key for the next opportunity for growth.

For example, business processes has matured over a decade into a management discipline that treats processes as assets that directly contribute to enterprise performance by driving operational excellence and business process agility. Today, business processes has become an essential source of performance that supports business success, some of which are:

- Optimizing the performance of end-to-end business processes that span functions as well as processes that might extend beyond the enterprise to include partners, suppliers, and customers (the value chain).
- Making the business processes visible (and thus explicit) to business and information technology (IT) constituents through business process modeling, monitoring, and optimization/simulation.
- Keeping the business process model in sync with process execution and empowering business users and analysts to use the model to improve process performance and outcomes.
- Enabling the effective integration of process activities, business measurements, rule management, content integration, and greater collaboration to set the base for continuous improvement.
- Enabling rapid iterations of processes and underlying systems for continuous process improvement and optimization.
- Delivering measurable improvement to enterprise performance that directly contributes to organizational success and competitive advantage.
- BPM is just one approach to the larger challenge known as business process improvement (BPI). Other approaches to BPI include business process re-engineering (BPR) and business process automation.

Hence, both executives and practitioners are focusing on process trends to gain a competitive advantage by being the early adopter. Our focus is on process mega trends and emerging trends as the driving force that will change how organizations work with and apply these trends successfully to their process landscape in order to gain a competitive advantage in the future.

MATURITY OF THE SUBJECT

The adoption of trends is tightly connected to the maturity. The rise of business process engineering and re-engineering results from a paradigm shift[2] that has already occurred by moving away from the previous function-oriented management practices. The new focus is now more towards practices that focus on customer value. The result of this shift also necessitated consideration of the

减少,甚至恒定或平稳,则存在一种趋势。各种规模的企业都使用这些数据来帮助预测未来或制定战略决策。

为什么趋势很重要?因为趋势有助于您为未来做好准备!从商业的角度来看,趋势有三种主要类型:新兴趋势、行业趋势和大趋势。如果一个组织忽略任何一种趋势,这个趋势或业务驱动因素最终可能演变为对其现有业务模型的直接威胁。如果拥抱趋势,就意味着他们抓住了下一个增长机会的关键所在。

业务流程在过去十年中日趋成熟并发展成为一种管理原则,它将流程视为直接促进企业绩效的资产,提高运营卓越性和业务流程敏捷性。如今,业务流程已成为支撑业务成功的重要因素,其中一些功能如下:

- 优化跨职能的端到端业务流程的性能,以及可扩展到企业范围外的流程的性能,使流程涵盖合作伙伴、供应商和客户(价值链);
- 通过业务流程建模、监控和优化/模拟,使业务流程对业务和IT参与者显而易见(从而清晰明确);
- 保持业务流程模型与流程执行同步,并授权业务用户和分析师使用该模型来改进流程性能和结果;
- 实现流程活动、业务度量、规则管理、内容集成和更密切合作的有效整合,为持续改进奠定基础;
- 实现流程和基础系统的快速迭代,以持续改进和优化流程;
- 为企业绩效提供可衡量的改善,直接促进组织的成功,增加竞争优势;
- BPM只是应对更大挑战的一种方法,称为BPI。BPI的其他方法包括BPR和业务流程自动化。

因此,企业高管和从业者都在关注流程趋势,并争取成为早期采用者来获得竞争优势。我们重点关注大趋势和新兴趋势,这些趋势作为驱动力将改变组织工作方式,并被应用于组织流程布局中,以期在未来获得竞争优势。

3.2.3 主体的成熟度

趋势的采纳与成熟度密切相关。业务流程工程和重建工程的兴起源于一种范式转变[2],以前的以职能为中心的管理模式已经逐步被抛弃。现在的重点集中在客户价值的实践上。这种转变还需要考虑企业战略、结构和文化的变革来支持新的基础设施。

enterprise strategy, structure, and culture that are required to support the new infrastructure.

According to a survey done by PriceWaterhouseCoopers AG,[3] organizations are critically aware of the importance of BPM to the future success of their business. To remain competitive, senior executives have identified the importance of the continuous optimization of business processes in terms of quality and efficiency for their administrative and production business processes[4] while retaining the differentiation among core competitive, core differentiating, and non-core processes. The survey found that many of these executives believed that their business would no longer exist in as little as 10 years if efforts to continuously improve and optimize their business processes were not pursued. In a similar tone, many of these executives saw that another key factor to the success of an organization is the collection and analysis of appropriate key process performance indicators.

Often when a trend emerges and the maturity is low, early adoption investors take advantage of the opportunity and develop unique leading solutions. Such practices from the leaders are called leading practices. Leading practices define and strengthen competitive advantage, innovation, and efficiency in the core differentiating competencies with a focus on the revenue model and value model. They are called the out-performers and are the first to take advantage of the new emerging trends and thereby outperform the market.

When a trend is in its early hype stages and become more mature, industry leaders adopt, invest, and develop industry practices to out-compete their peers. This is called industry adoption. Industry practices improve competitive parity and standardize core competitive competencies with a focus on performance models and service models. They are called industry leaders because they have the advantage of emerging trends and outperform the majority of the competition in their respective markets.

Finally, as the trend matures with wide adoption and years of experience it has becomes a standard or a best practice. The adoption becomes a best practice when organizations begin to improve and standardize their non-core competencies that focuses on the cost model and the operating model. Such organizations are considered followers who take advantage of best practices that are non-core to their business, while gaining the full advantage of trends with low risk and cost.

MEGA TRENDS

Mega trends are changes that are slow to form a tendency, but are likely to affect the future in all areas in the next 10–15 years, such as globalization, technology, economy, the workforce, demographics, politics, and the environment (Figure 1). Once in place, mega trends influence a wide range of activities, processes, and perceptions in businesses, governments, and societies, possibly for decades to come. They are the underlying forces that drive trends (i.e., aging population).

Process mega trends are already shaping the future. No one should dwell in neither the past nor the present, and there some trends that will definitely have a significant impact on how organizations apply and take advantage of processes in the next 10 years.

Both current and future process mega trends have enormous potential and will definitely change, improve, and revolutionize the future. Business process management

普华永道会计师事务所AG[3]的一项调查显示,组织清楚认识到BPM对其业务未来成功的重要性。为保持企业竞争力,企业高管已经意识到在质量和效率方面持续优化管理和生产业务流程[4]的重要性,同时要保持核心竞争力、核心差异化和非核心流程之间的差异。调查发现,许多高管认为,如果不持续努力改进和优化业务流程,他们的业务将在10年内不复存在。同样地,许多高管认为收集并分析适当的关键PPI是组织成功的另一个关键因素。

通常,当趋势出现但成熟度较低时,利用这种趋势的投资者会抓住机会开发独特领先的解决方案。这种来自领导者的实践叫做领导实践。领导实践在核心差异化能力中定义并加强竞争优势、创新和效率,重点关注收入模型和价值模型。领导实践被称为优胜者,是第一批利用新兴趋势从而超越市场的人。

当一种趋势处于早期的宣传阶段并变得比较成熟时,行业领导者会采用、投资和开发行业实践以超越同行,这通常被称为行业应用。行业实践通过关注绩效模型和服务模型来提高竞争公平性和标准化核心竞争能力。他们之所以被称为行业领导者,是因为他们利用新兴趋势的优势,并且在各自的市场中战胜了大多数竞争对手。

最后,随着趋势的成熟、广泛的采用和多年的经验积累,它已经成为一个标准或最佳实践。当组织开始专注于成本模型和操作模型来改进和标准化其非核心能力时,通常采用他们作为最佳实践。这些组织被称作最佳实践的追随者,虽然他们借鉴的最佳实践就其业务而言并非是核心的,但却能使他们在低风险、低成本的情况下充分利用趋势。

3.2.4　大趋势

大趋势是缓慢变化形成的一种趋势,可能会在接下来的10 ～ 15年内影响所有领域的未来,如全球化、技术、经济、劳动力、人口统计、政治和环境(图1)。大趋势一旦出现,可能在未来几十年影响到企业、政府和社会的各种活动、流程和观念。它们是推动趋势(如人口老龄化)的潜在力量。

流程大趋势正在塑造未来,任何人都不应该停留在过去和现在,有些趋势将会对组织在未来10年如何应用和利用流程产生重大影响。

当前和未来的流程大趋势都具有巨大的潜力,必将改变、改善和变革未来。

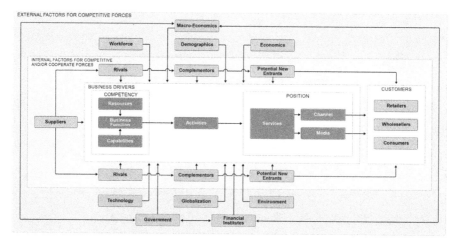

FIGURE 1

Megatrends as the driver of competitive forces.[5]

provides the context and best mechanism to achieve the full potential of technology trends for the next decade. BPM is at the inflection of underlying technologies and human participants, providing the perfect place to leverage technology trends while providing a business context.

EMERGING TRENDS

Emerging trends are maturing tendencies driven by mega trends that influence industry trends at different levels, such as process-driven case management, a technology mega trend and a trend in the insurance industry.

Emerging trends can be illustrated in many ways, e.g., hyper cycles (Gartner), radar systems (Forrester), mind-map footprints (Frost & Sullivan), usage curves and product lifecycles (Boston Consulting Group), and underground station lines. Common to all of them is an emphasis on a specific view that misleads the reader; the most popular one is the annually published Gartner Hyper-Cycle.

For those unfamiliar with these charts, the basic structure starts with a technology trigger near the origin of time and is visibly followed by a quick rise to the "peak of inflated expectations" that is often driven by a combination of unrealistic claims by proponents and the hopes of users desperate to believe those claims. The exaggerated peak of hype is inevitably followed by a crash of popularity into the so-called "trough of disillusionment." Many ideas just die here and drop off the curve, but for others a more realistic set of expectations develops as believers and early adopters begin to experience measurable benefits. It serves to push the idea (sometimes with changes) up the "slope of enlightenment."[6] This gradual advance passes an important point of inflection on the performance S curve known as the attitude confirmation. The next landmark is crossing a social chasm at another critical inflection point called the attitude plateau.[7] Once an idea successfully

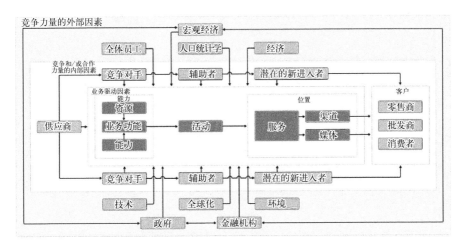

图1　作为竞争驱动力的大趋势[5]

BPM为发挥未来十年技术趋势的全部潜力提供了背景和最佳机制。BPM正处在底层技术和人工参与者的拐点上,在营造业务环境的同时,它也为利用技术趋势提供了完美的场所。

3.2.5　新兴趋势

新兴趋势是由那些在不同层次影响行业趋势的大趋势所驱动的正在走向成熟的趋势,如流程驱动的案例管理、技术大趋势和保险行业趋势。

新兴趋势可以通过多种方式加以说明,如技术成熟度曲线(Gartner)、雷达系统(Forrester)、思维图足迹(Frost & Sullivan)、各类曲线和产品生命周期(Boston Consulting Group)的使用以及地铁车站线。所有这些方法的共同点是强调一种误导读者的特定观点。最受欢迎的是每年发布的Gartner技术成熟度曲线。

对于那些不熟悉这些图表的人来说,基本结构从一个靠近时间起源的技术触发器开始,然后迅速上升到"高期望的峰值",这通常是由支持者不切实际的主张和用户不顾一切地相信这些主张共同驱动的。夸大的天花乱坠的宣传高潮过后,不可避免地出现人气暴跌进入所谓的"幻灭低谷"。许多想法都在这里消亡,并偏离了曲线,但对其他人来说,随着信奉者和早期采纳者开始体验到可衡量的利益,一套更现实的期望也随之形成。它有助于将思想(有时会随着变化)推上"启蒙的斜坡"[6]。这种渐进式的前进曲线与S曲线上的一个重要转折点相交,相交点叫做态度确认。下一个里程碑是跨越社会鸿沟与导入曲线相交的另一个关键的拐点,

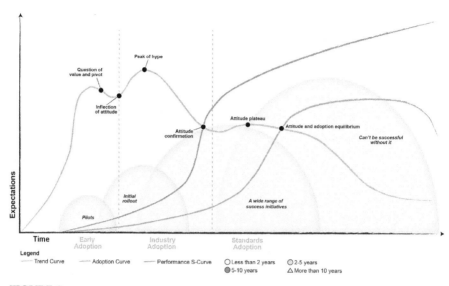

FIGURE 2

Process trends, which incorporates trends, actual adoption, performance, and the maturity life cycle.[8]

crosses the chasm, it plateaus as a generally recognized productivity concept for that industry. Some ideas fly quickly along these curves, passing older ideas that seem to just plod along at a much slower pace.

Hence, we have chosen to illustrate the emerging process trends in a hype trend model to give an independent and agnostic view of process trends, which incorporates trends, actual adoption, performance, and the maturity life cycle (Figure 2).

PROCESS TRENDS

Based on agnostic and vendor neutral research with the Global University Alliance and in consensus with their key authorities and leaders, we agreed on the following emerging process trends (Figure 3) that will influence the future of how organizations will adapt to, work with and apply processes.

EARLY ADOPTION

Trend phase: Pilots to initial rollout	*Market penetration*: Low	*Maturity*: Emerging
Benefit rating: Very high	*Investment required*: High	*Risk*: Very high

Characteristic: A potential trend breakthrough kicks things off. Early proof-of-concept stories and media interest trigger significant publicity. Often no usable

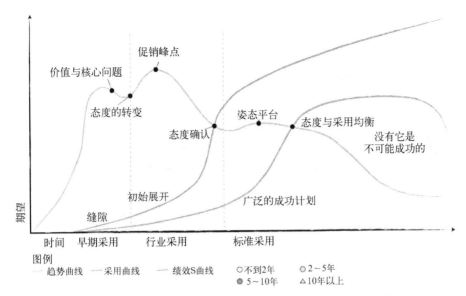

图2 过程趋势,包括趋势、实际采用、性能和成熟度生命周期[8]

称为态度平缓期[7]。一旦一个想法成功地跨越了鸿沟,它就会成为该行业公认的生产力概念。有些想法在这些曲线上快速传播,而那些陈旧的想法以一种慢得多的速度传播。

因此,我们选择在技术成熟度曲线模型中说明新兴流程趋势,以对流程趋势给出独立的、不可知的观点,其中包括趋势、实际采用、性能和成熟度生命周期(图2)。

3.2.6 流程趋势

基于与全球大学联盟的不可知论和供应商中立研究,并与他们的主要权威和领导人达成共识,我们赞同以下的新兴流程趋势(图3),这些趋势将对组织在未来如何适应、使用和应用流程产生影响。

3.2.7 早期采用

趋势阶段:试运行到初始启动	市场渗透率:低	成熟度:新兴
效益等级:非常高	所需投资:高	风险:非常高

特点:一个潜在的趋势突破开启了新时代。早期的概念验证故事和媒体兴趣引发了大量宣传。通常没有可用产品,商业可行性也未经证实。早期的宣传创作

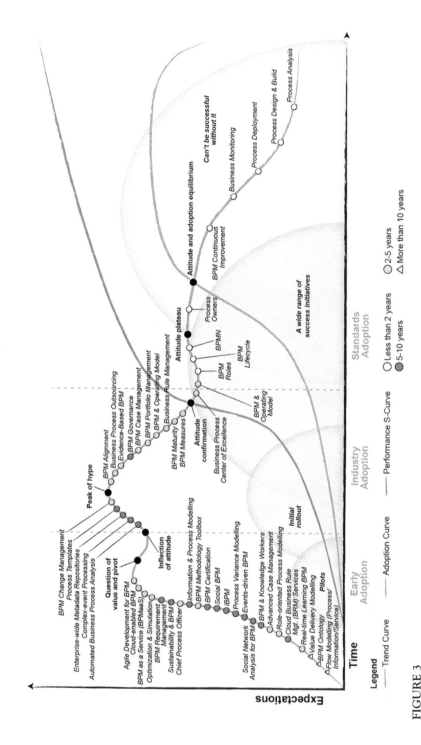

FIGURE 3

Business process trends.[9]

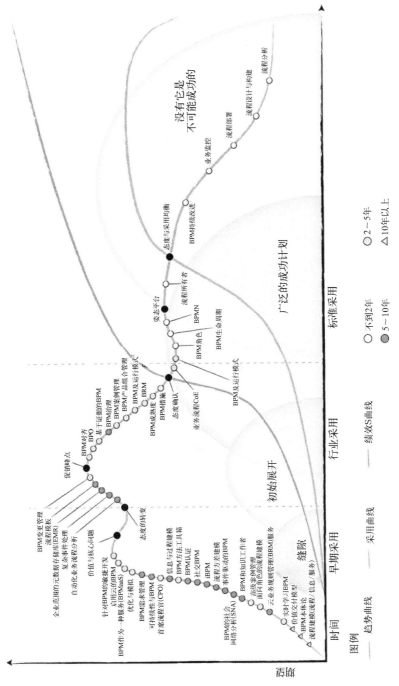

图 3　业务流程趋势[9]

products exist and commercial viability is unproven. Early publicity produces a number of success stories—often accompanied by scores of failures. Few organizations take action; many do not.

Trends are less than 5–10 years from mainstream adoption. It requires a high level of investment and high risk with the potential to deliver core differentiating aspects.

Business Performance Impact: Early adoption invests to take advantage of the opportunity and develop LEADing Practices. The LEADing Practices define and strengthen competitive advantage, innovation, and efficiency in the core differentiating competencies with focus on the revenue model and value model. They are called the outperformers and are the first to take advantage of the new emerging trends, and thereby outperform the market.

EARLY ADOPTER OF PROCESS TRENDS

1. **Extended Flow Modeling (a part of X-BPMN).** The next generation of BPM will benefit from the evolution of modeling approaches currently being advanced in architecture and engineering, enhancing a structured way of thinking, working, and modeling. Learning from other principles enables reuse of models as well as standardization of various concepts. What is especially relevant is the interlinkage between process flow, information flow, and service flow. Organizations realize the need to model the various flows both separately and together. We already see the first technology enabling such modeling; see example from iGrafx: www.igrafx.com/solutions/business-challenges/process-modeling

2. **BPM Ontology.** Many BPM and/or process frameworks, methods, and approaches such as Lean, Six Sigma, BPR (Business Process Reengineering), TQM (Total Quality Management), Zero Defect, BPMN (Business Process Modeling Notation), BPMS (Business Process Management Suite) have their own vocabulary. Each of these vocabularies has its own definition of terms such as business process, process step, process activity, events, process role, process owner, process measurement, and process rule. This variety of definitions might hamper communication. On one hand, the same word might have different meanings in different frameworks, methods, and approaches (i.e., homonymy). On the other hand, different words might have the same meaning in various frameworks, methods, and approaches (i.e., synonymy). When communicating, people are often unaware of homonymy and synonymy and expect the same words to have the same meaning and different words to have a different meaning, which might lead to miscommunication among people with different backgrounds (i.e., with training in a different framework, method, or approach). What is needed is a shared vocabulary (e.g., a folksonomy) that ensures a consistent use of terms. In a weak interpretation, such a folksonomy could be used as a central ontology to which all framework, method, and approach vocabularies are mapped to determine which words have the same and which have a different meaning in different frameworks, methods, and approaches. In a strong interpretation, such a central ontology

了许多成功的故事,往往伴随着大量的失败。少量组织采取行动,而大部分组织并不会采取行动。

从新兴趋势发展到主流应用通常不会超过5～10年。它不仅需要高水平的投资,而且在实现核心差异化过程中还可能存在高风险。

业务绩效影响:早期采用阶段投资要抓住机会并开发领导实践。领导实践在核心差异化能力中定义并加强了竞争优势、创新能力和效率,重点关注收益模式和价值模式。他们被称为优胜者,是第一批利用新兴趋势从而超越市场表现的人。

3.2.8　流程趋势的早期采用者

(1)扩展流程建模(extended flow modeling,X-BPMN的一部分)。下一代的BPM将受益于当前在架构和工程方面处于领先地位的建模方法的发展,这增强了结构化的思维、工作和建模方式。学习其他原理可以重用模型以及各种标准化概念。特别相关的是流程、信息流和服务流之间的互联。组织意识到需要对不同的流分别建模、共同建模。已经出现第一个支持这种建模的技术,请参阅来自iGrafx的示例:www.igrafx.com/solutions/business-challenges/process-modeling。

(2)BPM本体论。许多BPM或流程框架、方式方法,如精益、六西格玛、BPR、TQM、零缺陷、BPMN、BPMS都有自己的词汇表。每个词汇表都有自己的术语定义,如业务流程、流程步骤、流程活动、事件、流程角色、流程所有人、流程度量和流程规则。这种多样的定义可能会妨碍交流。一方面,同一个词在不同的框架、方式和方法中可能有不同的含义(即同音异义);另一方面,不同的词在不同的框架、方法和方法中可能具有相同的含义(即同义词)。在交流时,人们往往不知道同音异义,认为同一个词有相同的含义,不同的词有不同的含义,这可能导致不同背景的人之间的理解错误(即在不同的框架、方式或方法中进行培训)。所以需要一个共享词汇表(如一个大众分类表),以确保术语使用的一致性。在一个较弱的解释中,这样的大众分类可以作为一个中心本体论,所有框架、方式、方法词汇都映射到这个本体论中,以确定哪些词具有相同的含义,哪些词在不同的框架、方式和方法中具有不同的含义。在一个强有力的解释中,这样一个核心本体定义基本流程概念和它们之间的关系(如定义一系列流程步骤的能力),可以用作描述、记录和构

that defines fundamental process concepts and the relations between them (e.g., the ability to define a sequence of process steps) could be used as the reference vocabulary to describe, document, and structure process knowledge. Both interpretations would profit from a validated reference ontology. Hence, the need for a BPM ontology that can be applied within the areas of process modeling, process engineering, and process architecture is clear.

3. **Value-Oriented Process Modeling.** Often referred to as value-oriented process design or value delivery modeling (VDML), emerging in an era focused on the automation and optimization of business processes in the context of established business organizations. As such, they tended to focus on process flow within and between organizations, typically within individual lines of business. As the scope of automation expanded, processes were linked electronically but tended to preserve existing organizations and relationships, optimizing processes within lines of business, optimizing processes at an operating level. Value planning, value identification, value creation, and value realization are not really methods and approaches used by process teams today. However, advances in technology, global competition, and continuous business change have increased the need for business agility with a focus on the creation of customer value and optimization of business processes across the enterprise. This requires the ability of top management to analyze and guide the design of the business focusing on customer value, consolidating sharable capabilities, and linking business strategy to business transformation through a shared understanding of the desired business design and key objectives. In this area, we also see technology move in this direction, where VDML has been adopted as an object management group (OMG) modeling standard and is expected to be available in 2015.

4. **Real-time Learning BPM.** Organizations around the world struggle to crack the code for improving the effectiveness of managers, salespeople, scientists, and others whose jobs consist primarily of interactions with other employees, customers, and suppliers, and complex decision making based on knowledge and good judgment. As process and BPM adoption rise in organizations, enabling processes and continuous improvement around the knowledge workers and similar employees working in complex processes is a new challenge. Business processes are the heart of an organization and the support of the business processes by application systems is central to each organization. Introducing new applications requires employees to become trained and educated for them, often by multi-day presence trainings in advance.

When the software is rolled out organization-wide, it is expected that the cost and time savings will materialize in a short time. However, user errors slow down the efficiency of the new software and with it the execution of the connected business processes. Although they have been trained, employees are not able to use the new technology efficiently. Even though knowledge workers are often the core of many organizations, enablement of these employees with specific guidance at their point of need in a manner consistent with kaizen principles of quality and continuous improvement is frequently poor. In the following

造流程知识的参考词汇。这两种解释都将受益于经过验证的引用本体。因此,对可应用于流程建模、流程工程和流程体系结构领域的BPM本体的需求是显而易见的。

(3)价值导向的流程建模。通常被称为价值导向的流程设计或价值导向的流程建模(value delivery modeling, VDM),出现于专注在已有业务组织内实现业务流程自动化和优化的阶段。因此,该技术通常是在单个业务线内,倾向于关注组织内部和组织之间的流程。随着自动化范围的扩大,流程以电子化的方式连接,同时倾向于保留现有的组织和关系,优化多业务线内的流程和运营级别的流程。价值规划、价值识别、价值创造和价值实现不是当今流程团队真正使用的方式和方法。然而,技术进步、全球竞争和持续的业务变化增加了对业务敏捷性的需求,聚焦在整个企业范围内创造客户价值和优化业务流程。这就要求高级管理层能够分析并指导聚焦客户价值的业务设计,巩固可共享的能力,并通过共同理解所需业务设计和关键目标,将业务战略与业务转型联系起来。在这一领域,我们也看到技术朝着这个方向发展,其中VDML已被采用为OMG建模标准,预计将在2015年推出。

(4)实时学习BPM。世界各地的组织都在努力破解提高管理者、销售人员、科学家和其他人工作效率的密码,这些人的工作主要包括:与其他员工、客户和供应商的互动,以及基于知识和良好判断力的复杂决策。随着组织中流程和BPM采用率的提升,让知识工作者和在复杂流程中工作的类似员工支持流程和持续改进是一个新的挑战。业务流程是组织的核心,应用系统对业务流程的支持是每个组织的核心。引入新的应用程序需要员工接受培训和教育,通常需要提前进行多天的现场培训。

当软件在整个组织范围内推出时,预计成本和时间的节约将在短时间内实现。但是,用户使用错误将会降低新软件的效率,并随之降低所对应业务流程的执行速度。尽管他们接受过培训,但员工仍不能高效地使用新技术。尽管知识型员工通常是许多组织的核心,但在他们需要的时候,以符合质量改善原则和持续改进的方

section, we show how real-time learning based on business process guidance (BPG) can help employees to get along better with new processes. It is expected that real-time learning through BPG will grow in importance in the future.

a. More changes: Processes and applications will change even more frequently in the future, triggering a need for training and support among the employees using them.

b. More collections of applications: Instead of one large system installed and configured on premise, we will often see a collection of applications provided as a service out of the cloud. This asks for process guidance that works across applications and that can be configured and equipped with content by the user organization.

c. Social networks will be used more at work: We will also see more knowledge sharing and peer support using social network technologies at the workplace. Social BPG will provide users with access to social network communication channels and will help filter and display only messages that are relevant based on the process and application context of the user.

d. Users will influence provision of content: Statistics from software usage and user feedback will become an important source for content authors to provide additional content and improve the existing support content in the BPG system.

e. BPG will extend beyond the office: Mobile devices will bring process guidance to new areas such as repair and maintenance of machines. First prototypes are built in research projects where information and work instructions will be displayed with augmented reality techniques on top of live pictures taken through the built-in camera. Users can call experts that support them directly, seeing the machine in real time through the camera.

f. BPG is already a good concept supporting the introduction of new processes and applications and its potential will grow in the future as it enables the organization.

5. Cloud Business Rule Management (BRM) Services. Business rules are actionable elements of business policy; they are implicit and explicit business directives that define and describe guidance for taking a business action. Externalizing policies and rules create a need to manage them as an important business resource, and BRM has emerged as a structured discipline guiding business rule definition, categorization, governance, deployment, and use throughout the business life cycle. BRM is supported and enabled in this need to manage rules as an important business resource by two technology types: the business rule engine (BRE) and BRM system (BRMS). A BRE is core software that executes business rules that have been segregated from the rest of the application logic, matching a collection of rules (the rule set) against a set of given conditions to determine which rules apply. A BRMS is a comprehensive suite built around a BRE that facilitates the creation, registration, classification, verification, deployment, and execution of business rules. BRMS products constitute a modern incarnation of BRE products. A critical

式为他们提供的特定指导往往做得很糟糕。在下面的部分中,我们将展示基于业务流程指导(BPG)的实时学习如何帮助员工更好地与新流程相处。因此,通过BPG进行实时学习在未来将变得愈加重要。

a. 更多的变化:流程和应用程序在未来会更频繁地变化,从而产生对使用它们的员工进行培训和支持的需求。

b. 更多的应用程序集合:我们将常看到作为云服务提供的应用程序集合,而不是在本地安装和配置一个大型系统。这需要跨应用程序工作的流程指导,用户组织可以对内容进行设定和配置。

c. 社交网络将在工作中得到更多应用:我们还将在工作场所看到利用社交网络技术进行知识共享和同行支持。BPG将为用户提供访问社交网络通信渠道的权限,并基于对用户流程和应用环境的理解,帮助用户过滤并显示相关消息。

d. 用户会影响内容的提供:软件使用统计和用户反馈将成为作者提供额外内容和改进BPG系统现有支持内容的重要依据。

e. BPG将扩展到办公室之外:移动设备将把流程指导带入新的领域,如机器的维修和维护。第一个原型是在一个研究项目中构建的,信息和工作指令通过增强现实技术显示在内置摄像头拍摄的实况照片上。用户可以直接致电技术支持的专家团队,通过摄像头实时查看机器。

f. BPG已经是一个很好的概念,它支持新流程和应用程序的引入,并且随着它对组织的支持,其潜力将在未来不断增长。

(5)云业务规则管理(BRM)服务。业务规则是业务策略的可操作元素,它们是隐式和显式的业务指令,是对业务操作的指导的定义和描述。外部化的策略和规则使得它们被当作重要业务资源管理,而BRM已经成为在整个业务生命周期中指导业务规则定义、分类、治理、部署和使用的结构化规则。为了将规则作为重要的业务资源进行管理,BRM得到了两种技术类型的支持:业务规则引擎(BRE)和BRM系统(BRMS)。BRE是执行业务规则的核心软件,这些业务规则与应用程序逻辑的其余部分分离,根据一组给定的条件匹配一组规则(规则集),以确定应用哪些规则。BRMS是围绕BRE构建的综合套件,它有助于创建、注册、分类、验证、

distinction between a traditional BRE and a BRMS is that a BRMS incorporates support for seven capabilities: the execution engine (the BRE), repository, integrated development environment rule model simulation, monitoring and analysis management and administration rule templates. When BRMS or BRE functionality is provided as a core capability hosted in a cloud, it is called cloud BRM services. Cloud BRM services are a type of platform as a service (PaaS).Cloud BRM services can be obtained either as a separate offering or as a feature of a BPM PaaS. The primary business impact of cloud BRM services will derive from the business impact of BRM proper; cloud BRM services are just an alternative delivery vehicle for a concept BRM that can increase quality decision making when properly understood. Although BRM concepts have been prevalent in certain industries (for example, financial services) and in well-documented processes (for example, underwriting), there is no inherent limit to BRM's industry and process reach. Therefore, cloud BRM services have a similar potential reach, with an emphasis on "potential."

6. **Role-Oriented Process Modeling.** Traditional BPM and requirement concepts are insufficient for today's EA, business model, and value-driven approach to organization operational execution and strategic management, whereas requirements must support, link, and be decomposed from top objectives down to technology requirements. Consequently, business processes are architected and designed as a system of activities reflecting and supporting achievement of an organization's goals, strategies, and objectives. All of these can be classified as high-level business requirements. A role-oriented (people oriented) process modeling approach and discipline is required to create process-centric organizations as high level requirements must be decomposed, layered, and used to identify, model, architect, design, implement, and operate cross-functional process scenarios, each with a defined purpose, value–driven activity, and measurable outcomes (performance indicators) that relates directly to the desired business objectives (high-level requirements). All of these are functional capabilities, which are also requirements.

7. **Advanced Case Management.** This is at the nexus of BPM and enterprise content management (ECM) usage scenarios, and involves a mix of collaborative, unstructured, and structured processes. We see in multiple organizations requirements beyond traditional process modeling; among others, it is about empowering participants in a process by removing context tunneling and providing better support for exception handling, the ability to control flow and cross-flow information visibility. Organizations around the world have therefore started to invest in case management. The information model includes both data and documents, so changes in values, metadata, and life cycle state can all be used to model the case.

8. **BPM Knowledge Worker.** Introducing new applications requires employees to become trained and educated for them, often by multi-day presence trainings in advance. When the software is rolled out organization-wide, it is expected that the cost and time savings will materialize in a short time.

部署和执行业务规则。BRMS产品是BRE产品的现代化身。传统的BRE和BRMS之间的一个关键区别是，BRMS包含了对七种功能的支持：执行引擎（BRE）、存储库、集成开发环境、规则模型模拟、监视和分析管理以及管理规则模板。当BRM或BRE功能作为核心功能托管在云端时，称为云BRM服务。云BRM服务是平台即服务（platform as a service，PaaS）的一种类型。云BRM服务可以作为单独的产品或作为BPM PaaS的功能获得。云BRM服务的主要业务影响将源于BRM本身的业务影响。云BRM服务只是概念BRM的另一种搭载工具，完成了解后，它有助于提高质量决策。尽管BRM概念在某些行业（如金融服务）和记录完善的流程（如保险业）中很普遍，但BRM的行业和流程范围没有固有的限制。因此，云BRM服务具有类似的潜在影响力，并强调潜力。

（6）面向角色的流程建模。传统的BPM和需求概念不足以满足当今企业体系结构、业务模型和价值驱动的组织运营执行和战略管理方法，而需求必须支持、链接，将顶层目标分解为技术需求。因此，业务流程被设计成一个反映和支持组织目标、战略和目的实现的活动系统。所有这些都可以归类为高级业务需求。创建以流程为中心的组织需要以角色为导向（以人为本）的流程建模方法和规程，因为高级需求必须分解、分层，并用于识别、建模、构造、设计、实施和操作跨职能的流程场景，每个场景都有明确的目的、价值驱动活动和与期望的业务目标（高级需求）直接相关的可测量结果（绩效指标）。所有这些都是功能性能力，也是需求。

（7）先进的案例管理。这是BPM和企业内容管理（enterprise content management，ECM）使用场景之间的关系，涉及协作、非结构化和结构化流程的混合。我们在许多组织中看到了传统流程建模以外的需求。除此之外，它通过删除前后关联的暗箱为异常处理、控制流和跨流信息可视化能力提供更好的支持，来增强流程参与者的能力。因此，世界各地的组织开始投资案例管理。信息模型包括数据和文档，因此价值、元数据和生命周期状态的变化都可以用于建模案例。

（8）BPM知识工作者。引入新的应用程序需要员工接受培训和教育，通常需要提前进行多天的现场培训。当软件在整个组织范围内推出时，预计将在短时间

However, user errors slow down the efficiency of the new software and with it the execution of the connected business processes. Although they have been trained, employees are not able to use the new technology efficiently. Organizations are looking for better ways to provide the needed knowledge to their employees at the time of need. A new approach is real-time learning, in which information about the business process is presented to users automatically together with support on using software applications. It is gaining stronger acceptance in the market as a supplement or replacement to traditional software rollout training.

9. **Social Network Analysis for BPM.** Social network analysis (SNA) tools analyze patterns of relationships among people in groups. They are useful for examining the social structure and interdependencies (or work patterns) of individuals and organizations. SNA involves collecting data from multiple sources (such as surveys, e-mails, blogs, and other electronic artifacts), analyzing the data to identify relationships, and mining it for new information (such as the quality or effectiveness of a relationship). Organizational network analysis is a form of SNA that examines the information flow among individuals, and it depicts the informal social network, typically of groups working in the same enterprise. Value network analysis examines the deliverables exchanged among roles, typically groups of people from multiple organizations who need to work together. SNA scans social media to identify influential people, associations, or trends in the collective.

10. **Evidence-Based BPM.** As organizations gain awareness of the latent business value locked in their back-end systems' data stores, evidence-based BPM will become a day-to-day management tool rather than the subject of ad hoc initiatives triggered by punctual process performance issues. This shift will lead to the emergence of evidence-based process governance frameworks, allowing managers to effectively set up and steer long-term evidence-based BPM programs that deliver measurable value via continuous process improvement. In turn, increased evidence-based BPM maturity will spawn the deployment of real-time and predictive evidence-based BPM methods that will allow process stakeholders to respond to fine-grained process performance issues as they arise or even before they arise. In other words, evidence-based BPM methods will push the boundaries of contemporary business process monitoring practices by extending them with real-time predictive analytics. Evidence-based BPM will also enable continuous process auditing, whereby compliance violations are detected on a day-to-day basis, in contrast to contemporary postmortem process auditing approaches. Combined, these developments will bring BPM to the level of modern data-driven marketing approaches. Ultimately, every business process redesign decision will be made with data, backed by data, and continuously put into question based on data.

11. **Process Variance Modeling.** Business process variance should be seen as a viable way of allowing small differences in the way the core business functions are performed. It is advisable to introduce variation only in those business

内实现成本降低和时间节约。但是,用户操作错误会降低新软件的效率,并随之降低所连接业务流程的执行速度。尽管他们接受过培训,但员工不能有效地使用新技术。组织正在寻找更好的方法,以便在需要时向员工提供所需知识。这种新的方法叫实时学习,在这种方法中,有关业务流程的信息将自动提供给用户,同时提供软件应用的使用支持。作为传统软件推广培训的补充或替代,它在市场上得到了更广泛的认可。

(9)BPM的社会网络分析(SNA)。SNA工具分析群体中人与人之间的关系模式。这类工具对于检查个人和组织的社会结构和相互依赖性(或工作模式)很有用。SNA包括从多种来源(如调查、电子邮件、博客和其他电子产品)中收集数据、分析数据以确定关系,并挖掘数据以获取新信息(如质量或关系的有效性)。组织网络分析是SNA的一种形式,它检查个体之间的信息流,并描绘非正式的社会网络,通常是在同一企业中工作的群体。价值网络分析检查角色之间交换的可交付成果,通常是来自多个组织的需协同工作的人群。SNA扫描社交媒体,以确定集体中有影响力的人、团体或趋势。

(10)基于证据的BPM。随着组织认识到存储于系统后端的数据的潜在业务价值,基于证据的BPM将成为日常管理工具,而不是由流程时效问题触发的临时机动任务。这种转变将导致基于证据的流程治理框架的出现,使管理者能够有效地建立和指导长期的基于证据的BPM项目,而这些项目通过持续的流程改进提供可测量的价值。反过来,基于证据的BPM成熟度的提高将催生以实时和预测为目标的基于证据的BPM方法的部署,这些方法将使流程利益相关者在微小流程性能问题出现时甚至是出现前作出响应。换句话说,基于证据的BPM方法将通过使用实时预测分析扩展当代业务流程监控实践的边界。基于证据的BPM还将支持持续的流程审计,与当代的事后流程审计方法相比,通过这种方式可以在日常基础上检测到违反法规的行为。综合来看,这些发展将使BPM达到现代数据驱动营销方法的水平。最终,每一个业务流程重新设计决策都将以数据为基础,以数据为后盾,并根据数据不断提出问题。

(11)流程差异建模。业务流程差异应被视为允许在核心业务功能执行方式上存在细微差异的可行方法。建议只在代表组织核心差异化能力的业务流程中引入

processes that represent the core-differentiating competencies of the organization. This will allow an enterprise to develop its own practice and deliver unique value to clients and other stakeholders. For non-core and core-competitive competencies, best practice and best industry practice should suffice. Business process variance can be modeled in three different ways, depending on what is expected. If the aim is only to capture slight differences in the inputs, outputs, controls, and mechanisms of processes, it will be adequate to create only variances at the process activity or task levels. However, if the actual steps of the variant processes are different, true process variances can be used by presenting all the variances together in a single model or document, or a separate distinct process may even be developed. The modeling approach taken has a major impact on the management of the business processes and variances. When certain commonality between the master process and its variants is important, additional BPM techniques are necessary to maintain this traceability. This will require a great deal of attention to be given to establishing and maintaining the traceability links between the variants. Separate and distinctive processes introduce more process content, but standard BPM is applied because traceability to the master process is unnecessary. When introducing process variance, caution should be taken and the amount of variation should be minimized. If the development and modeling are not sufficiently controlled, the amount of additional and unnecessary content will quickly become unmanageable. However, if it is done well, it is an excellent way for organizations to acknowledge and embrace unique value enablers without losing out on the many benefits of business process modeling and management.

12. **Intelligent BPM (iBPM).** Recent evolution towards iBPM strategies and technology is the inclusion of more sophisticated reporting capabilities within the BPM environment itself. This is both enabled and in many ways necessitated by the greater flexibility of the architectures introduced with the BPM suites that provide BPM Phase 2 capabilities. With these environments, the ability to support non-sequential, goal-driven models is greatly increased, requiring more feedback (reporting) to enable successful execution of this type of less deterministic process models. With few exceptions, reporting on process events and business performance was previously done only after a process had executed, or otherwise within a separate environment disjointed from the process. This obviously prevented any opportunity to affect the direction or a process, but was based on a limitation of the management process as well as system and software architectures. Specifically with regard to BPM, process models were most commonly defined as proprietary structures, and in many cases compiled into software. Thus, changes either required bringing down and recompiling an application or were otherwise limited to discrete points in the process (such as exceptions and yes/no decision points).

13. **Social BPM.** This is a concept that describes collaboratively designed and iterated processes. These processes mirror the way work is performed from a doer's

变更。这将使企业能够开发自己的实践,并为客户和其他利益相关方提供独特的价值。对于非核心和核心竞争能力,最佳实践和行业最佳实践就足够了。业务流程差异可以以三种不同的方式建模,这取决于组织预期。如果目标仅仅是捕获流程的输入、输出、控制和机制的细微差异,那么只在流程活动或任务级别创建差异就足够了。但是,如果不同流程的实际步骤不同,那么可以通过在单个模型或文档中显示所有差异来使用真正的流程差异,甚至可以开发单个不同的流程。所采用的建模方法对业务流程和差异的管理有很大的影响。当主流程及其变体之间的某些共性非常重要时,就需要额外的BPM技术来维护这种可追溯性,这将需要大量的精力来建立和维护差异之间的可追溯链接。互相独立和互不相同的流程引入更多的流程内容,应用标准BPM是因为没有必要对主流程进行跟踪。在引入流程差异时,应谨慎行事,尽量减少变化量。如果开发和建模没有得到足够的控制,那么额外的和不必要的内容量将很快变得无法管理。相反,如果控制得很好,那么对于组织来说这是一种不损失业务流程建模和管理的诸多好处并可以承认和接受独特的价值推动者的好方法。

(12)iBPM。iBPM策略和技术的最新发展使BPM环境本身涵盖更复杂的报告功能。这是由具备"BPM阶段2"功能的BPM套件所引进的架构的更高灵活性所支持的,同时在许多方面也是必需的。在这些环境中,支持非顺序的、目标驱动的模型的能力得到了极大的增强,这需要更多的反馈(报告)来支持成功地执行这类不太确定的流程模型。除了少数例外,以前关于流程事件和业务性能的报告只在流程执行之后,或者在与流程脱节的独立环境中完成。这显然阻止了任何影响流程走向的机会,但这是基于管理流程以及系统和软件架构的限制。特别是在BPM方面,流程模型通常被定义为专有结构,并且在许多情况下被编译成软件。因此,所需的更改可能会导致应用程序停机和重新编译,或者仅限于流程中的离散点(如异常和是/否决策点)。

(13)社交BPM。这是一个描述协作设计和流程迭代的概念。这些流程反映了从实干者的角度执行工作的方式,以及从接收者的角度体验工作的方式。社交

perspective and experienced from a receiver's perspective. Social BPM is a concept that describes collaboratively designed and iterated processes. These processes mirror the way that work is performed from a doer's perspective and experienced from a receiver's perspective to harness the power of continuous learning. Social BPM resides at the intersection of process and collaborative activity. It is supported by BPM and social software that makes process design more visible and holistic. This includes the ability to support all process activities, such as collaboration, social networking, collective activities, and communications, that are a natural part of work to create a holistic process design that is open to influence and change from a variety of perspectives (for example, from customers, partners, suppliers, and employees). The value of social BPM is that it connects structured and unstructured knowledge-centric tasks by understanding the needs of each user (internal and external) and combines social technologies to achieve the process outcome. As such, social BPM moves BPM closer to design by doing. In practice, there are two distinct implementations of social BPM: one for process design and the other for process iteration. Social BPM design enables a group to collaboratively work on the design of a process. Social BPM iteration is the act of harnessing knowledge about how the process is experienced while it is being performed, and acting on this to change the process to better reflect preferences and shifts in the user experience. The business practice director will be the driving force to integrate social BPM techniques into process analysis and design.

14. **BPM Certification.** The need for skilled and experienced personnel to lead and participate in BPM activities is clear. The BPM profession requires a vendor neutral and agnostic Process eXpert and Process Architect certification with cross-disciplines, e.g., Business Process Principles (BPR, Six Sima, TQM, Lean, etc.), BPMN 2.0, eXtended BPMN, Process Monitoring, Value-Based Process Modeling, Continuous Improvement Approach, and Architectural Layer Modeling (Business, Application, and Technology). The eclectic nature of that skill and, by definition, the individuals who possess it, is also clear. Given the diversity of skills and experiences needed, would recruiters be better off looking for someone who is already certified in BPM? Certification in BPM, as discussed here, refers not only to certification in methodologies used in BPM (such as Six Sigma or IT Infrastructure Library) or a vendor-specific tool or methodology. Instead, we are referring to more generic, broadly scoped training in BPM as a discipline. There is growing interest in this type of certification, and a number of organizations have established their own distinct approaches to curricula, exams, assessments, and certifications for BPM.

15. **BPM Methodology or BPM Methodology toolbox.** These serve as solution accelerators and often feature commonly accepted practices for selected business processes. Process templates are becoming alternatives to traditional applications in certain process domains and industries, particularly when these process templates are based on an ICE, such as a BPMS. Today,

BPM位于流程和协作活动的交叉点。它得到了BPM和社交软件的支持,使流程设计更具可视性和整体性。这包括支持所有流程活动(如协作、社交网络、集体活动和沟通)的能力,这些活动是创建一个从各种角度(如从客户、合作伙伴、供应商和员工角度)影响和改变的整体流程设计的自然工作部分。社交BPM的价值在于它通过了解每个用户(内部和外部)的需求,将结构化和非结构化的以知识为中心的任务连接起来,并结合社会技术来实现流程结果。因此,社交化的BPM通过实践使BPM更接近设计。在实践中,有两种不同的社交BPM实现:一种用于流程设计,另一种用于流程迭代。社交化的BPM设计使一个团队能够协作地设计流程。社交BPM迭代是利用有关流程在执行流程中如何体验的知识,并以此改变流程以更好地反映用户体验中的偏好和变化的行为。业务实践总监的角色是将社交BPM技术集成到流程分析和设计中的驱动力。

(14)BPM认证。很明显,需要有技能和经验的人员来领导和参与BPM活动。BPM专业人员需要取得流程专家和流程架构师的跨学科技能认证,这些技能包括:业务流程原则(BPR、六西格玛、TQM、精益等)、BPMN 2.0、X-BPMN、流程监控、基于价值的流程建模、持续改进方法和架构层建模(业务、应用和技术)。考虑到所需的技能和经验的多样性,招聘人员如何寻找已经通过BPM认证的人是个关键。这里讨论的BPM认证不仅指在BPM中使用的方法(如六西格玛或IT基础结构库)或特定于供应商的工具或方法中的认证。相反,我们将BPM中更通用、范围更广的培训称为一种规程。人们对这种类型的认证越来越感兴趣,并且许多组织已经建立了自己独特的方法来进行课程设置、考试、评估和BPM认证。

(15)BPM方法或BPM方法工具箱。它们是解决方案的加速器,通常为选定的业务流程提供通用的实践。流程模板正在成为某些流程域和行业中传统应用程序的替代品,特别是当这些流程模板基于ICE(如BPMS)时。当前没有统一的

there is no unified BPM methodology. Instead, there are discrete methodologies that can be applied depending on the change or improvement being sought. The BPM methodologies apply across BPI (such as Six Sigma, Lean thinking, kaizen, Rummler-Brache, and business process re-engineering), application development (such as scrum, feature-driven development, and extreme programming), project management as well as implementation (PRINCE2, PMBOK, etc.), and change/transformation management. A growing number of BPM vendors provide methodologies that range from project implementation to broader BPI approaches. Consulting and system integration vendors are also incorporating BPM methods into their service delivery methodologies. However, choosing the right agnostic and vendor-neutral overachieving approach to methodologies that interconnect with all of them is required. Business process management methodologies initially operate with performing and driving business process intelligence (BPI) projects and rely on expert knowledge from seasoned BPI practitioners to be effectively used.

16. **Information and Process Modeling.** Also called anti-pattern information modeling, is the need to support information models with a more flexible process execution by avoiding well-known restrictions present in conventional BPM and workflow technology. The trend for information modeling in the market is about the challenge regarding process and information modeling and how one can produce adequate as-is and to-be process models that incorporate information models. Anti-pattern information modeling is often incorrectly understood to be concerned only with data modeling. The answer to this is not easy and is discussed in the X-BPMN chapter. One of the biggest challenges is the mistake that most BPM and BPMN concepts do not consider the process in its full context. A process always has a context; not considering its context in the purpose and goals perspective of the business is devastating. It keeps away the context that so many are looking for: the value perspective. Not considering the context to the business competencies can have the effect that nobody knows which processes are a part of the core differentiating competencies and which support the core competitive competencies of the organization. It does not matter how much we analyze the process itself; it cannot reveal this information. The same goes for services; whereas we all know that activities (processes) are needed to create services, most organizations do not know which processes creates what kind of service. Therefore, their process models do not consider the most vital aspects of the various value offerings to the consumers of the processes (e.g. employees and/or customers). Modeling the process without considering its relevant context results in process models that the executives and many others from the business or even architecture teams cannot use. We see too many BPM programs/projects within organizations that limit their as-is and to-be process models in this way. To structure the X-BPMN process groups, we categorize the relevant process context into layers.

BPM方法。相反,根据所寻求的变化或改进,可以应用离散的方法论。BPM方法适用于整个BPI[如六西格玛、精益思考、改善、Rummler-Brache(拉姆勒–布拉奇流程设计方法论)和BPR]、应用程序开发[如scrum(常用于敏捷软件开发)、功能驱动开发和极限编程]、项目管理和实施(Prince2、PMBOK等)以及变更/转换管理。越来越多的BPM供应商提供从项目实施到更广泛的BPI方法的各种方法。咨询和系统集成供应商也将BPM方法纳入其服务交付方法中。然而,对于与所有方法相互关联的方法,需要选择正确的不可知论和厂商中立的超常方法。BPM方法最初与执行和驱动项目一起运行,并依赖经验丰富的业务流程智能从业者提供的专家知识来有效使用。

(16)信息和流程建模。也称为反模式信息建模,是通过避免传统BPM和工作流技术中存在的众所周知的限制,以更灵活的流程执行来支持信息模型的需求。市场上信息建模的趋势是关于流程和信息建模的挑战,以及如何生成包含信息模型的适当的现状和未来流程模型。反模式信息建模通常被错误地理解为只涉及数据建模。这个问题的答案并不容易,我们将在X-BPMN一节中进行讨论。最大的挑战之一是大多数BPM和BPMN概念在其完整的环境中不考虑流程的错误。一个流程总是有一个背景,不考虑它在目的和目标方面的背景对于业务来说是毁灭性的。它避开了很多人正在寻找的背景:价值视角。不考虑业务能力的背景会产生这样的影响:没有人知道哪些流程是核心差异能力的一部分,哪些流程支持组织的核心竞争能力。不管我们如何分析这个流程本身,它都不能揭示这个信息,服务也是如此。虽然我们都知道创建服务需要活动(流程),但大多数组织不知道哪个流程创建什么样的服务。因此,他们的流程模型没有考虑提供给流程消费者(如员工和/或客户)各种价值产品的最重要方面。在不考虑其相关背景的情况下对流程进行建模会导致流程模型的产生,而这些模型是业务团队甚至架构团队的主管和许多其他人员无法使用的。我们看到组织中有太多的BPM程序/项目以这种方式限制它们的现状和未来流程模型。为了构造X-BPMN流程组,我们将相关的流程背景分类为层。

17. **Chief Process Officer.** We increasingly see in organizations a new top management position emerging, which we call the chief process officer (CPO). The CPO oversees the BPM-discipline of an organization, which creates significant value by moving business strategy systematically into people and IT-based execution at a pace and with certainty. The CPO works as a value scout across organizational boundaries, building an agility network for the organization. The need for this development is driven through digitalization in many companies. The CPO makes sure that IT is used in a way that produces the best business value. In a time when most technology moves to the cloud, business processes become a critical asset of an organization. The CPO manages these process assets using an outcome-driven process management discipline.

18. **Sustainability and BPM.** The management of organizations has experienced some interesting trends. The first one is a stronger focus on delivering value with a more comprehensive definition of it, encompassing not only financial aspects but also other stakeholder interests that put pressure on building more sustainable societies. Secondly, process management is a way of improving an organization's performance. And the third is the pervasiveness of IT as both a resource and an enabler. The need for organizations to become more sustainable from an economic, ecological, and social point of view through the management of processes and with a strong IT bend is clear. Therefore, a growing trend is to codify and guide through specific practices that, by linking strategy to operations, drive joint improvements in shareholder returns, the ecological footprint, and social impact, ideally from a life cycle point of view.

19. **BPM Requirements Management.** Whether for business innovation, transformation, or technology development, requirements management is the most widely used concept influencing design of anything in any industry. Consequently, it also influences design of business processes, both functional and end-to-end scenarios ("Our enterprise is our processes"). As a result, it impacts how well the organization operates.

    Today, BPM requirements management has become critical for any organization, heavily influencing the quality of its business designs and corporate results. The significance of the requirement concept to any organization lies in the fact that it is a key information carrier, interpreter, bridge to, and translator of desired enterprise goals with process and technology realization designs and performances using decomposition and mapping of high-level requirements into a network of more granular requirements. It applies throughout all pertinent types of enterprise layers (business, process, application, data, technology, organization, governance, etc.). In essence, requirements exist everywhere in any organization within each layer of its architecture and drives everything an organization does. Requirements are not stand-alone entities. They relate, decompose, or compose into other types of more granular requirements. Requirements are dynamic. They change, are impacted by changes to other requirements, or are added as new by business or technology. Requirements must therefore be continuously managed.

（17）首席流程官（CPO）。我们越来越多地看到组织中出现一个新的高层管理职位，我们称之为CPO。CPO的职责是监督组织的BPM规则，按照一定的效率和确定性将业务战略系统地移植到基于人员和IT的执行中，实现价值创造。CPO作为贯穿组织的"价值侦探"，负责为组织构建一个敏捷性网络。CPO确保IT的使用方式可以产生最佳的商业价值，这是由公司的数字化驱动的。在当今组织将大多数技术转移到云技术的时代，业务流程成为组织的关键资产，而CPO通过结果驱动的流程管理规程可以实现管理这些流程资产的功能。

（18）可持续性和BPM。组织管理经历了一些有趣的趋势。第一是更加注重交付价值，关于它更全面的定义，不仅包括财务方面，还包括其他利益相关者的利益，这些利益相关者的利益对建设更可持续的社会施加了压力；第二是流程管理是提高组织绩效的一种方法；第三是IT作为一种资源和启动程序无处不在。从经济、生态和社会的角度来看，组织需要通过流程管理和强大的IT应用变得更加可持续，这一点是显而易见的。因此，在理想情况下，从生命周期的角度来看，一个日益增长的趋势是对特定实践进行编纂和指导，该实践通过将战略与运营联系起来来推动股东回报、生态足迹和社会影响方面的共同改进。

（19）BPM需求管理。无论是业务创新、转型还是技术开发，需求管理都是影响任何行业、任何设计的最广泛使用的概念。因此，它还影响业务流程的设计，包括功能和端到端场景（"我们的企业就是我们的流程"）。因此，它会影响组织的运作。

如今，BPM需求管理已成为任何组织的关键，严重影响其业务设计和公司业绩的质量。需求概念对任何组织的重要性在于：它是所需企业目标的关键信息载体、解释者、桥梁和翻译器，通过将高级需求分解和映射到更细化的需求网络中实现流程和技术的设计和性能，从而实现所需的企业目标。本质上，需求存在于任何组织的体系结构的每一层中，并且驱动着一个组织所做的一切。需求不是独立的实体，它们关联、分解或组合成其他类型的更细粒度的需求。需求是动态的，它们受其他需求变化的影响会发生变化，或者被业务或技术添加为新需求。因此，必须对需求进行持续管理。

BPM requirements management requires a standardized terminology, builds common understanding, and makes available the standardized and integrated BPM requirement templates, enabling users of the BPM requirements managements body of knowledge to:

a. Identify the relevant objects to which the requirements have a relationship.

b. Decompose the business, application, and technology objects into the smallest parts that can, should, and need to be modeled, and then compose the detailed requirements to the objects' entities before building them (through mapping, simulation, and scenarios).

c. Visualize requirement relations to the specific object with the requirements templates/artifacts by using the requirements maps, matrices, and models.

d. Reduce and/or enhance complexity of requirements modeling, requirements engineering, and the use of requirements within architecture when applying the decomposition and composition standards.

e. Model the relevant requirements through the objects within the enterprise layers.

f. Add value perspective to requirements management.

g. Provide structured blueprinting and implementation that has specific phases for incorporating high-level and detailed business, application and technology requirements.

20. **Optimization and Simulation**—This enables organizations to experiment with a process, quickly determine process alternatives, and identify which alternatives are likely to produce the best outcomes under certain conditions. Optimization and simulation tools are useful technologies to, in essence, support process experimentation. These tools use a more scientific approach to process design and implementation. Optimization and simulation tools for BPM use an explicit process model (that is, an imitation of a business process) and enable the user (that is, the experimenter) to experiment with the process over time. Optimization and simulation allow the experimenter, perhaps a business process analyst, to see how the process holds up over time or in response to specific events. Does it bog down? Does the process break? What might we predict based on past behavior in production? Are there enough resources to handle all the calls, loans, claims, and other demands? Should you shift resources, and are they available? In other words, optimization and simulation allow you to run the process as if it were running in the real world. However, unlike processes running in the real world, if the optimized and simulated process breaks, no one gets hurt. It's all a simulation and it can be re-optimized and rerun. Using simulation and optimization tools, the assumptions, constraints, and scenarios of a process context can be verified with more certainty before the process model is actually deployed in the real world. Clearly, a prerequisite for performing business process optimization and simulation is that you must have an explicit business process model: the "imitation" mentioned in the definition. Business process modeling is a technique to graphically express how business processes and associated strategies are interrelated. Process modeling is used to better

BPM需求管理需要标准化术语、建立共识，并提供标准化和一体化的BPM需求模板，使BPM需求管理知识体系的用户能够：

a. 识别与需求相关的对象；

b. 将业务、应用程序和技术对象分解为可以、应该和需要建模的最小部分，然后在构建它们之前（通过映射、模拟和场景）将详细需求组合到对象的实体中；

c. 使用需求图、矩阵和模型，通过需求模板/制品来可视化特定对象的需求关系；

d. 在应用分解和组合标准时，降低或增强需求建模、需求工程和架构内需求使用的复杂性；

e. 通过企业层内的对象对相关需求进行建模；

f. 增加需求管理的价值视角；

g. 提供结构化的蓝图和实施，这些蓝图和实施具有特定的阶段，以结合高水平和详细的业务、应用和技术要求。

（20）优化和模拟：这使组织能够对流程进行试验、快速确定流程备选方案，并确定在特定条件下哪些备选方案可能产生最佳结果。用于BPM的优化和模拟工具使用显式流程模型（即业务流程的"仿制品"），并使用户（即试验者）能够随着时间的推移对流程进行试验。优化和模拟允许试验者（也许是业务流程分析师）查看流程如何随着时间的推移或对特定事件的响应而保持。优化和模拟允许试验人员（可能是业务流程分析人员）查看流程如何在一段时间内保持正常或对特定事件作出响应。它会停顿吗？流程是否中断？我们可以根据过去的生产行为预测什么？是否有足够的资源来处理所有的催款、贷款、索赔和其他需求？您是否应该转移资源？它们是否可用？换句话说，优化和模拟允许您像在现实世界中一样运行流程。然而，与现实世界中运行的流程不同，如果优化和模拟的流程中断，没有人会受到伤害。这都是可以重新优化和重新运行的模拟。使用模拟和优化工具，在实际部署流程模型之前，可以更加确定地验证流程背景的假设、约束和场景。显然，执行业务流程优化和模拟的前提条件是必须有一个明确的业务流程模型：定义中提到的"仿制品"。业务流程建模是一种以图形方式表示业务流程和相关策略如何相互关联的技术。流程建模用于更好地理解和诊断业务流程，以及流程中所有参与

understand and diagnose the business process, as well as the behavior of all the participating constituents within the process. Whereas process modeling is generally a static representation of the business process under study, simulation adds a dynamic component to this model. This technology profile specifically reflects the use of optimization and simulation tools when applied to designing and improving business processes by using explicit process models. It does not cover constraint-based optimization and simulation tools that are used for digital control systems, factory scheduling, transportation route scheduling, and other operations research and decision management applications that are not centered on process models.

21. **Business Process Modeling as a Service (BaaS).** BaaS gives you the opportunity to outsource your complete BPM so you can concentrate on your core business. The BPM as a service model (BaaS) is a service-oriented solution. Explained simply, BaaS is the outer shell of infrastructure as a service, BaaS, and software as a service: for example, combining all BPM services, from process analysis to real-time enterprise management, to integrated on-demand services: (1) automation of business processes, (2) process analysis and modeling with different specifications and scenarios, (3) process automation and process simulation using IBM BPM standard, (4) BPM and real-time enterprise management using new intelligence methods based on the BPM suite and BPM standard, (5) integration of technologies, e.g., RFID, and (6) integration of mobile devices (e.g., smartphones).

22. **Cloud-Enabled BPM (CE-BPM).** BPM technologies help manage the work of a single organization or multiple organizations. Business processes are the actual work of a single organization or multiple organizations. Business processes include formally defined activities as well as informal work practices. In addition, business processes may involve human and application activities, and they may be structured or unstructured. A CE-BPM platform is a platform for managing business processes in a private or public cloud. CE-BPM is often confused with BPMaa-S and BPMPaaS, which refers to the delivery of BPM technology functionality as a service by a cloud service provider, whereas CE-BPM refers to a cloud-enabled BPM technology product. CE-BPMs are typically purchased by enterprises to run shared business process service centers in a private cloud. A vendor may use the same technology in its BPMPaaS and its CE-BPM. The only difference is in the delivery model. BPM PaaS is delivered as a service; CE-BPM is delivered as a product and then is used to provide a public or private cloud service by an ESP or an internal IT organization. ESPs use CE-BPM as the underlying application infrastructure to deliver SaaS and business process utilities in the public cloud, as well as cloud-enabled outsourcing in community clouds. Providers of BPMPaaS may use their own or a third-party CE-BPM platform. A CE-BPM exhibits cloud-enabled application platform capabilities (see Gartner Reference Architecture for Cloud-Enabled Application Platforms). A CE-BPM must include at least one of the following BPM run-time capabilities: flow management, rule management, optimization and simulation, or BAM. It may

要素的行为。虽然流程建模通常是正在研究的业务流程的静态表示,但是模拟会向该模型添加一个动态组件。此技术概要文件具体反映了在使用显式流程模型设计和改进业务流程时使用优化和模拟工具的情况。它不包括用于数字控制系统、工厂调度、运输路线调度以及其他不以流程模型为中心的运筹学和决策管理应用程序的基于约束的优化和模拟工具。

(21)业务流程建模即服务(business process modeling as a service, BaaS)。BaaS为您提供了将完整的BPM进行外包的机会,因此您可以专注于核心业务。BaaS是一种面向服务的解决方案。简单解释一下,BaaS是基础设施即服务、BaaS和软件即服务的外壳:例如,将所有BPM服务结合起来,从流程分析到实时企业管理,再到集成的按需服务:① 业务流程的自动化;② 具有不同规范和场景的流程分析和建模;③ 使用IBM BPM标准的流程自动化和流程模拟;④ 使用基于BPM套件和BPM标准的新智能方法的BPM和实时企业管理;⑤ 技术整合,如RFID;⑥ 移动设备整合(如智能手机)。

(22)应用云的BPM(CE-BPM)。BPM技术有助于管理单个组织或多个组织的工作。业务流程是单个组织或多个组织的实际工作。业务流程包括正式定义的活动以及非正式的工作实践。此外,业务流程可能涉及人工和应用程序活动,它们可能是结构化的或非结构化的。CE-BPM平台是用于管理私有云端或公共云端中的业务流程的平台。CE-BPM经常与BPMaa-S和BPMPaaS混淆,后者指的是云服务提供商将BPM技术功能作为服务提供,而CE-BPM指的是支持云端的BPM技术产品。CE-BPM通常由企业购买,以在私有云中运行共享业务流程服务中心。供应商可以在其BPMPaaS及其CE-BPM中使用相同的技术。唯一的区别在于交付模式,BPMPaaS作为服务提供;CE-BPM作为产品提供,然后用于由ESP或内部IT组织提供公共或私有云服务。ESP使用CE-BPM作为底层应用程序基础架构,在公共云中提供SaaS和业务流程实用程序,以及在社区云中实现云端外包。BPMPaaS提供商可以使用自己的或第三方CE-BPM平台。CE-BPM展示了支持云的应用程序平台功能(请参阅适用于云的应用程序平台的Gartner参考架构)。CE-BPM必须至少包含一种以下BPM运行时的功能:流管理、规则管理、优化和模拟或业务活动监控(BAM)。它可以选择性地包括各种设计时的BPM功能,如业务

optionally include a variety of design-time BPM capabilities, such as business process modeling and automated business process discovery.

23. **Agile Development for BPM**. This represents a development methodology that is a highly accelerated, incremental approach aimed at delivering high-priority, demonstrable business value. Agile development for BPM combines management disciplines as well as agile software development methods. The nature of agile BPM means process improvement or physical process implementation starts before the models are fully complete, avoiding the big design up-front problem, which delays benefits realization. Agile development for BPM methods is defined in terms of values, principles, and best practices rather than overly prescriptive plan-driven processes. Lean and agile practices of collaboration, customer focus, short cycles, and value delivery are applied to BPM suites (BPMSs) and BPM technologies, as well as the BPM (the process of process improvement) cycle. Agile BPM builds on the growing trends of social BPM and business process analysis (BPA) for the masses, both of which increase user involvement in process discovery, modeling, and implementation. Agile BPM methods attempt to establish a high level of collaboration among business process owners, architects, and the IT organization. They also attempt to flatten the project and organizational structure, often through self-organizing teams. Agile BPM methods are based on empirical process control, which accepts requirements changes and validates project direction with short, business-focused delivery cycles. Use of agile BPM is most necessary in situations requiring frequent process change, and is particularly important for continuous process improvement use scenarios.

INDUSTRY ADOPTION

Trend phase: "Initial roll out" to "a wide range of successful initiatives"	
Market penetration: Medium–Low *Benefit rating*: High	*Maturity*: Medium *Investment required*: Medium–High

Characteristic: Early publicity produces a number of success stories, often accompanied by scores of failures. Some companies take action; many do not. More instances of how the technology can benefit the enterprise start to crystallize and become more widely understood. Second- and third-generation products appear from technology providers. More enterprises fund pilots; conservative companies remain cautious.

Trends are less than 5 years from mainstream adoption. They require a medium level of investment and medium risk with the potential to deliver industry competitive advantage.

流程建模和自动业务流程发现。

（23）BPM的敏捷开发。这代表了一种开发方法，这种方法是一种高度循序渐进的方法，旨在提供高优先级、可证明的业务价值。BPM的敏捷开发结合了规则管理和软件敏捷开发方法。敏捷BPM的本质意味着在模型完全完成之前开始流程改进或物理流程实施，进而避免出现前期大的设计问题，造成延迟利益实现的后果。BPM方法的敏捷开发是根据价值观、原则和最佳实践来定义的，而不是过度的规范计划驱动流程。协作、客户关注、短周期和价值交付的精益和敏捷实践适用于BPM套件和BPM技术，以及BPM周期。敏捷BPM建立在大众社交BPM和BPA不断增长的趋势之上，这两者都增加了用户对流程发现、建模和实施的参与。敏捷BPM方法尝试在业务流程所有人、架构师和IT组织之间建立高级别的协作。他们还试图通过自组织团队来平整项目和组织结构。敏捷BPM方法基于经验流程控制，该流程控制接受需求变更并通过简短的、以业务为中心的交付周期来验证项目方向。在需要频繁更改流程的情况下，最需要使用敏捷BPM，这对于持续的流程改进使用方案尤为重要。

3.2.9　行业应用

趋势阶段："初步推出"到"广泛的成功举措"	
市场渗透：中—低	成熟度：中等
效益等级：高	所需投资：中—高

特点：伴随着几十次失败，早期的宣传产生了许多成功的案例。一些公司采取了行动，而许多公司则没有。有关技术如何使企业受益的实例开始受到肯定，并越来越被广泛理解。第二代和第三代产品来自技术提供商。更多的企业为试点提供资金，保守的公司则仍然保持谨慎。

趋势距离主流采用还不到5年。它们需要中等水平的投资和中等风险，以提供行业竞争优势。

Performance impact: Industry leaders adopt, invest, and develop industry practices to outcompete their peers. Industry practices improve their competitive parity and standardize core competitive competencies with a focus on the performance model and service model.

They are called the industry leaders because they outperform their peers with their advantage of emerging trends.

Industry Adoption of Process Trends

1. **Automated Business Process Analysis (BPA).** (BPA for the masses) This provides a simpler modeling approach tailored to business roles rather than technical roles, enabling BPA tools to become popular among businesspeople. The resulting benefits will include faster realization of the desired business performance improvements and better ability to meet time and budget targets, owing to better process understanding as well as extra insight into process impacts to avoid unpleasant surprises. Business process analysis for the masses is a developing trend toward a simpler modeling approach tailored to business roles, rather than a technical or BPA expert. It is simpler in that it uses familiar business terms, with attention to business goals and outcomes, and less inclusion of technical terms to support implementations. Because this trend will enable BPA tools to become popular among business people, Gartner referred to it as "BPA for the masses." The traditional BPA tool category has focused on the need of business architects and analysts to collaborate with others, requiring more robust methods and tooling than many business process modelers care to deal with. However, BPA for the masses will be targeted directly at business staff regardless of position or role, to provide them with easy-to-grasp insights into their own business processes. The goal is to capture the informal shadow process, concepts, and information often missing in more formal in-process modeling and user requirements definitions. BPA for the masses tooling allows for collaboration around communities of interest to develop peer interactions, knowledge exchanges, and consensus building. Harvesting information from common formats such as Microsoft's Excel, Word, PowerPoint (EVP) and Visio is a key requirement, as is the ability to communicate either with BPA traditional models or with common business formats. We expect BPA for the masses to be increasingly delivered via thin clients on-premise or the cloud through SaaS, because this will allow for communities to grow unimpeded. We see increased use of mobile technology for capturing at the source process-related information, which allows for BPA where needed.

2. **Complex-Event Processing (CEP).** CEP is the basis for many pattern-based strategies, particularly those that require continuous intelligence. When combined with BPM, CEP not only helps detect patterns, it also allows an organization to quickly act on those patterns through executable business processes. CEP is a style of computing that is implemented by event-driven, continuous intelligence systems. CEP differs from other kinds of computing in that insight is

绩效影响：行业领导者采用、投资和开发行业实践以超越同行。行业实践通过关注绩效模型和服务模型来提高其竞争均等性并标准化核心竞争能力。

他们被称为行业领导者，因为他们凭借新兴趋势超越同行。

1. 流程趋势的行业应用

（1）自动化BPA。这（面向大众的BPA）提供了一种针对业务角色而非技术角色的简单建模方法，使BPA工具在业务人员中变得流行。由此带来的好处将包括更快地实现预期的业务绩效改进和更好地满足时间和预算目标的能力，因为对流程更好地理解以及对流程影响的额外洞察将会避免令人不快的意外。面向大众的BPA是一种发展趋势，它朝着更简单的、针对业务角色而定制的建模方向发展，而不是技术或BPA专家。更简单的是，它使用熟悉的业务术语，关注业务目标和结果，很少包含技术术语来支持执行。由于这一趋势将使BPA工具在业务人员中变得流行，Gartner将其称为面向大众的BPA。传统的BPA工具类别侧重于业务架构师和分析师与其他人协作的需求，需要比许多业务流程建模师所关心的更强大的方法和工具。然而，面向大众的BPA将直接针对业务人员，无论其职位或角色如何，为他们提供易于理解的领悟自身业务流程的方法。目标是得到非正式的影子流程（shadow process）、概念和信息，这些信息常在更正式的流程内建模和用户需求定义中丢失。面向大众的BPA工具允许围绕感兴趣的团体进行协作，以开发对等交互、知识交流和共识构建。从Microsoft Excel、Word、PowerPoint和Visio等常见格式中收集信息是一项关键要求，与BPA传统模型或通用业务格式进行通信的能力也是如此。我们期望通过精简客户端内部部署或通过SaaS云端来为大众提供BPA，因为这将允许团体畅通无阻地发展。我们看到越来越多地使用移动技术来捕获源流程相关的信息，这都需要使用BPA。

（2）复杂事件处理（CEP）。CEP是许多基于模式的策略（特别是那些需要持续智能化的策略）的基础。与BPM结合使用时，CEP不仅可以帮助检测模式的类型，还可以让组织通过可执行的业务流程快速处理这些模式。CEP是一种由事件驱动的连续智能系统实现的计算方式。CEP与其他类型的计算不同之处在于：通过组合来自多个数据点（事件对象）的信息来获得洞察力。CEP系统使用算法和规则

derived by combining information from multiple data points (event objects). A CEP system uses algorithms and rules to process streams of event data that it receives from one or more sources. It generates new summary-level facts (called complex events) and puts them in context to identify threat and opportunity situations. This information is then used to guide the response in sense-and-respond business activities. CEP is event-driven because the computation is triggered by the receipt of event data. CEP systems run continuously, so they are available to act as soon as the data arrive. Data is processed immediately upon arrival. In contrast, time-driven and request-driven IT systems store the data when it arrives, and processing is triggered later by a clock (in a time-driven system) or by a request from a person or computer program (in a request-driven system). One can produce complex events in a scheduled computation (time-driven processing) or in response to an ad hoc user query or method call (request-driven processing). However, the term "CEP" is generally only applied to event-driven processing.

Here, we focus on general-purpose, reusable event-processing software platforms that are customized at development time to implement CEP applications. The core of these platforms is a software engine that runs the CEP algorithms and rules. Commercial event-processing platform products typically include development and administrative tools; other tools to implement graphical business dashboards and alert end users; and adapters for various input event data sources and output devices.

3. **Enterprise-wide Metadata Repositories**. Metadata is defined as "information that describes various facets of an information asset to improve its usability throughout its life cycle". Generally speaking, the more valuable the information asset, the more critical managing the metadata about it becomes because the contextual definition of metadata provides the understanding that unlocks the value of the data. Examples of metadata are abstracted levels of information about the characteristics of an information asset, such as its name, location, perceived importance, quality, or value to the organization, as well as its relationship to other information assets. Metadata can be stored as artifacts in metadata repositories in the form of digital data about information assets that the enterprise wants to manage. Metadata repositories are used to document and manage metadata (in terms of governance, compliance, security, and collaborative sharing) and to perform analysis (such as change impact analysis and gap analysis) using the metadata. Metadata repositories can also be used to publish reusable assets (such as application and data services) and browse metadata during life cycle activities (design, testing, release management, and so on) in the common sources of metadata, should meet enterprise-wide metadata management needs. These include several categories of metadata repositories, such as those used in support of tool suites (tool suite repositories), project-level initiatives and programs (community-based repositories), and those used to federate and consolidate metadata from multiple sources (enterprise repositories) to manage metadata in

来处理从一个或多个源接收的事件数据流。它生成新的汇总级事实(称为复杂事件)并将其置于背景中以识别威胁和机会情况。然后,此信息用于指导和响应业务活动中的反应。CEP是事件驱动的,因为计算是由事件数据的接收触发的。CEP系统持续运行,因此可以在数据到达时立即执行和处理数据。相比之下,时间驱动和请求驱动的IT系统在数据到达时存储数据,稍后由时钟(在时间驱动的系统中)或通过人/计算机程序(在请求驱动系统中)的请求触发处理。复杂事件可能产生于计划的计算(时间驱动的处理)中,或对临时用户查询的响应中,或方法调用(请求驱动的处理)中。但是,术语CEP通常仅适用于事件驱动的处理。

在这里,我们将重点放在通用的、可重用的事件处理软件平台上,这些平台在开发时定制以实现。这些平台的核心是运行CEP算法和规则的软件引擎。商业事件处理平台产品通常包括:开发和管理工具;用于实现图形业务仪表板和提醒终端用户的其他工具;适用于各种输入事件数据源和输出设备的适配器。

(3)企业范围的元数据存储库(EMR)。元数据被定义为"描述信息资产的各个方面的信息,目标是提高信息资产在整个生命周期中的可用性"。一般而言,信息资产越有价值,管理元数据的关键性就越大,因为元数据的背景对数据价值的理解提供了帮助。元数据的示例是关于信息资产特征的抽象级别的信息,如其名称、位置、感知的重要性、质量或对组织的价值,以及其与其他信息资产的关系。元数据可以以关于企业想要管理的信息资产的信息化数据的形式作为制品存储在元数据存储库中。元数据存储库用于记录和管理元数据(在治理、合规性、安全性和协作共享方面),并使用元数据执行分析(如变更影响分析和差距分析)。元数据存储库还可用于发布可重用资产(如应用程序和数据服务),并在元数据的公共源生命周期活动(设计、测试、发布管理等)中浏览元数据,同时应满足企业范围的元数据管理需求。这些包括多个元数据存储库类别,如用于支持工具套件(工具套件存储库)、项目级计划和程序(基于社区的存储库)的存储库,以及用于联合和合并多个源(企业存储库)中的元数据以在更大的企业范围内管理元数据的存储库。

a more enterprise-wide fashion. Here, we focus on the state of the repository markets—because there are now many sub-markets—in terms of this need to federate and consolidate metadata in an enterprise-wide manner. We are seeing more and more organizations, even those that already own enterprise repositories, acquiring several other best-of-breed repositories, each focused on different communities of users in projects and programs involving data warehousing, master data management, business process modeling and analysis, service-oriented architecture (SOA), and data integration - just to name a few - types of communities. In each case, these community-focused repositories have shown benefits in improved quality and productivity through an improved understanding of the artifacts, the impact queries, and the reuse of assets, such as data and process artifacts, services, and components. This has resulted in the subsetting of what once was the enterprise repository market into smaller communities of interest, using solutions that are less expensive and easier to manage. However, attempting to federate metadata across multiple repositories to provide an enterprise-wide view of metadata is no simple task, but rather a cornerstone of advanced process modeling.

4. **Process Templates.** "Process templates" is an overarching term that describes pre-built business process design, execution, and management artifacts that accelerate time to solution. They are also known by various names such as "solution frameworks," "solution templates," "solution kits," "starter kits," "process accelerators," and "process pods." Process templates should be agnostic and vendor neutral. Typically, process templates are graphical and are based on process flows, rules or SOA. The contents vary dramatically among vendors or providers. Some offer simple visual process models that are useful in jump-starting discussions about target processes for improvement. Others provide pre-built detailed process models, technical reference models, candidate service definitions, technical service libraries, rule sets, user interface templates, simulation scenarios, recommended governance policies, delivery and deployment guides, and process improvement methodologies. Some vendors sell process templates as products, whereas others treat them as software assets primarily intended for use in professional service engagements. Process templates are not intended to deliver 100% of a solution. Instead, they are meant to be changed by an implementer. A process template can be extended (that is, the implementer can add capability beyond what was provided by the original assets). It can also be adjusted or configured to accommodate the unique requirements of a process. In many cases, process templates are designed to allow business stakeholders to extend the solution, not just IT personnel.

Process templates use models to manipulate one or more aspects of the process. Some templates are broad (including activities, rules, work flows and user interfaces) and some are narrow, such as a rule set only. Nevertheless, in the BPM market, model-driven pre-built solution content is typically referred to as process templates.

在这里,我们关注存储库市场的状态,因为现在有许多子市场需要以企业范围的方式联合和整合元数据。我们看到越来越多的组织(甚至那些已经拥有企业存储库的组织)也采购了其他几个最佳存储库,每个存储库分别关注在项目和项目组中不同用户社群,这些项目或项目组包括数据仓库、主数据管理、业务流程建模和分析、面向服务的体系结构(SOA),以及数据整合中的几种社区。在每种情况下,这些以社区为中心的存储库通过改进对制品、影响查询和资产重用(如数据和流程制品、服务和组件)的理解,在提高质量和生产力方面显示出了优势。这促使曾经的企业存储库市场被细分为更小的兴趣社区,使用更便宜、更容易管理的解决方案。然而,尝试跨多个存储库联合元数据以提供企业范围的元数据视图不是一项简单的任务,而是高级流程建模的基础。

(4)流程模板。流程模板是一个总体术语,它描述了预先构建的业务流程设计、执行和管理制品,从而加快了解决方案的速度。它们也有不同的名称,如"解决方案框架""解决方案模板""解决方案工具包""初学者工具包""流程加速器"和"流程包"。流程模板应该是不可知的,并且与供应商无关。通常,流程模板是图形化的,基于流程活动、规则或SOA。不同的供应商之间的内容差异很大。有些提供了简单的可视化流程模型,这些模型在启动有关目标流程的讨论以进行改进时非常有用。其他则提供预先构建的详细流程模型、技术参考模型、候选服务定义、技术服务库、规则集、用户界面模板、模拟场景、推荐治理策略、交付和部署指南以及流程改进方法。一些供应商将流程模板作为产品销售,而其他供应商则将其视为主要用于专业服务约定的软件资产。流程模板不能提供100%的解决方案。相反,它们是由执行者进行更改的,包括:可以扩展流程模板(也就是说,执行者可以在原始资产提供的功能之外添加功能);还可以调整或配置它以适应流程的独特需求。在许多情况下,流程模板的设计允许业务涉众扩展解决方案,而不仅仅是IT人员。

流程模板使用模型来操作流程的一个或多个方面。有些模板很宽泛(包括活动、规则、工作流和UI),有些模板很狭窄,如仅限规则集。然而,在BPM市场中,模型驱动的预先构建解决方案内容通常被称为流程模板。使用流程模板,生成的应

With process templates, the resulting application is driven by the metadata reflected in the process model. This means that the application's behavior is determined by direct manipulation of the explicit process model, rather than through the setting of parameters or by writing code. Instead of parameters, which restrict application behavior to predetermined options only, a process orchestration engine reads the explicit business process model and directly executes it.

5. **BPM Change Management.** The implementation of a BPM change management program demands a whole new way of working in an organization, and also implies looking differently at your organization. This is something that many organizations underestimate. Old, existing ways of working and managing/directing people must be changed. This fact alone begs for a clear change at the management level, but it also requires change at lower organization levels. This new way of working should be accepted before working in a process-oriented manner can become successful. When organizations decide to implement process improvements and/or BPM, they must not only pay attention to the new possibilities and the factors that stimulate successful implementation; they must also be aware of the restrictions. These restrictions or barriers are often bound to the organization culture, to the comfort one obtains from holding a certain position, and to power and status. Management must deal with these barriers and actively deal with the factors that stimulate implementation as well. Clear and accurate communication is important for successful change management. This implies a need to build integrity and trust, which will have implications for the specific tactics that will be adopted in implementing the changes required. There are many tactics that can be selected from the tool kit for each area, and the actual tactics adopted will need to match the particular business, but if you have a framework from which to select, the likely success of your BPM change management project is increased.

6. **BPM Alignment.** BPM alignment focused on reusability and accelerated automation needs an understanding of what alignment is, how to develop an alignment competency, and what considerations should be made by organizations to ensure alignment is adequately adopted. Alignment of BPM provides for the policy or strategy of the organization to drive the alignment of BPM portfolios, programs, and projects that require the relevant stakeholders (business process owners) to develop a common understanding of their business process so that there is a transformation of business process from as-is through to-be. The to-be business processes that have been aligned can then be used in enterprise transformation and innovation to enable improved financial measures of performance and replication of the same success across project, portfolio, and programs. The BPM alignment objective is combined with BI MDM, SOA, and/or the cloud; the strategic value and the effect on the organizational performance are significant.

用程序由流程模型中反映的元数据驱动。这意味着应用程序的行为是通过直接操作显式流程模型而不是通过设置参数或编写代码来确定的。流程编排引擎读取显式业务流程模型并直接执行它,而不是仅将应用程序行为限制为预定选项的参数。

(5) BPM变更管理。BPM变更管理计划的实施需要一种全新的组织工作方式,同时也意味着要以不同的方式看待组织,这被许多组织所低估,我们必须改变原有的工作和管理/指导人员的方式。这一事实本身就要求在管理层进行明确的变革,但也要求在较低的组织层进行变革。在以面向流程的方式工作获得成功之前,应该接受这种新的工作方式。当组织决定实施BPM流程改进时,他们不仅必须关注新的可能性和激励成功实施的因素,还必须意识到这些限制。这些限制或障碍往往与组织文化、从担任某一职务中获得的舒适感以及权力和地位有关。管理层必须应对这些障碍,并积极应对阻碍实施的因素。清晰准确的沟通对于成功的变更管理至关重要,这意味着需要建立完整性和信任,这将对实施所需更改采用的特定策略产生影响。可以从每个领域的工具包中选择许多策略,并且所采用的实际策略将需要与特定的业务相匹配,如果您有一个可从中选择的框架,那么您的BPM变更管理项目成功的可能性将增加。

(6) BPM协调。针对可重用性和加速自动化的BPM调整需要了解什么是协调性、如何开发协调能力,以及组织应该考虑什么以确保协调性得到充分采用。BPM的协调性为组织的政策或战略提供了驱动BPM组合、计划和项目的协调性,这些项目需要相关的利益相关者(业务流程所有人)对其业务流程进行共同理解,从而实现业务流程从现有到将来的转变。然后,可以在企业转型和创新中使用调整后的未来业务流程,以改进财务指标,从而在方案、投资组合和项目中实现成功的复制。BPM协调目标与商务智能主数据管理(business intelligence master date management,BI MDM)、SOA或云端相结合,其战略价值和对组织绩效的影响是显著的。

7. **Business Process Outsourcing (BPO).** This is likely to yield high benefits to BPO providers as well as buyers. To gain maximum benefit, a BPO program should go through a formal close-down. There is no point in arguing lost causes once irrevocable decisions have been taken. Staff and companies alike need to accept the new situation and move forward. However, there will be a lot of information generated during the life of the program, and this will have been stored with varying degrees of formality by the team members. This information needs to be formally filed away for future reference. In this light, there are no simple criteria to conduct an outsourcing versus in-house analysis. The benefits associated with outsourcing are numerous, and one should consider each project on its individual merits. Ongoing operational costs that may be avoided by outsourcing are also a consideration. In a nutshell, outsourcing allows organizations to be more efficient, flexible, and effective, while often reducing costs.

8. **Evidence-Based BPM.** As organizations gain awareness of the latent business value locked in their back-end systems' data stores, evidence-based BPM will become a day-to-day management tool rather than the subject of ad hoc initiatives triggered by punctual process performance issues. This shift will lead to the emergence of evidence-based process governance frameworks, allowing managers to effectively set up and steer long-term evidence-based BPM programs that deliver measurable value via continuous process improvement. In turn, increased evidence-based BPM maturity will spawn the deployment of real-time and predictive evidence-based BPM methods that will allow process stakeholders to respond to fine-grained process performance issues as they arise or even before they arise. In other words, evidence-based BPM methods will push the boundaries of contemporary business process monitoring practices by extending them with real-time predictive analytics. Evidence-based BPM will also enable continuous process auditing, whereby compliance violations are detected on a day-to-day basis, in contrast to contemporary postmortem process auditing approaches. Combined, these developments will bring BPM to the level of modern data-driven marketing approaches. Ultimately, every business process redesign decision will be made with data, backed by data, and continuously put into question based on data.

9. **BPM Governance.** Governance in organizations is not a new trend; as a matter of fact, few industries are not demanded to prove compliance in multiple areas. Governance in terms of monitoring, evaluation, and audits are part of all organizations, daily tasks. The trend we have seen for years and now with the advanced abilities of process intelligence, evidence-based process mining, rules modeling and performance management. BPM governance become a part most organizations apply. By tackling compliance as well as continuous improvement via BPM governance, the organizations have an agile way to more easily respond to regulatory change, enable faster decision-making, and link it to the continuous improvement loop.

（7）BPO。这可能会给BPO供应商和买家带来高收益。为了获得最大的收益，BPO计划应该有正式的结束措施，一旦做出了不可挽回的决定，就没有必要争论失败的原因。员工和公司都需要接受新形势并向前发展。但是，在项目的生命周期中会产生大量的信息，并且团队成员将以不同程度的形式存储这些信息。这些信息需要正式归档以备将来参考。因此，没有简单的标准来进行外包与内部分析。与外包相关的好处是很多的，我们应该根据每个项目的优点来考虑每个项目。外包可能避免的持续运营成本也是一个考虑因素。简而言之，外包可以使组织更高效、灵活和有效，同时通常也能降低成本。

（8）基于证据的BPM。随着组织认识到其后端系统数据存储中所存在的潜在业务价值，基于证据的BPM将成为日常管理工具，而不是由于突发情况而采取的措施。这种转变导致基于证据的流程治理框架的出现，使管理者能够有效地建立和指导长期的基于证据的BPM项目，这些项目通过持续的流程改进提供可测量的价值。反过来，基于证据的BPM成熟度的提高将催生实时和预测的基于证据的BPM方法的部署，这些方法将允许流程利益相关者对出现的甚至之前出现的细粒度流程性能问题作出响应。换句话说，基于证据的BPM方法将通过使用实时预测分析来扩展它们，从而推动当代业务流程监控实践的边界。基于证据的BPM还将支持持续的流程审计，与当代的事后流程审计方法相比，通过这种方式，可以在日常基础上检测到违反法规的行为。结合起来，这些发展将使BPM达到现代数据驱动营销方法的水平。最终，每一个业务流程重新设计决策都将以数据为基础，以数据为后盾，并根据数据不断提出问题。

（9）BPM治理。组织治理并不是一种新趋势。事实上，很少有行业不需要证明在多个领域的合规性。监督、评估和审计方面的治理是所有组织日常任务的一部分。多年来，随着流程智能化、基于证据的流程挖掘、规则建模和性能管理能力的提高，这一趋势越来越显著，因此BPM治理成为大多数组织应用的一部分。通过BPM治理解决合规性和持续改进，组织拥有了一种更容易响应法规变更、支持更快决策制定并将其链接到持续改进循环的敏捷方法。

10. **BPM and Enterprise Architecture.** BPM and enterprise architecture (EA) should be an integrated part of the enterprise modeling, engineering, and architecture concepts. There are multiple benefits and different ways to combine the disciplines to create the needed business transformation and innovation, that could achieve the quality and longevity for enterprises. The key distinction for BPM as a discipline is added focus on flexible and dynamic process design and process orchestration and automation through enabling architecture. In addition to reduced costs through continued improvement and automation, BPM provides the foundation for converged and agile business and IT responsiveness and is the key to applying the principles. The success of interlinking BPM with EA derives from the proper coordination between planning and execution of the overlapping principles in the approaches. This, in turn, requires a company's understanding of EA and the process life cycles of the enterprise and the establishment of appropriate collaboration between EA and BPM governance approaches to ensure interlinking of the described approaches.

Whereas value management, BPM, and EA each have value on their own, we have described how they are naturally synergetic and work best when used together for better business performance and value outcomes and strategic alignment of business and IT. When these approaches are used together, performance drivers and operational excellence and thereby possible improvement areas are provided by the BPM context that outlines where to change the input–output model and provides an understanding of where to create the value and how and where to measure performance. Business architecture provides the design principles for solution transformation, and the rest of EA provides the discipline for translating business vision and strategy into architectural change. Although governance principles can apply the needed standards and rules, all are required for sustainable continuous improvement, optimization, and innovation. It is important to realize the value of direct collaboration across the described boundaries. Only when supported by appropriate collaboration and governance processes can BPM and EA roles work effectively together toward the common goals of the enterprise. The key to business–IT alignment and what glues it all together is the processes and activities. The notion of having business process optimization and integration of approaches has been around for a long time. Yet, around the same time that EA and governance became a mainstream topic in the context of business and IT alignment, the focus in many process optimization communities shifted subtly to BPM to go beyond an optimization approach.

11. **BPM Case Management.** In the past few years, the ECM and BPM markets have converged into a common use case called case handling, case management, or adaptive case management. The goal of case management is to make knowledge workers more productive by empowering them with control over the process outcome; providing them with full visibility and ability to manipulate all process data; and allowing them to collaborate to manage and evolve to complete each process instance. This trend is an evolution of the document-centric BPM that is motivating vendors to provide a deeper integration

（10）BPM和企业架构。BPM和企业结构（enterprise architecture，EA）应该是企业建模、工程和体系结构概念的集成部分。现实中，将各学科结合起来有多种好处和不同的方法，创造所需的业务转型和创新，从而实现企业的质量和寿命提升。BPM作为一门学科的关键区别在于，通过启用体系结构，将重点放在灵活和动态的流程设计、流程协调和自动化上。BPM除了通过持续改进和自动化来降低成本，还提供融合和敏捷业务的基础和IT响应性，是应用这些原则的关键。将BPM与EA连接起来的成功，源于方法中重叠原则的规划和执行之间的适当协调。反过来，这要求公司了解企业的EA和流程生命周期，并在EA和BPM治理方法之间建立适当的协作，以确保所述方法的相互关联。

尽管价值管理、BPM和EA各自都有自己的价值，但我们已经描述了它们如何自然地协同工作，并在一起使用时发挥最佳效果以实现更好的业务绩效和价值结果，以及业务和IT的战略协调。当这些方法一起使用时，绩效驱动因素和运营卓越性以及可能的改进领域由BPM所处背景提供，该背景概述了在何处更改输入/输出模型，并提供了对在何处创建价值以及如何/何处进行绩效度量的理解。业务体系结构提供了解决方案转换的设计原则，而EA的其余部分提供了将业务远景和策略转换为体系结构更改的规程。尽管治理原则可以应用所需的标准和规则，但所有这些都是持续改进、优化和创新所必需的。重要的是要实现跨越所述边界的直接协作的价值。只有在适当的协作和治理流程的支持下，BPM和EA角色才能有效地共同实现企业的共同目标。业务与IT协调以及将它们黏合在一起的关键是流程和活动。业务流程优化和方法集成的概念已经存在了很长时间，然而，在EA和治理成为业务和IT一致性背景下的主流话题的同时，许多流程优化社区的焦点都超越了优化方法并巧妙地转向BPM。

（11）BPM案例管理。在过去的几年中，ECM和BPM市场已经融合到一个称为案例处理、案例管理或适应性案例管理的通用用例中。案例管理的目标是通过赋予知识工作者对流程结果的控制权，使他们具有充分的可视性和操作所有流程数据的能力，并允许他们协作管理和发展以完成每个流程实例，从而提高知识工作者的工作效率。这种趋势是以文档为中心的BPM的发展，它促使供应商在ECM和BPM技术之间提供更深入的集成。供应商正在为知识工作者整合协作技术，以

between ECM and BPM technology. Vendors are incorporating collaboration technology for knowledge workers to manage the data and outcome of each process instance. The result is that BPM products are becoming more flexible, are better integrated with ECM technology, and provide better collaboration environments for knowledge workers.

STANDARD ADOPTION

Trend phase: "A wide range of successful initiatives" to "Cannot be successful without"	
Market penetration: High *Benefit rating:* Low–Medium	*Maturity:* High *Investment required:* Low

Characteristic: Mainstream adoption starts to take off. Criteria for assessing provider viability are more clearly defined and become a standard of whom to apply it to. The technology's broad market applicability and relevance are clearly paying off.

Trends are widely mainstream adopted and deliver out-of-the-box functionality.

Performance Impact: Wide adoption and years of experience have become standard or best practice. Organizations adopt to best practices to improve and standardize the non-core competencies with focus on the cost model and operating model. Organizations such as these are referred to as followers, and take advantage of best practices that are non-core to their business while gaining full advantage of trends with low risk and cost.

STANDARDS ADOPTION OF PROCESS TRENDS

1. **BPM and Operating Model.** Is an abstract representation of how an organization operates across resource, process, organization, and technology domains to accomplish its functions. This includes decision as to how a company wants to operate with regard to standardizing and integrating processes across various organizational domains (e.g. business units, geographies, product lines, franchises). It facilitates discussions as to how a company wants to pursue its business model. The purpose of an operating model is to categorize the organization into groups of how it operates, to increase understanding and suggest opportunities for improvement. In the context of BPM, an operating model can be used in various ways:
   a. A BPM operating model refers to both the level of integration and standardization of the BPM concepts, the BPM team, a shared facility, and how it operates in enabling process innovation and transformation for the organization.
   b. The BPM Center of Excellence (COE) acts both as the initiation point and the organization's custodian to the point of accountability of its processes, tasked with ensuring sustainability, maturity, governance, alignment, as well as the measurements and reporting that makes BPM successful.

管理每个流程实例的数据和结果。其结果是，BPM产品变得更加灵活，更好地与ECM技术集成，并为知识工作者提供更好的协作环境。

3.2.10　标准应用

趋势阶段："一系列成功的举措" 到 "没有了就不能成功"

市场渗透率：高	成熟度：高
效益等级：低—中	所需投资：低

特点：主流应用开始兴起。评估提供者生存能力的标准被更清楚地定义，并成为应用它的标准。这项技术的广泛市场适用性和相关性显然得到了回报。

趋势被广泛采用并提供开箱即用的功能。

性能影响：经过广泛地采用和多年的经验积累，趋势已成为标准或最佳实践。组织采用最佳实践来改进和标准化非核心能力，重点关注成本模型和操作模型。像这样的组织被称为最佳实践的追随者，他们将最佳实践用于非核心业务，以在低风险和低成本的环境下充分利用趋势。

3.2.11　流程趋势的标准应用

（1）BPM和操作模型。这是组织如何跨资源、流程、组织和技术域运行以实现其功能的抽象表示。这包括决定公司希望如何在不同组织领域（如业务部门、地理位置、产品线、特许经营）标准化和一体化流程方面进行运营。它有助于讨论一家公司如何追求其商业模式。操作模型的目的是将组织分为不同的组，对其如何操作进行分类，以增加理解并提出改进的机会。在BPM的使用中，操作模型可以以各种方式使用。

a. BPM操作模型指的是BPM概念、BPM团队、共享设施的一体化和标准化水平，以及如何运作以实现组织的流程创新和转型；

b. BPM CoE既是启动点，也是组织流程责任点的管理者，其任务是确保可持续性、成熟度、治理、协调，以及使BPM成功的度量和报告；

    c. BPM portfolio management also needs to consider the organizational operating model as well as the level of process standardization and integration across the BPM portfolio.

    d. For BPM alignment management, it is essential to define the level of process standardization and integration for any alignment initiative.

    e. The BPM teams develop process templates that need to be integrated and standardized across various operating components.

2. **Business Rule Management (BRM).** BRM guides business rule definitions, categorizations, governance, deployments, and use throughout the business life cycle. When combined with other BPMTs, BRMs simplify process change and accelerate process agility. Two level-setting definitions are required before discussing BRMs: (1) Business rules: implicit and explicit business directives that define and describe guidance for taking a business action (a decision, constraint, option, or mandate—for example, if an applicant wants more than $1 million in insurance coverage and he has high blood pressure, he will be charged at a higher rate). (2) BRM: a structured discipline guiding business rule definition, categorization, governance, deployment, and use throughout the business life cycle. BRM is defined as a comprehensive business rule offering that facilitates the creation, registration, classification, verification, deployment, and execution of business rules in the support of BRM. A BRM is the next-generation evolution of the more-mature technological foundation known as a business rule engine (BRE). A critical distinction between a traditional BRE (execution engine only) and a BRM is that the latter is much more than an execution engine and development environment. A BRM goes well beyond a BRE and broadens the historical technology ecosystem to incorporate rich support for seven key component areas: execution engine, repository, integrated development environment rule model simulation, monitoring and analysis management and administration, and rule template.

3. **BPM Center of Excellence (BPM CoE).** A BPM CoE represents an internal consultancy and promoter of BPM, including training and awareness, and offers a "one-stop shop" that provides services to multiple BPM projects, programs, and initiatives. The BPM CoE implements, chooses, and supports the guidelines, standards, and tools, and offers services that enable the enterprise to progress with and adopt BPM. A well-planned process improvement strategy includes a BPM CoE model that best fits an organization's needs as it starts up or grows its BPM program. A BPM CoE is essential for BPM to become institutionalized within an organization. The BPM CoE acts as an internal consultancy and promoter for BPM, including training and awareness, and offers a "one-stop shop" that provides services to multiple BPM projects, programs, and initiatives. The BPM CoE implements, chooses, and supports the guidelines, standards, and tools, and offers services that enable the enterprise to progress with and adopt BPM. A well-planned process improvement strategy includes a BPM CoE model that best fits an organization's needs as it starts up or grows its BPM program. A BPM CoE guides process improvement projects by applying

　　c. BPM产品组合管理还需要考虑组织操作模型以及整个BPM产品组合的流程标准化和一体化级别；

　　d. 对于BPM协调性管理，非常重要的一点是必须为任何协调计划定义流程标准化和一体化的级别；

　　e. BPM团队开发流程模板，需要在不同的操作组件之间进行一体化和标准化。

　　（2）BRM。BRM在整个业务生命周期中指导业务规则定义、分类、治理、部署和使用。当与其他业务流程建模工具（business process modeling tools，BPMTs）结合时，BRMs简化了流程更改并加速了流程敏捷性。在讨论BRMs之前，需要两个级别设置定义。① 业务规则：隐式和显式业务指令用于定义和描述采取业务操作的指导（一种决策、约束、选择或授权，例如，如果申请人想要超过100万亿美元的保险范围，并且患有高血压，他将被收取更高的费用）；② BRM：在整个业务生命周期中指导业务规则定义、分类、治理、部署和使用的结构化规则。BRM是一种全面的业务规则产品，它有助于创建、注册、分类、验证、部署和执行业务规则，以支持BRM。BRM是更成熟的技术基础，即BRE的下一代演进。传统的BRE（仅执行引擎）和BRM之间的一个关键区别是后者远不止是执行引擎和开发环境。BRM远远超出了BRE，并扩展了历史技术生态系统，以包含对七个关键组件领域的丰富支持：执行引擎、存储库、集成开发环境、规则模型模拟、监控和分析管理以及规则模板。

　　（3）BPM CoE。BPM CoE代表BPM的内部咨询和推动者，包括培训和意识，为多个BPM项目、计划和程序提供"一站式服务"。BPM CoE实施、选择和支持指导方针、标准和工具，并提供使企业能够推进并采用BPM的服务。精心策划的流程改进策略包括BPM CoE模型，该模型最适合组织启动或扩展其BPM计划的需求。BPM CoE对于BPM在组织内制度化至关重要。BPM CoE通过应用标准和经

standards and proven techniques to ensure that they deliver business value and can be leveraged for future efforts supporting business agility. It delivers a standard methodology (or methodology toolbox where multiple approaches are required), repository, and best practices for engaging in process redesign and transformation activities. These may cover the disciplines of modeling, real-time measurement, and content management rules. The center offers multidisciplined senior process improvement staff who support work ranging from consulting on small projects to turnkey program management for large and complex transformation efforts.

The implementation models for a BPM CoE will vary. The BPM CoE can be centralized or federated, and may report into the business, IT, or a blended relationship. Business process directors who understand the necessary capabilities to guide the enterprise's BPM efforts and what capabilities already exist in the organization may choose to incorporate the needed components into existing competency centers or governance groups to achieve the same outcomes. The BPM CoE is not just for large enterprises; mid-sized and small organizations can have a fully functional BPM CoE. The requirement is not staffing, but functionality. One person can manage this function and establish a BPM program. Some organizations use alternative naming for the BPM CoE: for example, "business process CoC" or "process and service improvement group."

4. **BPM Roles.** To succeed with any business process initiative today, it is crucial to understand the BPM roles, features of a role, motivation, measurements, and challenges faced in identifying and using business process-centric roles today. Clarity shed regarding the BPM roles will provide accountability. The concept of "role" is separate and distinct from the persons or things that access the rights of and perform its responsibilities. Given the rights granted in a business context and supported by the skills and knowledge needed to exercise those rights, a role can be treated and managed as a conceptual thing of significance in the design of enterprises and organizations. Therefore, a standard way of thinking, working, modeling, and governing is applied to exploration of the nature of roles to create a standardized and repeatable method for identifying, characterizing, and documenting roles and then this approach is applied to finding and describing the roles needed within the BPM COE.

5. **BPM Life Cycle.** The organizational requirements of implementing a usable and effective BPM life cycle in any organization is a demanding task in itself; even more difficult is the need to structure the life cycle in a way that fully and in a detailed and explicit manner revolves around accomplishing not only process-related goals, but more importantly, business objectives, goals, and strategy. In a nutshell, processes are essentially a sequenced flow of steps and activities that have been specifically designed to achieve a defined business objective and eventually allow for the fulfillment of a strategy on behalf of the organization. Thus, processes act as a chain reaction of actions that are indirectly responsible

过验证的技术来指导流程改进项目,以确保它们提供业务价值,并可用于支持业务敏捷性的未来工作。它提供了标准方法(或需要多种方法的方法工具箱)、存储库以及参与流程重新设计和转换活动的最佳实践。这些可能涵盖建模、实时测量和内容管理规则。该中心提供多学科的高级流程改进人员,他们支持从小型项目咨询到大型复杂转型工作的总包项目管理等各种工作。

BPM CoE的实施模型会有所不同。BPM CoE可以是集中式或联合的,可以向业务、IT或混合关系报告。了解指导企业BPM工作的必要能力以及组织中已存在的功能的业务流程主管可以选择将所需组件合并到现有的能力中心或治理组中,以实现相同的结果。BPM CoE不仅适用于大型企业,中型和小型组织也可以拥有功能齐全的BPM CoE。其要求不是人员的配置,而是功能型的。一个人可以管理此功能并建立BPM计划。一些组织使用BPM CoE的替代命名:如"业务流程CoC"或"流程和服务改进小组"。

(4)BPM角色。要在今天成功地实施任何业务流程计划,必须了解BPM角色、角色特征、动机、度量以及在当今识别和使用以业务流程为中心的角色时所面临的挑战。关于BPM角色的明确性将提供责任。角色的概念是独立的,并与获得权利并履行其职责的人或物相区别。鉴于在商业环境中授予的权利以及行使这些权利所需的技能和知识的支持,可以将角色视为在企业和组织设计中具有重要意义的概念性事物来对待和管理。因此,标准的思考、工作、建模和管理方法应用于探索角色的性质,以创建一种标准化和可重复的方法来识别、描述和记录角色,然后将这种方法应用于寻找和描述BPM CoE中所需的角色。

(5)BPM生命周期。在任何组织中实施可用且有效的BPM生命周期本身就是一项艰巨的任务,更加困难的是,需要以一种完全且详细和明确的方式构建生命周期,这种方式不仅要实现与流程相关的目标,更重要的是实现业务目的、目标和战略。简而言之,流程本质上是一个有序的步骤和活动流程,可专门用于实现既定的业务目标,并最终允许代表组织实现战略。因此,流程充当了间接负责实现组织战略的行动的链式反应,这就是最终目标。大多数BPM和流程生命周期几乎全部

for fulfilling the strategy of an organization; and ultimately, that is the goal. Most BPM and process life cycles focus almost exclusively on process-oriented solutions and goals and more or less circle around technical problems and other non–business related challenges. That is where many organizations go about it the wrong way. As mentioned earlier, processes are but a tool to fulfill the goals of the business. With that in mind, it is important to maintain a strong focus on business objectives and goals when designing the structure and the steps involved with the BPM life cycle. Process goals have to serve the needs of the business, and designing a tight collaboration between process objectives and goals and business objectives and goals is of the utmost importance.

6. **Business Process Model and Notation** (BPMN). BPMN is a standard for business process modeling that provides a graphical notation for specifying business processes in a Business Process Diagram, based on a flowcharting technique similar to activity diagrams from a unified modeling language. The objective of BPMN is to support BPM for both technical users and business users by providing a notation that is intuitive to business users, yet able to represent complex process semantics. The BPMN specification also provides mapping between the graphics of the notation and the underlying constructs of execution languages, particularly business process execution language. The primary goal of BPMN is to provide a standard notation readily understandable by all business stakeholders. These include the business analysts who create and refine the processes, the technical developers responsible for implementing them, and the business managers who monitor and manage them. Consequently, BPMN serves as a common language bridging the communication gap that frequently occurs between business process design and implementation. Currently there are several competing standards for business process modeling languages used by modeling tools and processes. Widespread adoption of the BPMN will help unify the expression of basic business process concepts (e.g., public and private processes, choreographies), as well as advanced process concepts (e.g., exception handling, transaction compensation). BPM initiative developed BPMN, which has been maintained by the OMG.

7. **BPM Continuous Improvement.** Organizations continue to invest massive amounts of money and time in improving their business processes. Why? Because of the imperative to optimize their business operations for their markets, even as their markets shift with changing customer expectations and the accelerating drumbeat of competition. The pull away from structured processes to more ad hoc and exception management, with a higher degree of process flexibility, as well as the need to support mobile solutions, are the core drivers of the current BPM landscape[10]. BPM is charging toward a market opportunity of 6.6 billion USD and will be the basis of the next generation of packaged apps. This BPM landscape report lays out the path to the future state and describes the impact of this shifting landscape on business, customers, and partners.

专注于面向流程的解决方案和目标,或多或少地围绕技术问题和其他与业务相关的挑战,这就是许多组织以错误的方式去做的事情。如前所述,流程只是实现业务目标的工具。考虑到这一点,在设计BPM生命周期所涉及的结构和步骤时,必须始终关注业务目的和目标。流程目标必须满足业务需求,并且在流程目的和目标与业务目的和目标之间设计紧密协作至关重要。

(6)BPMN。BPMN是业务流程建模的标准,它基于类似于统一建模语言的活动图的流程图技术,提供用于在业务流程图中指定业务流程的图形符号。BPMN的目的是为技术用户和业务用户提供一种对业务用户直观但能够表示复杂流程语义的符号,从而支持BPM。BPMN规范还提供符号图形与执行语言(特别是业务流程执行语言)的底层构造之间的映射。BPMN的主要目标是提供一个标准符号,供所有业务利益相关者容易理解,其中包括创建和完善流程的业务分析师、负责实施流程的技术开发人员以及监控和管理流程的业务经理。因此,BPMN作为一种通用语言,弥合了业务流程设计和执行之间经常出现的通信鸿沟。目前,建模工具和流程使用的业务流程建模语言有几个互为竞争关系的标准。广泛采用BPMN将有助于统一基本业务流程概念(如公共和私有流程、编排)以及高级流程概念(如异常处理、事务补偿)的表达。BPMN由BPM发展而来,并由对象管理组织OMG进行维护。

(7)BPM持续改进。企业继续投入大量的资金和时间来改进其业务流程。为什么?因为必须为他们的市场优化业务运营,即使他们的市场随着不断变化的客户期望和竞争的鼓点而变化。从结构化流程转向特殊和异常管理,具有更高的流程灵活性,同时支持移动解决方案的需求,是当前BPM环境[10]的核心驱动因素。BPM正朝着66亿美元的市场机遇收费,并将成为下一代打包应用程序的基础。此BPM环境报告列出了通向未来状态的路径,并描述了这种变化环境对业务、客户和合作伙伴的影响。

8. **Business Monitoring or Business Activity Monitoring (BAM)**—This describes the processes and technologies that provide real-time situation awareness, as well as access to and analysis of critical business performance indicators, based on event-driven sources of data. BAM is used to improve the speed and effectiveness of business operations by keeping track of what is happening now and raising awareness of issues as soon as they can be detected. BAM describes the processes and technologies that provide real-time situation awareness, as well as access to and analysis of critical business performance indicators, based on event-driven sources of data. BAM is used to improve the speed and effectiveness of business operations by keeping track of what is happening now and raising awareness of issues as soon as they can be detected. BAM applications may emit alerts about a business opportunity or problem, drive a dashboard with metrics or status, make use of predictive and historical information, display an event log, and offer drill-down features. Events from a BAM system may trigger another application or service, communicated via a messaging system. The processing logic of a BAM system may use query, simple-stream or CEP.

9. **Process Development.** Process deployment is where the organization launches, implements, executes, deploys, activates, completes, concludes and transitions the processes to execution (go live). The process release and deployment management in the BPM life cycle aims to plan, schedule, and control the movement of releases to test in live environments. The primary goal of release management and deployment management is to ensure that the integrity of the live environment is protected and that the correct components are released on time and without errors.

10. **Process Design and Build.** Business process design is the concept by which an organization understands and defines and designs the business activities that enable it to function. Process design is concerned with designing the processes of a business to ensure that they are optimized and effective, meet customer requirements and demands, and support and sustain organizational development and growth. A well-designed process will improve efficiency, deliver greater productivity, and create more business value. The most common initiatives behind business process design projects are:
    a. Customer and supply chain management
    b. Operational performance improvement
    c. Business process integration, standardization, and automation
    d. Cost reduction
    e. Creating new business opportunities

Business process design typically occurs as an early, critical phase in BPM projects rather than as an end in itself.

11. **Process Analysis.** Process analysis is a standard practice in the market that helps managers improve the performance of their business activities. The ultimate goal when organizations model business processes is to describe what

（8）业务监控或BAM。这描述了基于事件驱动的数据源提供实时态势感知、访问和分析关键业务性能指标的流程和技术。通过跟踪当前正在发生的事情并在发现问题后立即提高对问题的认识，BAM可用于提高业务运营的速度和效率。BAM应用程序可能发出有关业务机遇或问题的警报，使用指标或状态驱动仪表板，利用预测性和历史信息，显示事件日志，并提供向下钻取功能。来自BAM系统的事件可能触发另一个应用程序或服务，通过消息传递系统进行通信。BAM系统的处理逻辑可以使用查询、简化流或CEP。

（9）流程开发。流程部署是组织启动、实施、执行、部署、激活、完成、结束流程并将其转换为执行（上线）的环节。BPM生命周期中的流程发布和部署管理旨在计划、调度和控制发布的运转，以便在实际环境中进行测试。发布管理和部署管理的主要目标是确保实时环境的完整性得到保护，并确保正确的组件按时发布且没有错误。

（10）流程设计和构建。业务流程设计是组织理解、定义和设计使其能够运行的业务活动的概念。流程设计涉及设计业务流程以确保其优化和有效，满足客户要求和需求，以及支持和维持组织的发展和增长。一个设计良好的流程将提高效率、提高生产力并创造更多业务价值。业务流程设计项目背后最常见的举措是：

a. 客户与供应链管理；

b. 运营绩效改进；

c. 业务流程一体化、标准化和自动化；

d. 降低成本；

e. 创造新的商业机会。

业务流程设计通常在BPM项目中作为一个早期的、关键的阶段出现，而不是作为其本身的终点。

（11）流程分析。流程分析是市场上的一种标准实践，可帮助业务经理提高其业务活动的能力。当组织为业务流程建模时，最终目标是以详细的层次结构描述业务所做的工作，从高层到业务流程可见的层次。在这个内容中，流程分析是对

the business does in a hierarchy of detail from a high level down to the level where processes of the business become visible. In this content process analysis is a step-by-step breakdown of all the relevant process aspects, including the inputs, outputs, and the BPM COE operations that take place during the phase. It can be a milestone in continuous improvement. The process analysis approach consists of the following steps: (1) definition of the scope and the objectives of the study, (2) documentation of the status quo and definition of performance measures, (3) assessment and performance evaluation, and (4) development of recommendations.

CONCLUSION

In this chapter we have focused on the current and developing process trends. We have given a perspective on emerging trends, industry trends, and mega trends and have detailed and explained the trends for organizations to be able to learn from others to see what new trends are emerging and what others do successfully. Our recommendation is clear: Executives and process practitioners should make adoption decisions based on the ability to learn from others. Our experience has shown that it has a much higher benefit–risk ratio and performance curve.

End Notes

1. *The Oxford English Dictionary* (published by the Oxford University Press, 2014).
2. van Rensburg A., "A framework for business process management," *Computers & Industrial Engineering* 35, (1998): 217–220.
3. Müller T. Zukunftsthema geschäftsprozessmanagement. 'Zukunftsthema Geschäftsprozessmanagement.'pwc-PricewaterhouseCoopers. Februar 2011. PricewaterhouseCoopers AG (2012).
4. Business Process Intelligence, Daniela Grigori et al., Computers in Industry Vol. 53, Elsevier B.V., 2004.
5. LEADing Practice Driver & Forces Modelling Reference Content #LEAD-ES20001PG.
6. Everett M, *Diffusion of innovations: Technology & Engineering* (Rogers: Press of Glencoe, 1962).
7. Geoffrey A. Moore, *Crossing the Chasm* (Publisher Capstone, 2000).
8. LEADing Practice Business Process Reference Content #LEAD-ES20005BP.
9. Ibid.
10. Prepare For 2013's Shifting BPM Landscape, Craig Le Clair, Alex Cullen, Julian Keenan, Forester Media, INC., 2013.

所有相关流程方面的逐步分解，包括在阶段中发生的输入、输出和BPM CoE操作。它可以成为持续改进的里程碑。流程分析方法包括以下步骤：① 确定研究范围和目标；② 记录现状和确定绩效指标；③ 评估和绩效评估；④ 制定建议。

3.2.12　结论

在本节中，我们重点介绍了当前和发展的流程趋势。我们对新兴趋势、行业趋势和大趋势进行了展望，并详细说明了这些趋势，使组织能够向他人学习，了解正在出现的新趋势以及其他组织成功完成的工作。我们强烈建议：管理人员和流程实践者应该根据向他人学习的能力来决策采纳或不采纳。我们的经验表明，它具有更高的效益风险比和绩效曲线。

参考文献

［ 1 ］ The Oxford English Dictionary (published by the Oxford University Press, 2014).
［ 2 ］ van Rensburg A., "A framework for business process management," Computers & Industrial Engineering 35, (1998): 217−220.
［ 3 ］ Müller T. Zukunftsthema geschäftsprozessmanagement. 'Zukunftsthema Geschäftsprozessmanagement.' pwc-PricewaterhouseCoopers. February 2011. PricewaterhouseCoopers AG (2012).
［ 4 ］ Business Process Intelligence, Daniela Grigori et al., Computers in Industry Vol. 53, Elsevier B.V., 2004.
［ 5 ］ LEADing Practice Driver & Forces Modelling Reference Content #LEAD-ES20001PG.
［ 6 ］ Everett M, Diffusion of Innovations: Technology & Engineering (Rogers: Press of Glencoe, 1962).
［ 7 ］ Geoffrey A. Moore, Crossing the Chasm (Publisher Capstone, 2000).
［ 8 ］ LEADing Practice Business Process Reference Content #LEAD-ES20005BP.
［ 9 ］ Ibid.
［10］ Prepare For 2013's Shifting BPM Landscape, Craig Le Clair, Alex Cullen, Julian Keenan, Forester Media, INC., 2013.

BPM Center of Excellence

Mark von Rosing, Maria Hove, Henrik von Scheel, Rogan Morrison

INTRODUCTION

To achieve the organizational objectives of business process management (BPM), the implementation of a BPM office or center of excellence (CoE) is a key component of success in the management of business processes. BPM is not a discipline that can be done on one day and then disregarded the next; it must be considered as a consistent business stream that flows through an organization, preventing the organization and its processes from slipping back, making ongoing improvements to organizational processes, and providing the means to execute the strategic objectives of the various stakeholders. This business stream—the stewardship or management of business processes—BPM, is similar to other business disciplines that require an initiation point and a means to ensure ongoing sustainability as the organization builds its maturity. The BPM CoE acts as both the initiation point and the organization's custodian and point of accountability of its processes, tasked with ensuring sustainability, maturity, governance, alignment, as well as the measurements, reporting, templates, and artifacts that makes BPM successful.

THE CHALLENGE BPM CoE FACES

For most organizations, a number of pain points[1] exist with respect to implementing and sustaining a successful BPM CoE. As they do this, they need to be aware of what happens when a BPM CoE is not in place to guide and govern the overall BPM. For the executive, the reasons as well as the consequences of not implementing a successful and ongoing BPM within the organization are important.

- **Not managing the processes across all areas**: Within the organization, only some of the parts have completed detailed process mapping, orchestration, and optimization. This leaves process integration across the entire enterprise structure disjointed and unmanaged. A process will simply stop at the end of a department, line of business, or other business unit's logical boundary. The result is the hampering of service and information flows and duplication of roles, resulting in inefficiency.
- **No mandate to make the necessary changes, i.e., no business process charter**: The business process charter defines the mandate for initiating improvement targets.[2] It gives process owners direction and approval for their initial focus on BPM improvements.[3] These targets should tell the process owner the areas in which executives want to see improvements, e.g.,
  - Standardize the budget process, i.e., "Reduce the 15 different budget processes we currently have to one."

The Complete Business Process Handbook. http://dx.doi.org/10.1016/B978-0-12-799959-3.00012-4

3.3　BPM CoE

Mark von Rosing, Maria Hove, Henrik von Scheel, Rogan Morrison

3.3.1　介绍

组织为了实现BPM的目标,最关键组成部分是实施BPM办公或CoE。BPM不是一个可以在一天内完成然后下一天忽略的学科,它必须被视为一个贯穿组织的连续业务流,防止组织及其流程后退,持续改进组织流程,并为各利益相关者提供落实战略目标的手段。此业务流(业务流程管理)与其他业务规程类似,都需要一个起始点和一种方法,以确保在组织构建其成熟度时持续的可持续性。BPM CoE既是启动点,又是组织的托管人,也是流程的责任点,其任务是确保BPM成功的可持续性、成熟度、治理、一致性,以及度量、报告、模板和组件。

3.3.2　BPM CoE面临的挑战

对于大多数组织来说,在实施和维持一个成功的BPM CoE方面存在许多痛点[1]。当他们这样做时,他们需要知道当一个BPM CoE没能指导和管理整个BPM时会发生什么。对于业务主管来说,了解在组织内没有成功实施BPM和持续执行BPM的原因及结果都是很重要的。

- 没有跨所有领域管理流程:在组织内部,如果只是部分完成了详细的流程映射、编排和优化,这将致使整个企业组织中的流程体系脱节且无法管理。流程将停止在部门、业务线或其他业务单元的末端,其结果是服务和信息流受到阻碍、角色重复、导致效率低下。

- 没有对必要的更改进行授权,即没有业务流程章程:业务流程章程定义了启动改进目标的授权[2]。它为流程所有人提供了他们最初关注BPM改进的指导和批准[3]。这些目标应该告诉流程所有人主管希望看到改进的领域,举例如下:

  - 规范预算流程,即"将我们目前拥有的15个不同的预算流程减少到1个";

- Increase the accuracy of the data entry inputs from 65% to 95%.
- Reduce the total cycle time from ordering to delivery from 18 h to 2 h
- **No vision for the future**: BPM should never be content with always delivering the as-is processes. BPM that does not have a future vision of the ever-changing business will forever lag behind the business changes and hence deliver a much lower value than what is expected. At executive forums, the Return on Investment (ROI) will always be questioned.
- **Lack of BPM budget**: As a resulting lack of budget in the BPM CoE, only a percentage of the processes are converted from as-is to to-be. Because only processes in the to-be state are a true representation of the business, the organization is left in a state of permanent under-delivery and underperformance. The old way of doing things creeps back in and the result is a step forward but two steps backward. Confusion is created and the results for an organization are self-limiting.
- **Lack of clear rules for participation in BPM within the organization**: A clear lack of defined roles and responsibilities[4] of who should work with BPM creates divisions, and a lack of accountability and responsibility pervades the organization. This results in missed improvement targets and an organization whose business processes stagnate, adding little value even in cases where much potential exists.
- **Not enough skilled resources in the BPM CoE to fulfill the mandate**: A lack of skilled resources such as business analysts available for each project results in a distinct lack of quality and causes major delays in projects such as Enterprise Resource Planning (ERP) implementations that are reliant on business processes. This adds to project costs and untimely delays.
- **Lack of benchmarking and best practice**: No standard or point of reference guidelines have been set for business process modeling, orchestration, or optimization. This leads to significant variations in BPM quality and often results in having the record of the business processes in separate BPM repositories and tools that do not communicate with one another. This leads to a disjointed set of business processes across the entire organization.
- **BPM governance**: The cost and complexity of adhering to compliance requirements are a drain on resources and a significant pain point. In many cases, particularly where the compliance relates to maintaining BPM (e.g., in keeping process changes and the record of these changes updated), BPM CoE support is needed to satisfy these requirements effectively. The requirements are the same across the enterprises, yet each line of business or department addresses them individually, and this is expensive. Furthermore, enterprises devote resources to compliance unnecessarily. Because the requirements are so complex and their resources are limited, enterprises may not meet the requirements to the extent that they would wish.
- **BPM alignment**: Although the information technology (IT) marketplace may convey the impression that BPM is easy and simple, the reality is that BPM is not easy. The complexities of a BPM implementation have their roots in the

- 将数据输入的准确性从65%提高到95%；
- 将从订购到交付的总周期时间从18小时缩短为2小时。

- 没有对未来的愿景：BPM永远不应该满足于始终提供as-is（当前）流程。对不断变化的业务没有未来愿景的BPM将永远落后于业务变化，因此交付的价值比预期的要低得多。如果此情况发生在高管论坛上，企业的ROI将始终受到质疑。

- 缺乏BPM预算：由于BPM CoE中预算不足，只有一部分流程从as-is（当前）转换为to-be（未来）。因为只有to-be状态的流程才是业务的真实表述，否则组织将长期处于交付不足和绩效不佳的状态，旧的做事方法慢慢地又回来了，结果是前进了一步，却后退了两步。混乱由此产生，结果是组织受到自我限制。

- 组织内缺乏参与BPM的明确规则：显然，谁应该与BPM合作缺乏明确的角色和责任[4]，这就造成了部门隔阂，组织内部将普遍缺乏责任制和责任感。这会导致组织错过改进目标和业务流程停滞，因此即使在存在很大潜力的情况下也不会增加价值。

- BPM CoE中没有足够的熟练资源来完成任务：缺乏熟练资源，如每个项目的业务分析师，会导致质量明显不足，并使依赖于业务流程的企业资源规划（ERP）实施等项目出现重大延误。这增加了项目成本和延误。

- 缺乏标准和实践经验：没有为业务流程的建模、协调或优化设置标准或参考指南。这导致了BPM质量的显著变化，并且常常导致业务流程记录在不同的BPM存储库和工具中，而这些存储库和工具彼此之间没有通信。这会导致整个组织中出现一组不连贯的业务流程。

- BPM治理：遵守合规要求的成本和复杂性是对资源的消耗和一个重大的痛点。在许多情况下，特别是在与维护BPM相关的合规性时（例如，在保持流程更改和更新这些更改的记录时），需要BPM CoE支持来有效地满足这些需求。在整个企业中，需求是相同的，但是每一个业务线或部门都分别处理它们，这是昂贵的。此外，企业不必要地将资源投入到合规性上。由于要求如此复杂且资源有限，企业可能无法达到他们希望的程度。

BPM一致性：尽管IT市场可能会给人一种BPM简单的印象，但现实是BPM并不简单。BPM实现的复杂性根源于构建复合应用程序中的架构挑战（跨业务、应用程序和技术层的一致性）。这些包括UI的编制、与记录系统的集成、管理非功

architectural challenges (alignment across the layers of business, application, and technology) in building composite applications. These include orchestration of user interfaces, integration with systems of record, managing nonfunctional aspects such as transactional requirements, availability, governance, performance, and scalability, and so on.[5] BPM development is certainly not easy and therefore the alignment of its construct into an enterprise can be seen as a major pain point. In addition, the visibility of BPM solutions to end users makes this architectural alignment focus even more important.

- **Performance management**: Performance management of business processes enables executives to be informed as significant events and trends occur in their businesses. It assists managers in aligning performance goals and identifies areas of opportunity and concern related to BPM investment and results. Current approaches to BPM performance management are cumbersome and inflexible and solve only part of the problem. They do not extend from the organizational objectives down to the level of the individual manager or employee and they do not provide a link between business consulting and the systems required to monitor and manage BPM performance. As a result, employees do not work at optimal productivity, which negatively influences performance and employee satisfaction and reduces customer retention.[6]

- **Value management**: BPM value management is the pursuit and stewardship of the anticipated benefits of processes that are of worth, importance, and significance to a specific stakeholder or group of stakeholders. In many organizations no relationship between BPM and value exists. Because of this, the nature of BPM and its contribution to stakeholder value is not anticipated in the organizational strategy. This means that BPM investments that can otherwise improve the worth of the organization are isolated and have little or no influence on providing the means to achieve strategic ends. The final results is an organization that has little consideration for the way in which its information, service, and data flows operate and which people are involved in these aspects, leading to duplications of task, ineffectual data services, and poor information within reporting and measurements.

- **BPM maturity**: BPM maturity relates to the degree of formality and optimization of process management. Many organizations only adopt a single view of BPM maturity; they consider only one of the following three possible maturities:
  - Maturity in developing BPM solutions
  - Maturity in adopting BPM solutions
  - Maturity of governance through BPM solutions

For organizations that take one of these narrow views, they are left at a disadvantage in their global competitiveness. Maturity cannot be considered in only one realm; it needs to have a holistic approach, and hence the maturity against which BPM performance is measured needs the same perspective.

- **Lack of an adequate change management approach within the BPM CoE**: Managing change is tough, but part of the problem is that there is little agreement regarding what factors most influence transformation and innovation

能方面,如事务需求、可用性、治理、性能和可伸缩性等[5]。BPM开发当然不容易,因此其构造与企业的一致性可以视为一个主要的痛点。此外,BPM解决方案对最终用户的可见性使得这种体系结构一致性更加重要。

- 绩效管理:业务流程的绩效管理使业务主管能够在其业务中发生重大事件和出现趋势时得到通知。它帮助管理者调整绩效目标,并确定与BPM投资和结果相关的机会和关注领域。当前的BPM绩效管理方法既繁琐又不灵活,只能解决部分问题。它们不会从组织目标扩展到单个经理或员工的级别,也不会在业务咨询、监控和管理BPM绩效所需的系统之间提供链接。因此,员工不能以最佳生产力工作,这会对绩效和员工满意度产生负面影响,并降低客户保留率[6]。

- 价值管理:BPM价值管理是对流程预期收益的追求和管理,这些流程对特定的利益相关人员或组织具有价值和重要性。在许多组织中,BPM和价值之间不存在任何关系。因此,BPM的本质及其对利益相关者价值的贡献在组织策略中没有被预期。这意味着BPM投资可以通过其他方式提高组织的价值,但这些投资是孤立的,对提供实现战略目标的方法几乎没有影响。最终的结果是,组织很少考虑其信息、服务和数据流的操作方式,以及哪些人参与了这些方面,从而导致任务的重复、无效的数据服务以及报告和度量中的不良信息。

- BPM成熟度:BPM成熟度与流程管理的形式化和优化程度有关。许多组织只采用单一的BPM成熟度标准,它们只考虑以下三种可能的成熟度之一:
  - BPM解决方案的开发成熟度;
  - BPM解决方案的应用成熟度;
  - BPM解决方案的治理成熟度。

对于采取这些狭隘观点之一的组织来说,它们在全球竞争力方面处于劣势。成熟度不能只在一个领域中考虑,它需要有一个整体的方法,因此衡量BPM绩效所依据的成熟度需要相同的视角。

- 在BPM CoE中缺乏适当的变更管理方法:管理变更很困难,但部分问题是,对于哪些因素最能影响变革和创新举措几乎没有一致意见。如果请五

initiatives. Ask five executives to name the one factor critical to the success of these portfolios and you are likely to get five different answers.[7] What is missing, we believe, is a focus on the hard factors. These factors bear three distinct characteristics:

- Organizations are able to measure the change in direct or indirect ways
- Organizations can easily communicate their importance, both within and outside organizations
- Businesses are capable of influencing those elements quickly

Some of the hard factors that affect a transformation and innovation initiatives are:

- The time necessary to complete the initiative
- The number of people required to execute it
- The financial results that intended actions are expected to achieve

WHAT HAPPENS WITHOUT A BPM CoE?

What happens if an organization does not have a sound BPM CoE in place? Some of the symptoms attributed to organizations not properly analyzing, selecting, and managing its BPM improvements are related to the process as described in Figure 1 below:

In Figure 1, we see business events, business processes, and business decisions each as separate. Connecting them, through BPM, to a common view that is maintained and managed via BPM allows insight into the business events to be applied, through business decisions, to affect the business processes, and vice versa.

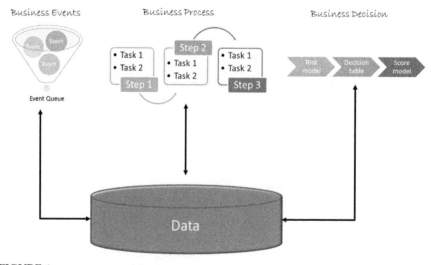

FIGURE 1

Business events driving business decisions through business process management.

位高管指出对这些投资组合成功至关重要的一个因素,您可能会得到五个不同的答案[7]。我们认为,缺少的是对困难因素的关注。这些因素具有三个显著特征:

- 组织能够以直接或间接的方式衡量变化;
- 组织可以轻松地在组织内部和外部传达其重要性;
- 企业能够快速影响这些要素。

影响变革和创新举措的一些有力因素包括:

- 完成计划所需的时间;
- 执行所需的人数;
- 预期行动将实现的财务结果。

3.3.3　没有BPM CoE会发生什么?

如果组织没有健全的 BPM CoE 会发生什么? 由组织没有正确地分析、选择和管理其BPM改进导致的一些症状与下面图1描述的流程相关。

在图1中,我们将业务事件、业务流程和业务决策视为独立的。通过BPM将它们连接到通过BPM维护和管理的公共视图,可以通过业务决策深入了解要应用的业务事件,从而影响业务流程,反之亦然。

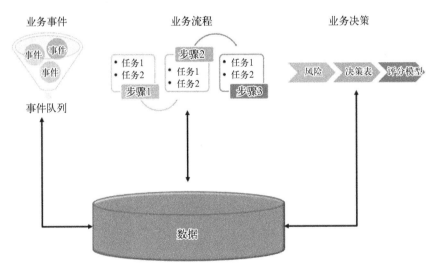

图1　通过BPM驱动业务决策的业务事件

- **Lack of business agility to react to business events**: Events range from those that take place in the course of normal operation or that may occur as planned or unplanned occurrences that drive responses and processes internal to a business and result in a business decision: for example, the introduction of mobile technology as a service. BPM applies designs to improve the detection, identification, and response to associated business events with efficiency and effectiveness.

Why it is important: The increasing speed of commerce is introducing more events per day, per minute, and per second for many businesses, increasing the backlog of challenges in the event queue. Hence, efficient detection and response to just a fraction of these events can result in millions of dollars gained or saved per year. It makes sense that businesses are focusing on improving their performance in identifying events or event patterns, as well as generating the right response, at the right time, and for the right reason.[8]

- **Inability to implement BPM technology**: BPM technologies have become cornerstone products for enterprise software vendors. BPM is solidly rooted as a core element of their product stack and continues to drive complementary software development, including B2B integration, content management, and decision‑making.[9] Lack of a BPM CoE that helps select the correct BPM technology suited to the organization and works to ensure that its implementation as part of the wider ERP implementation is effective leads to:
  - Ineffective use of organization resources
  - Inability to adjust with agility and flexibility to meet changing business requirements and customer expectations
  - Inefficient response time to internal and external events
  - Limited speed at which an organization can bring new products or services to market
  - Reduction of customer satisfaction
  - Reduction of competitive advantage and positioning

Why it is important: Consider a manufacturing organization that loses millions of dollars because of an underperforming logistics operation. When orders are unable to be shipped, the root cause may be lack of consistent and efficient processes that create challenges to compensate and adjust priorities, resources, and tasks.

High-performing organizations invest in BPM approaches and technologies to better address business challenges that cause inefficiencies, lost revenue, and lower customer satisfaction. Improving the way business processes are managed is a critical component for improving operational performance.

- **Inability to assist decision‑making**: Decision management helps organizations extract strategic decision-making logic from traditional programmed solutions into a central repository maintainable by business experts. It improves operational performance by facilitating faster time to market for decision-making changes and enabling organizations to react better to market opportunities.

- 缺乏对业务事件作出反应的业务敏捷性：事件的范围从在正常运营流程中发生的事件，或可能作为计划内或计划外发生的事件，这些事件将驱动企业内部的响应和流程，到导致业务决策：例如，将移动技术作为服务方式引入。BPM应用设计来提高对相关业务事件的检测、识别和响应的效率和有效性。

为什么这一点很重要：商业增长的速度为许多企业每天、每分钟和每秒钟都产生更多的事件，从而增加了事件队列的积压的挑战。因此，对这些事件的一小部分进行有效的检测和响应，每年可以获得或节省数百万美元。在识别事件或事件模式以及在正确的时间出于正确的原因生成正确的响应方面，企业关注于提高绩效是有意义的[8]。

- 无法实施BPM技术：BPM技术已成为企业软件供应商的基础产品。BPM作为其产品组合的核心元素根深蒂固，并继续推动互补的软件开发，包括B2B整合、内容管理和决策[9]。缺少能够帮助选择和纠正适合组织和工作的BPM技术的BPM CoE，并确保其作为更广泛ERP执行的一部分的实施将有效导致：
  - 组织资源的无效使用；
  - 无法灵活调整以满足不断变化的业务需求和客户期望；
  - 对内部和外部事件的响应时间效率低下；
  - 限制组织将新产品或服务推向市场的速度；
  - 降低客户满意度；
  - 降低竞争优势和定位。

为什么这一点很重要：如果我们考虑一个由于物流运作不佳而损失数百万美元的制造企业，订单无法发货的根本原因可能是缺乏一致和高效的流程，这些流程会给补偿、调整优先级、资源和任务带来挑战。

高绩效组织通过投资BPM方法和技术以更好地应对导致效率低下、收入损失和客户满意度降低的业务挑战。改进业务流程的管理方式是提高运营绩效的关键组成部分。

- 无法支持决策：决策管理帮助组织从传统的编程解决方案中提取战略决策逻辑，并将其放入业务专家可维护的集中存储库中。它通过促进更快地将决策变更推向市场和使组织能够更好地对市场机会作出反应来改进业务性能。

Why it is important: Staying competitive in a fast-moving market requires agility, flexibility, and precision. Decision-making has created value for organizations across many industries with diverse business challenges, and creates a decision-making platform allowing organizations to align business policies to market challenges and opportunities.

- **Cross-functional, resource, and cause and effect challenges**: This point highlights several areas in which a poor or nonexistent BPM CoE leads to business challenges in which the following symptoms exist:
  - Project and functional managers often clash over scarce BPM resources
  - Priorities of BPM initiatives frequently change, with resources constantly reassigned
  - BPM initiatives begin as soon as approved by senior managers, irrespective of the resources availability
  - Even if the strategic BPM initiative is implemented, the organization frequently does not achieve the desired improvement, because it is not measured against a benchmark or baseline.
  - There is no comprehensive document (BPM charter) that links all of the organization's BPM undertakings to the strategic plan
  - The list of BPM initiatives is not properly prioritized. Therefore, it is presumed that all ideas should be implemented simultaneously.

CAUSE AND EFFECT MATRIX

The symptoms associated with a lack of the abilities described in the previous section and not fulfilled through a BPM CoE unfortunately represent only half of the story. The additional consequences are portrayed in Figure 2. These effects include a reluctance to eliminate or align BPM initiatives and the need to engage in the maintenance of an ever-widening tunnel, rather than support a decreasing funnel of BPM pipeline portfolio process projects, which then lead to a chronic lack of resources, poor quality, missed deadlines, and inconsistent process mapping, orchestration, and optimization, and eventually to shortfalls within strategic objectives.

A logical outcome of such an approach is the apparent lack of real to-be processes that capture the essence of an organization's way of working, standard operating procedures or policies are misunderstood between departments, and alignment between other disciplines is missing.[10]

The answer to the pain points and the symptoms highlighted above is the development of a BPM CoE. The BPM CoE is discussed in some detail in the next section with reference to a portfolio management approach.

LESSONS LEARNED REGARDING BPM CoE

In the following section, our aim is to communicate some of the knowledge or understanding gained about what a BPM CoE could and should look like in the form of the specific construct, participating roles, the work, the perspective it brings to the portfolio of work, and the services delivered.

为什么这一点很重要：在快速发展的市场中保持竞争力需要敏捷性、灵活性和精确性。决策为许多面临各种业务挑战的行业的组织创造了价值，并创建了一个决策平台，使组织能够根据市场挑战和机遇调整业务政策。

- 跨职能、资源和因果挑战：这一点突出了几个领域，其中较差或不存在的 BPM CoE 导致业务挑战，其中存在以下症状。
  - 项目经理和职能经理经常因缺乏 BPM 资源而发生冲突。
  - BPM 计划的优先级经常发生变化，资源不断重新分配。
  - 无论资源可用性如何，BPM 计划一经高级经理批准就立即开始。
  - 即使实施了战略 BPM 计划，组织也经常无法实现预期的改进，因为它不是根据基准或基线来衡量的。
  - 没有将组织的所有 BPM 承诺与战略计划相关联的全面文档（BPM 章程）。
  - 未正确确定 BPM 计划的优先级。因此，我们假定所有的思想都应该同时实现。

3.3.4　鱼骨图

不幸的是，与 3.3.3 小节中描述的能力不足以及没有通过 BPM CoE 实现相关的症状仅代表了一半的情况，其他后果如图 2 所示。这些影响包括：不愿意消除或调整 BPM 计划以及需要维护不断扩大的隧道，而不是支持减少 BPM 路径组合项目漏斗，从而导致长期缺乏资源、质量差、错过最后期限、流程图、协调和优化不一致，最终导致战略目标中的不足。

这种方法的一个合乎逻辑的结果是，明显缺乏获得组织工作方式本质的实际 to-be 流程，部门之间误解了标准操作程序或政策，并且缺少其他学科之间的一致性[10]。

对上述痛点和症状的回答是开发一个 BPM CoE。3.3.5 小节将参考投资组合管理方法详细讨论 BPM CoE。

3.3.5　关于 BPM CoE 的经验教训

在下面的部分中，我们的目标是以特定的结构、参与角色、工作、它对工作组合带来的视角以及所提供的服务的形式，传达关于 BPM CoE 可以和应该是什么样子的一些知识或理解。

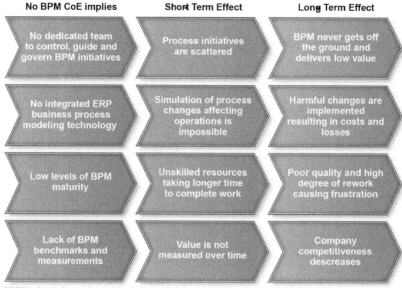

| No BPM CoE implies | Short Term Effect | Long Term Effect |

(©)LEADing Practice Business Process Reference Content [#LEAD-ES20005BP]

FIGURE 2

Short- and long-term effects of limited to no business process management (BPM) center of excellence (CoE).[11]

One of our biggest lessons learned is the importance of viewing the BPM CoE from a portfolio approach and thereby interlinking BPM alignment, BPM governance, and BPM change management: the many programs and projects the BPM CoE needs to govern and align. Such considerations will shape the specifics of the BPM CoE approach[12] and the services offered to the larger organization. On the other side, this defines what kind of roles need to be involved in the BPM CoE as well as how they will work together with multiple other teams, e.g., business architects, transformation experts, change managers, application architects. After all, the work that the BPM CoE will do or support will involve parts or full aspects of the process life cycle[13] and all its roles, tasks, and deliverables in such a manner that the lines of governance are clear and that portfolio, program, and project management team should have a clear line of alignment and integration. The clear suggestion is therefore that your BPM CoE construct include:

1. The definition of which work needs to be done and how (the work of the BPM CoE)
2. The roles needed in the BPM CoE
3. Portfolio process management
4. All execution is done through and managed within the process life cycle
5. A clear understanding and approach to BPM governance

领导实践业务流程参考内容

图2　受限于非BPM CoE的短期和长期影响[11]

　　我们学到的最大经验教训之一是,从组合方法中查看BPM CoE的重要性,从而将BPM调整、BPM治理和BPM变更管理相互关联:BPM CoE需要管理和调整的许多计划和项目。这些考虑将形成BPM CoE方法[12]的细节以及向更大的组织提供的服务。另一方面,这定义了在BPM CoE中需要涉及的角色类型,以及它们将如何与多个其他团队(如业务架构师、转换专家、变更管理者、应用程序架构师)协作。毕竟,BPM CoE将要做或支持的工作将涉及流程生命周期的部分或全部方面[13],以及流程生命周期的所有角色、任务和可交付成果,其方式应确保治理线清晰,并且组合、项目和项目管理团队应具有明确的一致性和整合线。因此,明确建议您的BPM CoE结构包括:

　　(1)需要完成哪些工作以及如何完成(BPM CoE的工作)的定义;

　　(2)BPM CoE中需要的角色;

　　(3)流程组合管理;

　　(4)所有执行都是在流程生命周期内完成和管理的;

　　(5)对BPM治理的清晰理解和方法;

6. A BPM maturity holistic view in the BPM CoE
7. BPM performance management described for BPM executed within the BPM CoE
8. BPM alignment to value so that the strategic intent is woven into the BPM CoE
9. Alignment of the BPM CoE to existing enterprise standards, enterprise architecture, and enterprise modeling and other IT disciplines
10. Enabled business decisions through evidence-based BPM
11. Continuous improvement in terms of change management in the BPM CoE

We will cover point 1 and 2 in this chapter; the rest of the points are covered in other chapters.

Work of a BPM CoE

A BPM CoE refers to a team, a shared facility, and an entity that provides leadership, evangelization, best practices, research, support, and/or training for the management of business processes. A BPM CoE is a tool that can be used to revitalize stalled BPM initiatives.[14] Its mandate is to govern, steward, and control a process life cycle across a set of roles, tasks, and activities. The BPM CoE integrates its function into the specific strategic, tactical, and operational layers within the enterprise. Furthermore, a BPM CoE will work closely with portfolios, programs, and projects, especially those that are involved with the implementation of applications and technology, and where they involve complex ERP solutions.

To position the BPM CoE services accurately, it must have the authority to oversee and consider the link between the process initiatives within its mandate of execution and all the enterprise layers, e.g., business, application, and technology. Furthermore, the scope of any initiative must be done within a portfolio approach.[15] For a portfolio approach to work and to result in the alignment of multiple initiatives necessitates an understanding of the existing and targeted process architecture. This is depicted in Figure 3.[16] For most process practitioners, introducing a portfolio approach requires an understanding of where program and project levels fit in against the process scope. It further requires they understand how the process portfolio, program, and project levels interact across the enterprise layers and against aspects of time and detail.

As with any other CoE, the BPM CoE does not simply develop on its own; it is not self-creating and it needs to be initiated, integrated, and aligned to the greater enterprise structure. This is the main failing in the current development approach taken with many BPM CoEs. In most instances, organizations attempt to take a shortcut and develop their BPM CoE by leveraging its creation on the back of the services it is to deliver, instead of initiating and sustaining it with consideration of an enterprise portfolio perspective. Taking an enterprise portfolio approach aids strategic alignment, collaboration between process and

（6）BPM CoE中的BPM成熟度整体视图；

（7）针对在BPM CoE内执行的BPM描述的BPM性能管理；

（8）BPM与价值的一致性，使战略意图融入BPM CoE；

（9）将BPM CoE与现有企业标准、企业体系结构、企业建模和其他IT规程相融合；

（10）通过基于证据的BPM实现业务决策；

（11）在BPM CoE的变更管理方面持续改进。

我们将在本节中介绍第（1）点和第（2）点；其余的要点将在其他章节中介绍。

1. BPM CoE 的工作

BPM CoE指的是一个团队、一个共享设施和一个实体，该实体为业务流程的管理提供领导、传播、最佳实践、研究、支持或培训。BPM CoE同时是一种工具，可用于恢复停滞的BPM计划[14]。它的任务是在一组角色、任务和活动中支配、管理和控制流程生命周期。BPM CoE将其功能整合到企业内的特定战略、战术和运营层。此外，BPM CoE将与产品组合、与程序和项目紧密合作，尤其是那些涉及应用程序和技术实施的项目，以及涉及复杂ERP解决方案的项目。

为了准确定位BPM CoE服务，它必须有权监督和考虑其执行任务中的流程计划与所有企业层（如业务、应用程序和技术）之间的链接。此外，任何举措的范围必须在产品组合方法中完成[15]。对于组合方法的工作并导致多个计划的一致性，需要了解现有的和有针对性的流程架构，如图3所示[16]。对于大多数流程从业者而言，引入组合方法需要了解程序和项目级别与流程范围相适应的位置。它还要求他们了解流程组合、计划和项目级别如何跨企业层以及时间和细节方面进行交互。

与任何其他CoE一样，BPM CoE不仅仅是单独开发，它不是自我创建的，它需要启动、整合并与更大的企业结构相融合。这是当前许多BPM CoE采用的开发方法中的主要失败之处。在大多数情况下，组织试图通过在其提供的服务的支持下利用其创建来获取捷径并开发其BPM CoE，而不是从企业组合的角度来启动和维持它。采用企业组合方法有助于战略调整、流程与架构工作之间的协作，以及在

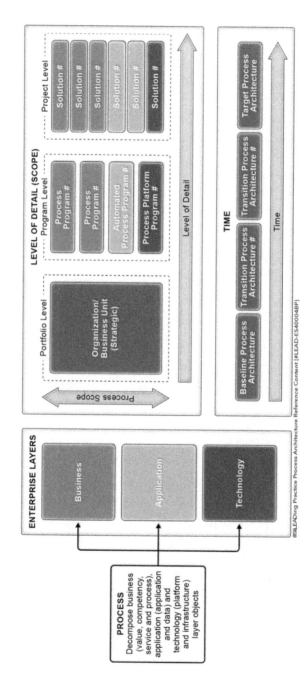

FIGURE 3

Business process management centre of excellence management of the enterprise layers process decomposition and composition.[17]

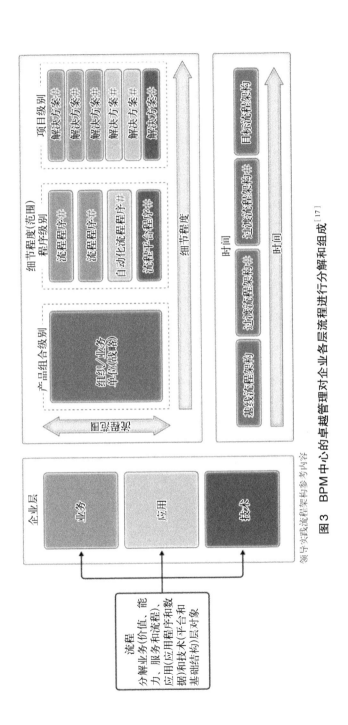

图3 BPM中心的卓越管理对企业各层流程进行分解和组成 [17]

architectural work, and the creation of a direct relationship between BPM initiatives and business innovation. Where business innovation is one of the driving forces for change and therefore identifying and creating value, the BPM CoE created in this manner offers a vastly improved way of implementing changes and sustaining results.

The enterprise portfolio approach deals with the various levels within the enterprise structure (Figure 3). The enterprise layers (business, application, and technology) have processes that require decomposition and composition. The BPM CoE is the responsible party that should take control of the process decomposition and composition.

The highest level, the portfolio, deals with the high-level scope initiatives and is able to differentiate the process into strategic aspects as well as define the process automation requirements. The program levels deal with the tactical scope initiatives of the business, application, and technology as the level of scope detail increases; furthermore, it develops the process automation within the tactical level. The third level is the project level, which deals with the operational scope initiatives and detailed process automation across the operation level. At this level, the business, application, and technology business process requirements are considered in detail.

The BPM CoE development should consider the technology landscape, which it simply has not done in the past. As is represented in Figure 3, technology process development is described over time to ensure adequate integration and alignment within the enterprise portfolio approach. The technology process architecture similarly develops over time from a baseline toward a target (to-be) process architecture through transitional phases of process architecture. This ensures that the technology process architecture can meet the application and business demands and requirements enabling business innovation.

Typical BPM CoE Roles

One of the primary areas of responsibility for the delivery element of a BPM CoE is to provide the staffing, expertise, and experience required to execute the pipeline of programs and projects within the portfolio. This requires a staffing model and a resource pool, aligned with roles in the BPM CoE.

Any staffing model for BPM projects should take into account a need for scalability to keep pace with BPM adoption across the enterprise. This scalability can be enabled by understanding the specialization of various roles involved in successfully implementing a BPM project, and the recommended level of involvement by these roles in each project. These resources are likely to start being used for multiple projects as resource demands grow for the BPM CoE.[18]

The BPM CoE roles are highlighted in Figure 4.[19,20]

Such roles within the BPM CoE ensure that there is an accountability framework for creating decisions and determining the services, architecture, standards, and policies for continuous management of business processes. Over all, they ensure

BPM计划与业务创新之间建立直接关系。如果业务创新是变革从而识别和创造价值的驱动力之一,那么以这种方式创建的BPM CoE提供了一种大大改进的实施变革和维持结果的方式。

企业组合方法处理企业结构中的各个级别(图3)。企业层(业务、应用程序和技术)具有需要分解和组合的流程。BPM CoE是控制流程分解和组成的负责方。

最高级别的产品组合处理高级别的范围方案,能够将流程区分为战略方面,并定义流程自动化需求。随着范围细节级别的增加,项目级别处理业务、应用程序和技术的战术范围计划;此外,它在战术级别开发流程自动化。第三个级别是项目级别,它处理操作范围计划和整个操作级别的详细流程自动化。在这个层次上,将详细考虑业务、应用程序和技术业务流程需求。

BPM CoE的开发应该考虑到技术环境,这在过去根本没有做过。如图3所示,技术流程开发随着时间的推移而进行描述,以确保在企业组合方法中进行充分的整合和协调。通过流程体系结构的过渡阶段,技术流程体系结构也随着时间的推移从基线发展到目标(to-be)流程体系结构。这确保了技术流程体系结构能够满足应用程序和业务需求以及支持业务创新的需求。

2. 典型的 BPM CoE 角色

BPM CoE交付要素的主要职责之一是提供执行组合中项目和项目通道所需的人员、专业知识和经验。这需要一个配置模型的人员和一个资源池,并与BPM CoE中的角色保持一致。

针对BPM项目的任何人员配置模型都应考虑到可扩展性的需求,以跟上企业对BPM的采用。这种可扩展性可以通过理解成功实现BPM项目所涉及的各种角色的专门化以及这些角色在每个项目中的推荐参与级别来实现。随着对BPM CoE资源需求的增长,这些资源可能开始用于多个项目[18]。

图4突出显示了BPM CoE角色[19,20]。

BPM CoE中的此类角色确保存在用于创建决策和确定业务流程持续管理的服务、体系结构、标准和策略的问责框架。总而言之,他们确保流程的可持续性、成

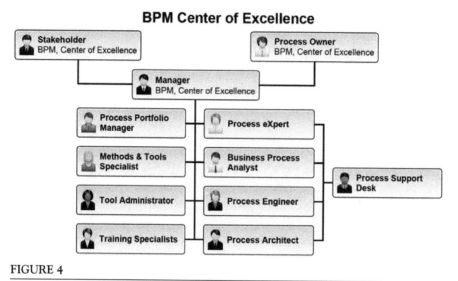

FIGURE 4

Typical bussiness process management centre of excellence roles.

the sustainability, maturity, governance, and alignment, of processes both horizontally from end to end across the enterprise, but in a manner that ensures their connection and contribution to strategy.

BPM CoE Process Life Cycle

The BPM CoE roles work in various programs and projects where the specific tasks always relate to a specific phase of the process life cycle; therefore, the roles, tasks, and deliverable part of the portfolio, program, and project can be managed within the process life cycle. This is done through managing the deliverables/artifacts assigned to the specific roles and their tasks within the process life cycle[21] (Figure 5). Because the process management life cycle has already been aligned to the project management life cycle, the key touch points can be highlighted for specific focus on governance and value delivery.

The process life cycle is controlled, directed, and governed by the BPM CoE across all areas of the business. The BPM CoE requires a specific team of skilled individuals to accomplish this.

BPM CoE Portfolio Process Management

Another key feature of the BPM CoE is the alignment of the portfolio inputs processes to outputs flow. It describes how a BPM CoE would integrate a portfolio working model into its current state of working. It helps define how the daily operation of a

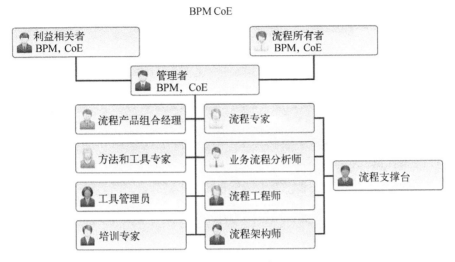

图4　典型的BPM CoE的角色体系

熟度、治理和一致性,从整个企业的端到端,但是要以一种能够确保他们与战略相关联且有所贡献的方式进行。

3. BPM CoE 流程生命周期

BPM CoE 角色在各种程序和项目中工作,其中特定任务总是与流程生命周期的特定阶段相关。因此,可以在流程生命周期内管理组合、程序和项目的角色、任务和可交付部分。这是通过管理分配给流程生命周期中特定角色及其任务的可交付结果/制品来完成的[21](图5)。流程管理生命周期已经与项目管理生命周期相协调,因此可以突出关键接触点,以具体关注治理和价值交付。

流程生命周期由业务的所有领域的BPM CoE控制、指导和管理。BPM CoE 需要一个由专门的技术人员组成的团队来完成这一任务。

4. BPM CoE 组合流程管理

BPM CoE的另一个关键特征是投资组合输入流程与输出流程的一致性。它描述了一个BPM CoE如何将一个组合产品工作模型整合到当前的工作状态中。它有助于定义BPM CoE的日常操作将如何基于输入标准运行,这将导致启动一组

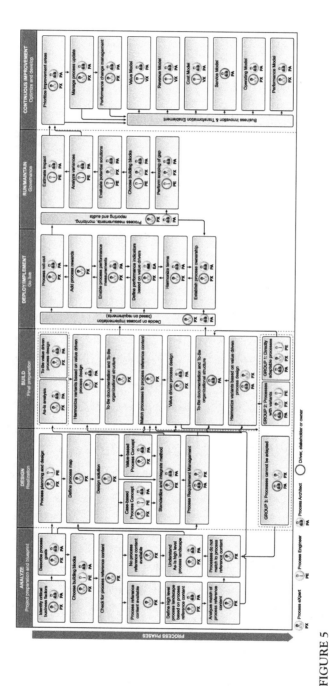

FIGURE 5

Process management life cycle.[22]

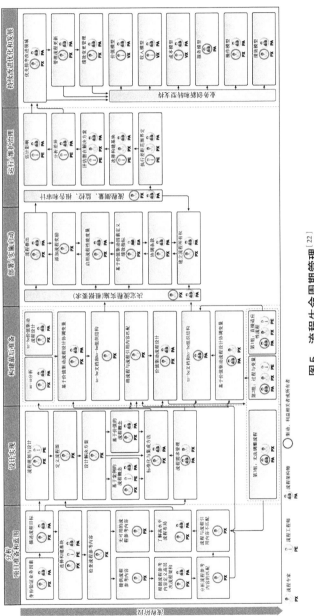

图 5 流程生命周期管理[22]

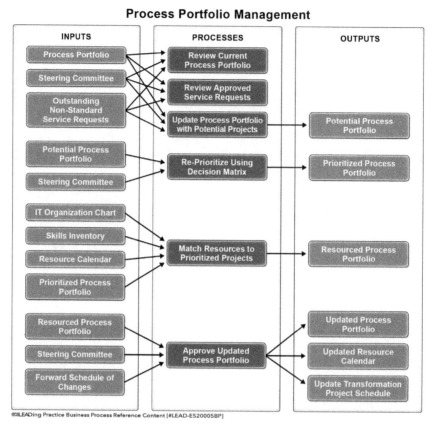

FIGURE 6

Business process management centre of excellence operational portfolio management expected inputs processes–outputs/deliverables.

BPM CoE would function based on input criteria, which would result in the initiation of a defined set of processes and respective expected outputs from these processes. This is shown in Figure 6.[23,24]

The outputs of the processes are in essence the deliverables that a BPM CoE should consider enabling through its ongoing operation.

A Clear Understanding and Approach to BPM Governance

Governance must deal with two aspects: governed guidelines, and governing guidelines. The Open Group's SOA Governance Framework[25] calls these aspects governed and governing, respectively. Any of the first type prescribes activities to be done within the core discipline (in this case, the core discipline is BPM), whereas the second type makes sure that all that needs to be done is actually completed.

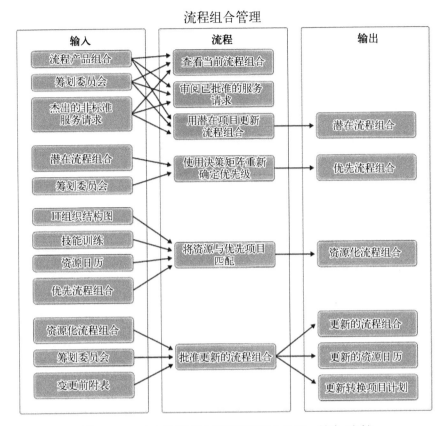

图6　BPM CoE运营组合管理预期输入流程—输出/交付

已定义的流程以及这些流程各自的预期输出,如图6所示[23,24]。

　　流程的输出本质上是一个BPM CoE应该考虑通过其正在进行的操作实现的可交付结果。

5. 对BPM治理的清晰理解和方法

　　治理必须涉及两个方面:治理指导方针和治理指南。开放组织的SOA治理框架[25]分别将这些方面称为治理和管理。第一种类型中的任何一种都规定了要在核心规程中完成的活动(在本例中,核心规程是BPM),而第二种类型确保需要完成的所有工作都是实际完成的。

Research suggests that BPM governance, by providing guidance on models, met-rics, and management accountability, is the way to reduce chances for failing large-scale BPM initiatives.[26] To achieve this the BPM CoE needs to define the approach to governance. Part of this work is to ensure that there is a good understanding across all layers of the organization of what BPM governance is. This means that the executives, managers, and individual contributors each understand their role in adhering to the BPM governance laid out by the BPM CoE.

A strong governance framework for BPM needs to be used to help guide the BPM approach. For more information, see the BPM governance chapter. A high-level overview of relevant topics is presented in Figure 7.

The alignment category (guidelines ensuring strategic and tactical alignment of BPM with enterprise business objectives, investment policies, other initiatives, and stakeholders) includes the following guidelines:

- BPM guiding principles
- High-level business process architecture
- Business process portfolio management (prioritizing BP for redesign/automation)
- BPM investment policies
  - Project planning, approval, and funding
  - BPM budget access and transparency policies

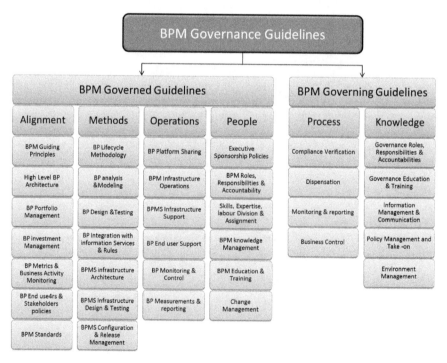

FIGURE 7

Business process management (BPM) center of excellence governance guidelines.

　　研究表明,通过对模型、指标和管理问责制提供指导,BPM治理是减少大规模BPM计划失败的有效方法[26]。为了实现这一点,BPM CoE需要定义治理方法,这项工作的一部分是确保在组织的所有层中都能很好地理解什么是BPM治理。这意味着高管、经理和个人贡献者都了解他们在遵守BPM CoE规定的BPM治理方面的作用。

　　需要使用强大的BPM治理框架来帮助指导BPM方法。有关更多信息,请参阅5.4 BPM治理一节。相关主题的高级概述如图7所示。

　　一致性类别(指导方针,确保BPM与企业业务目标、投资政策、其他计划和利益相关者的战略和战术协调)包括以下指导方针。

- BPM指导原则。
- 高级业务流程架构。
- 业务流程组合管理(BPPM,优先考虑业务流程的重新设计/自动化)。
- BPM投资政策:
  - 项目规划、批准和融资;
  - BPM预算访问和透明度政策。

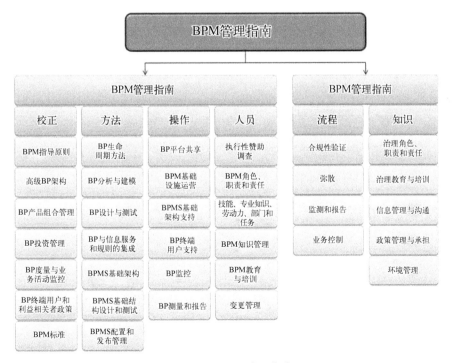

图7　BPM CoE治理指南

- Business process metrics, Key Performance Indicators (KPI's) and business activity monitoring
- Business process end user and stakeholders policies
- BPM standards

The methods category (guidelines prescribing methodologies, best practices, and standards for modeling, implementing, and commissioning business processes) includes the following guidelines:

- Business process life cycle methodology
- Business process analysis and modeling
- Business process design and testing
- Business process integration with information, services, and rules

The operations category (guidelines prescribing operational procedures and best practices for BPM, BPMS, and underlying infrastructure) includes the following guidelines:

- Business process platform sharing
- BPMS infrastructure operations
- BPMS infrastructure support
- Business process end user support
- Business process monitoring and control
- Business process measurements and reporting

The people category (guidelines for sponsorship, roles, and organizations, resources management and training, knowledge management, and communication) includes the following guidelines:

- Executive sponsorship policies
- BPM roles, responsibilities, and accountability
- Skills, expertise, labor division, and assignment
- BPM Knowledge Management
- BPM education and training
- Collaboration and communication

The chart in Figure 7, breaks BPM governing guidelines into two categories.

The processes category (guidelines prescribing the steps for governing processes) includes the following guidelines:

- Compliance verification
- Dispensation
- Monitoring and reporting
- Business control

The knowledge category (guidelines for managing and communicating content, resources, and facilities involved in the planning, execution, measurement, and analysis of the governing processes) includes the following guidelines:

- Governance roles, responsibilities, and accountability
- Governance education and training

- 业务流程指标、KPI 和业务活动监控。
- 业务流程最终用户和利益相关者政策。
- BPM 标准。

方法类别（用于建模、实施和调试业务流程的指导方针、最佳实践和标准）包括以下指导方针：

- 业务流程生命周期方法；
- 业务流程分析和建模；
- 业务流程设计和测试；
- 业务流程与信息、服务和规则的整合。

运营类别（规定 BPM、BPMS 和基础设施的运营程序和最佳实践的指南）包括以下指南：

- 业务流程共享平台；
- BPMS 基础设施运营；
- BPMS 基础设施支持；
- 业务流程最终用户支持；
- 业务流程监控；
- 业务流程度量和报告。

人员类别（赞助、角色和组织、资源管理和培训、知识管理和沟通指南）包括以下指南：

- 高管赞助政策；
- BPM 角色、职责和责任；
- 技能、专业知识、分工和分配；
- BPM 知识管理；
- BPM 教育和培训；
- 协作和沟通。

图 7 中的图表将 BPM 管理指南分为两类。

流程类别（规定流程管理步骤的指南）包括以下指南：

- 合规性验证；
- 分配；
- 监控和报告；
- 业务控制。

知识类别（管理和交流管理流程规划、执行、测量和分析中涉及的内容、资源和设施的指南）包括以下指南：

- Information management and communication
- Policy management and take-on
- Environment management

A BPM Maturity Holistic View in BPM CoE

As discussed in the BPM maturity chapter, the term "maturity" relates to the degree of formality and ripeness of practices within a specific filed, from ad hoc practices to formally defined steps, managed result metrics, active optimization, and improvement of the processes. Maturity models have the aim to improve existing practices from and can be viewed as a set of structured levels that describe how well the behaviors and practices of an organization can reliably and sustainably produce required outcomes. A maturity model can be described as a structured collection of elements that describe certain aspects of maturity in an organization. A maturity model in the context of a BPM CoE may, for example:

- Provide a quick scan of the status quo
- Identify where there is low maturity and therefore a place to start
- Provide a way to define what improvement the BPM CoE will focus on
- Provide an approach for prioritizing actions
- Provide the benefit of a community's prior experiences
- Provide a common language and a shared vision

A maturity model can be used as a benchmark for comparison and as an aid to understanding. In the BPM maturity chapter, we elaborated on the need to view the maturity of the process in its context of:

- The purpose and goal of the process, including aspects such as link to strategic business objectives, goals, critical success factors, value drivers, performance indicators, requirements, etc.
- The relationship to the organizational aspects such as business areas and groups, business competencies, business capabilities, etc.
- Which business, information and / or data objects are involved in terms of input, output, storage, etc.
- Its flows and interlink to services flow, information flow, and even value flow.
- The owners involved, which can be the process owners, but also business area owners, services owners, or data owners
- The various roles involved, which span from traditional process roles to application roles and service roles
- The various rules and compliance aspects involved, e.g., business rules, process rules, service rules, and security and data rules that influence and direct the various aspects of the process
- Automation in terms of application modules involved, and application tasks as well as application services
- Of the various measurements in terms of KPIs, Process Performance Indicators (PPI's), and Service Performance Indicators (SPI's) that are part of reporting and real-time decision-making

- 治理角色、责任和问责；
- 治理教育和培训；
- 信息管理和沟通；
- 政策管理和承担；
- 环境管理。

6. BPM CoE 中的 BPM 成熟度整体视图

正如在BPM成熟度一节中所讨论的，术语成熟度与特定领域内实践的正式程度和成熟度有关，从临时的实践到正式定义的步骤、管理结果度量、主动优化和流程改进。成熟度模型的目标是从现有的实践中改进，并且可以看作一组结构化的层次，描述组织的行为和实践如何能够可靠和可持续地产生所需的结果。成熟度模型可以叙述为描述组织中成熟度的某些方面的元素的结构化集合。在BPM CoE环境中的成熟度模型可以，例如：

- 快速浏览现状；
- 确定哪些地方成熟度较低，因此有一个开始的地方；
- 提供一种定义BPM CoE将关注哪些改进的方法；
- 提供优先行动的方法；
- 提供社区以往经验的好处；
- 提供共同语言和共同愿景。

成熟度模型可以作为比较的基准，也可以作为理解的辅助。在BPM成熟度一节中，我们阐述了在以下情况下看待流程成熟度的必要性：

- 流程的目的和目标，包括与战略业务目的、目标、关键成功因素、价值驱动因素、绩效指标、要求等的链接；
- 与组织方面的关系，如业务领域和集团、业务能力、业务潜在能力等；
- 在输入、输出、存储等方面涉及哪些业务、信息和/或数据对象；
- 服务流、信息流甚至价值流的流动和互联；
- 所涉及的所有者，可以是流程所有人，也可以是业务领域所有者、服务所有者或数据所有者；
- 涉及的各种角色，从传统流程角色到应用程序角色和服务角色；
- 涉及的各种规则和合规性方面，如业务规则、流程规则、服务规则以及影响和指导流程各个方面的安全和数据规则；
- 自动化所涉及的应用程序模块、应用程序任务以及应用程序服务；
- 在KPI、PPI和SPI方面的各种度量中，这些指标是报告和实时决策的一部分；

- Data aspects involved, including the data components, data entities, data objects, and possibly the data services and data tables
- Which media and channels are involved
- Technological aspects from platforms aspects such as platform services and platform devices (mobile devices) to infrastructure aspects
- Services that are created within the process as well as the service flow that goes across the various processes

In a BPM CoE context, however, the challenge with existing BPM maturity models is that the BPM CoE maturity is not really considered through the multiple perspectives in which it need to be considered. A BPM CoE maturity model needs to covers at least the following perspectives:

1. The maturity of developing the BPM standards
2. The BPM services offered
3. The BPM governance set in place
4. Instituting BPM security
5. Organizing change management around BPM initiatives
6. Ensuring latest BPM development is aligned
7. Executing BPM transformation
8. Adopting BPM solutions

An overview of the additional BPM CoE maturity aspects is illustrated in Figure 8.

BPM Performance Management Is Executed Within the BPM CoE

For BPM adoption, governance, and related value aspects to be successfully implemented, ongoing BPM performance management must be included in the overall concept to be managed by the BPM CoE for performance management to provide a complete solution designed to assist the business in identifying the appropriate metrics to maximize value for the business, translate them into objectives for managers and employees throughout the organization, and to provide employees at all levels of the organization the capabilities and feedback they need to optimize their performance.

Implementing such a BPM performance management methodology can provide significant bottom line business benefits, and provide growth in annual savings. Benefits include:

- Increasing employee productivity
- Increasing customer satisfaction and retention
- Increasing revenue
- Reducing employee attrition and costs associated with it
- Reducing performance management workload
- Reducing hiring and training costs
- Reducing learning curve for new hires
- Reducing IT workload

- 涉及的数据方面,包括数据组件、数据实体、数据对象,以及可能的数据服务和数据表;
- 涉及哪些媒体和渠道;
- 从平台服务和平台设备(移动设备)等平台方面到基础设施方面的技术方面;
- 在流程中创建的服务,以及贯穿各个流程的服务流。

然而,在一个BPM CoE环境中,现有的BPM成熟度模型面临的挑战是,没有真正通过考虑多个视角来考虑BPM CoE成熟度。BPM CoE成熟度模型至少需要涵盖以下方面:

(1)制定BPM标准的成熟度;

(2)提供的BPM服务;

(3)BPM治理已经就位;

(4)建立BPM安全性;

(5)围绕BPM计划组织变更管理;

(6)确保最新的BPM发展一致;

(7)执行BPM转换;

(8)采用BPM解决方案。

附加的BPM CoE成熟度方面的概述如图8所示。

7. 在BPM CoE中执行BPM绩效管理

对于要成功实施的BPM采用、治理和相关价值方面,必须将持续进行的BPM绩效管理纳入由绩效管理的BPM CoE管理的总体概念中,以提供一个完整的解决方案,旨在帮助业务部门确定适当的指标,以最大化业务部门的价值。将其转化为整个组织的经理和员工的目标,并为组织的各级员工提供优化绩效所需的能力和反馈。

实施这样的BPM绩效管理方法可以提供显著的业务收益,并提供每年节省的增长。好处包括:

- 提高员工生产力;
- 提高客户满意度和保持率;
- 增加收入;
- 减少员工流失和与之相关的成本;
- 减少绩效管理工作量;
- 降低招聘和培训成本;
- 降低新员工的学习曲线;
- 减少IT工作量。

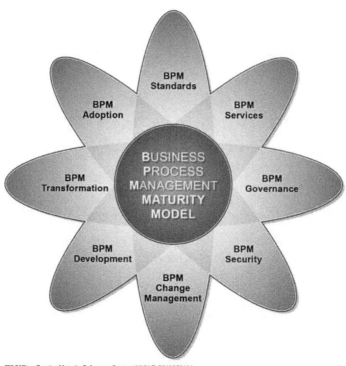

IIXILEADing Practice Maturity Reference Content [#LEAD-ES60003AL]

FIGURE 8

Considering the additional business process management (BPM) centre of excellence maturity aspects.[27]

Furthermore, such a performance management system can be used within the BPM CoE to measure and optimize any business process, including the performance management process itself. This can result in a self-perpetuating virtual cycle of continued performance optimization, leading to exponential improvement over time.

BPM Alignment to Value So That the Strategic Intent Is Woven Into the BPM CoE

As already discussed in the value-driven process design chapter, one of the most challenging aspects in business is to get the agreement of value aspects from a group of people (employees, managers, and executives) who have come from different backgrounds and cultures and have different competing forces and drivers. Added to this mix is the requirement to align the BPM to the defined value within the organization so that the strategic intent of the business operations is captured as a

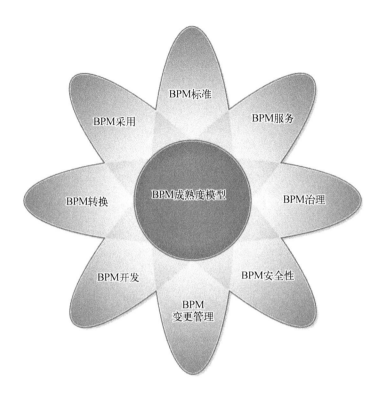

图8 考虑附加的BPM CoE成熟度方面[27]

此外，这样的绩效管理系统可以在BPM CoE中用于度量和优化包括绩效管理流程本身的任何业务流程，这会导致持续性能优化的自始至终的虚拟循环，并随着时间的推移导致指数级的改进。

8. BPM与价值的一致性，使战略意图融入BPM CoE

正如价值驱动流程设计一节中已经讨论过的，商业中最具挑战性的方面之一就是从一组来自不同背景和文化、具有不同竞争力量和驱动力的人（员工、经理和高管）那里获得价值方面的一致性。此外，还需要将BPM与组织内的定义值保持一致，以便将业务运营的战略意图捕获为一个BPM CoE可以将其整合的公共基

common thread which a BPM CoE can integrate. The capturing of value aspects is seen as vital as it sets the direction for the various BPM CoE projects and the process execution. Contrary to ordinary BPM approaches, BPM alignment understands that an organization can simultaneously pursue multiple strategies and goals and needs to align the underlying execution. For example, an organization can pursue both high growth and profits by defining unique critical success factors that break the conventional value–cost tradeoff by simultaneously pursuing both differentiation in the market and low cost in its operations. Although these would be two different projects, they can have lots of common steps, tasks, and maybe even artifacts. This is where BPM alignment to value is relevant; it captures value elements and aligns them and the underlying BPM initiatives.

Alignment of BPM CoE to Existing Enterprise Standards, Enterprise Architecture, Enterprise Modeling, and Other IT Disciplines

A BPM CoE cannot and should not exist in isolation within any organization; this principle drives the alignment of BPM to enterprise standards, enterprise architecture, enterprise modeling, and other disciplines. Furthermore, process is one of the four key pillars of enterprise management and is therefore a fundamental integration stream indicated in Figure 9.

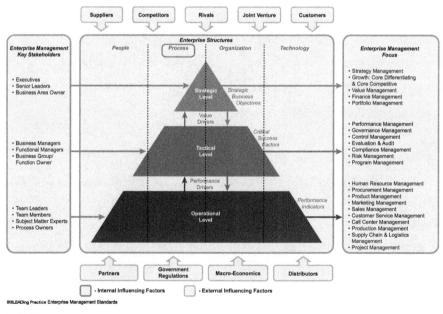

FIGURE 9

Business process management integration across the enterprise through process.[28]

础。价值方面的捕获被认为是至关重要的,因为它为各种BPM CoE项目和流程执行设定了方向。与常规的BPM方法相反,BPM协调理解组织可以同时追求多个策略和目标,并且需要协调底层的执行。例如,一个组织可以通过定义独特的关键成功因素来追求高增长和利润,在打破传统的价值–成本权衡的同时追求市场差异化和低成本运营。尽管这将是两个不同的项目,但它们可以有许多常见的步骤、任务,甚至可能是制品。这就是BPM与价值的一致性相关的地方:它捕获价值元素,并将它们与基础的BPM计划相协调。

9. 将BPM CoE与现有的企业标准、企业体系结构、企业建模和其他IT规程进行结合

BPM CoE不能也不应该单独存在于任何组织中。这一原则推动了BPM与企业标准、企业架构、企业建模和其他规程的结合。此外,流程是企业管理的四个关键支柱之一,因此图9所示的是基本集成流。

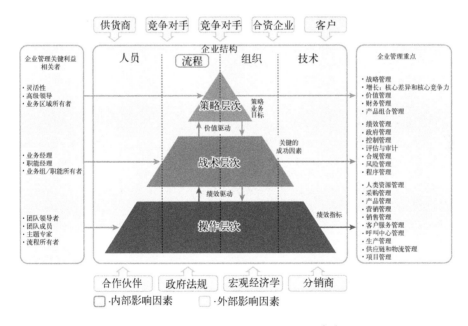

图9　BPM通过流程整合到整个企业[28]

The BPM CoE should thus have an influence across all major levels in the organization. The CoE should be an integral part of an alignment management framework that the business sets up, ensuring that participation in daily operational activities as well as in the portfolio, program, and project management delivery is attained, sustained, and optimized.

Continuous Improvement and BPM CoE Change Management

A challenge faced by both newly formed and well-matured BPM CoEs is the continuous expectations to make ongoing improvement in response to changing needs, wants, performance, and/or value concerns/expectations.[29] Change is not something that could or might happen to the modern enterprise; it is happening. It is a fact of any BPM initiative or BPM portfolio, and thus any BPM CoE. The issue of change therefore requires considerable consideration.

Change management is a process and the use of tools and techniques to manage the people side of change processes to achieve the required outcomes, and to realize the change effectively within the individual change agent, the inner team, and the wider system[30]: in, this case, the BPM CoE and the complete enterprise organization. Change scope encompasses the addition, modification, and removal of anything that could have an effect on the BPM portfolio, BPM programs, BPM projects, BPM services, or specific processes within the various organizational flows. Consequently, for any BPM CoE, BPM change management is a critical and core competency that must be the center around which the changes will be happening. In the modern BPM CoE organization, we believe that they will realize that this competency becomes a core differentiator and a link to stakeholder value and performance expectations.

BPM change management has linkages to many other key BPM services such as changes to the process itself, process architectures, procedures, systems, process measurements, performance metrics, process monitoring, process reporting, process compliance, and process documentation, as well as changes to BPM services, BPM projects, and other items. The way in which this is accomplished is by building change interventions at regular intervals throughout the process life cycle and then using these change interventions to ensure communication/feedback to relevant stakeholders (Figure 10). The feedback loop occurs only during the running/maintaining

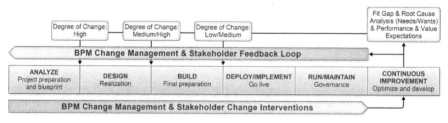

FIGURE 10

Process life cycle with business process management (BPM) change management interventions and change feedback loop.[31]

因此,BPM CoE应该在组织中的所有主要级别上具有影响力。CoE应是业务建立的协调管理框架的一个组成部分,因此,要确保CoE参与日常运营活动以及组合产品、计划和项目管理交付,并得到持续和优化。

10. 持续改进和BPM CoE变更管理

新形成的和成熟的BPM CoE都面临着一个挑战,那就是对不断变化的需要、需求、绩效或价值关注/期望做出持续的改进[29]。变更不是现代企业可以或可能发生的事情,而是正在发生的,这是任何BPM计划或BPM组合的一个事实。因此,任何BPM CoE都是如此,变更的问题需要仔细考虑。

变更管理是一个流程,使用工具和技术来管理变更流程中的人员以实现所需的结果,并在单个变更代理、内部团队和更广泛的系统中有效地实现变更[30]:在这种情况下,是BPM CoE和完整的企业组织。变更范围包括添加、修改和删除任何可能对BPM组合、BPM程序、BPM项目、BPM服务或各种组织流程中的特定流程产生影响的内容。因此,对于任何BPM CoE,BPM变更管理都是一项关键的核心能力,必须是变更发生的中心。在现代BPM CoE组织中,我们相信他们会意识到这种能力成为核心差异化因素,并与利益相关者的价值和绩效期望挂钩。

BPM变更管理与许多其他关键的BPM服务有联系,如流程本身的变更、流程架构、流程、系统、流程度量、绩效度量、流程监控、流程报告、流程合规性和流程文档,以及对BPM服务、BPM项目和其他项目的变更。实现这一目标的方法是在整个流程生命周期中定期建立变更干预措施,然后使用这些变更干预措施确保与相关利益相关者的沟通/反馈(图10)。反馈循环仅在运行/维护和持续改进阶段发生,它不可能发生在之前的阶段,因为它们已经作为项目/投资组合的一部分被关

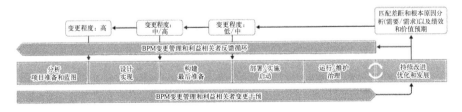

图10 流程生命周期与BPM变更管理干预和变更反馈循环[31]

and continuous improvement phases. It cannot occur during the previous phases because they have been closed as part of the project/portfolio.

- Degree of change—low: changes that can be achieved with low work amounts are referred to by different names in the various BPM CoEs: fast changes, quick changes, or even standard changes. All have one thing in common: they have a low degree of change and are for the most part preauthorized because they have low risk, are relatively common, and follow a known procedure or work flow.
- Degree of change—medium: Changes that can be achieved with medium work amount are for the most part referred to as normal change requests.
- Degree of change—high: Changes that include a high degree of change can also have different names. Some of the names are big change, strategic change, major incident change, and emergency change. All not only have a high degree of change in common, they also have a certain relationship to performance and value creation, and therefore must be implemented as soon as possible.

Change interventions do not just occur in the process life cycle; they have to be planned into specific points in the process phases. The degree of change determines the amount of change intervention required. This is detailed in Figure 11, which defines what tools should be used against what degree of change and the degree of active involvement required by the stakeholders in adopting the change.

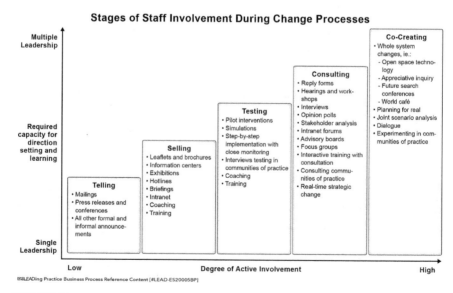

FIGURE 11

Stages stakeholder involvement.[32]

闭了。

- 变更程度——低：在不同的BPM CoE中，可以通过不同的名称来引用低工作量实现的变化：快速变化、快速变化，甚至是标准变化。它们都有一个共同点：它们的变化程度很低，而且大部分都是预先授权的，因为它们的风险很低，比较常见，并且遵循已知的程序或工作流程。
- 变更程度——中：可通过中等工作量实现的变更是指最常见的变更请求。
- 变更程度——高：包含高度变化的变更也可以有不同的名称，一些名称是重大变化、战略变化、重大事件变化和紧急变化。所有这些都不仅具有高度共同的变化，它们还与绩效和价值创造有一定的关系，因此必须尽快实施。

变更干预不仅发生在流程生命周期中，还必须在流程阶段的特定点进行规划。变更程度决定了所需的变更干预量。这在图11中有详细说明，它定义了什么样的工具应该用来衡量变更的程度，以及利益相关者在实施变更时所需要的积极参与程度，以及应该使用哪些工具。

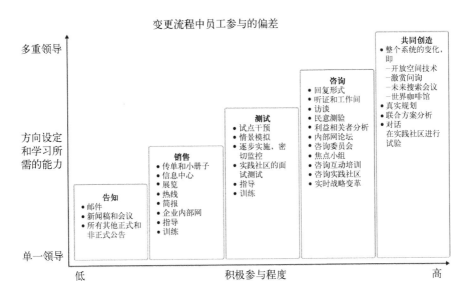

图11　利益相关者参与的阶段[32]

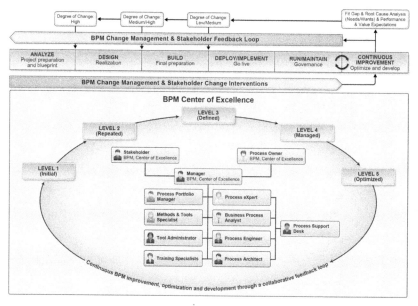

FIGURE 12

Consolidated business process management (BPM) center of excellence model.[33]

CONCLUSION

The key to successful BPM is a BPM CoE.[34] The BPM CoE helps establish and improve BPM maturity and ensures a consistent and cost-effective way of offering BPM services. This chapter elaborated on the need to manage BPM initiatives with allocated roles and responsibilities through a BPM CoE, managing the portfolio, programs, and projects through their specific process life cycle and the changes accruing. In this chapter we therefore illustrated the key aspects of a BPM CoE, as shown in Figure 12: the life cycles, roles, and process portfolio, program, projects, and changes that occur as a part of continuous improvement of the organization. This is combined with a maturity view on BPM through five levels, all of which need to be balanced together to enable value identification, creation, and realization.

End Notes

1. Pain points are areas that organizations can make improvements to.
2. Sandy Kemsley, *Business Process Discovery* (2011), www.tibco.com.
3. Gabriel Kemeny and Micheal Reame, *Creating a Charter for Your Process Improvement Project* (2010), http://www.processgps.com/.
4. Dan Morris, *7 Key Roles You Need in Your BPM Center of Excellence* (2013), http://www.processexcellencenetwork.com/.
5. S. Simmons and D. Wakeman, "BPM Voices: Don't forget the basics of software design when architecting BPM solutions." *IBM Business Process Management Journal* (2014).

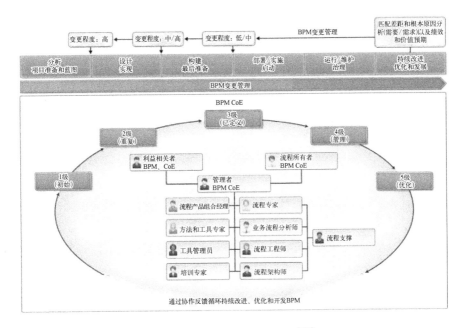

图12　综合BPM CoE模型[33]

3.3.6　结论

成功的BPM的关键是BPM CoE[34]。BPM CoE有助于建立和改进BPM成熟度,并确保提供BPM服务的一致性和经济高效的方式。本节阐述了通过一个BPM CoE来管理具有分配角色和职责的BPM计划的必要性,通过特定的流程生命周期和累积的变更来管理产品组合、程序和项目。因此,在本节中,我们说明了BPM CoE的关键方面,如图12所示:作为组织持续改进的一部分而发生的生命周期、角色和流程组合、计划、项目和变更。这是通过五个级别与BPM的成熟度视图结合在一起,所有这些都需要平衡在一起,以实现价值识别、创建和实现。

参考文献

[1] Pain points are areas that organizations can make improvements to.

[2] Sandy Kemsley, Business Process Discovery (2011), www.tibco.com.

[3] Gabriel Kemeny and Micheal Reame, Creating a Charter for Your Process Improvement Project (2010), http://www.processgps.com/.

[4] Dan Morris, 7 Key Roles You Need in Your BPM Center of Excellence (2013), http://www.processgps.com/.

[5] S. Simmons and D. Wakeman, "BPM Voices: Don't forget the basics of software design when architecting BPM solutions." IBM Business Process Management Journal (2014).

6. K. Forbes, *Driving Business Value Through Performance Management* (2002) http://www.crmxchange.com/whitepapers/pdf/bluepumpkin-WP_PerformanceManagement.pdf.

7. Harold L. Sirkin, P. Keenan, and A. Jackson, "The hard side of change management." *Harvard Business Review* (2005).

8. Dyer L. and Ericksen J, "Complexity-based agile enterprises: putting self-organizing emergence to work", in *The Sage Handbook of Human Resource Management*, ed. A. Wilkinson et al (London, 2009).

9. Unlocking the Potential of Business Process Management", *iGrafx* (2013), www.igrafx.com.

10. LEADing Practice Business Process Reference Content #LEAD-ES20005BP.

11. Portfolio Management Reference Content LEAD-ES10019AL.

12. Portfolio Management Reference Content LEAD-ES10019AL.

13. Business Process Reference Content LEAD-ES20005BP.

14. M.O. George, *The lean six sigma guide to doing more with less* (John Wiley and Sons, 2010):261, ISBN: 9780470539576.

15. Taken from the Portfolio Management Reference Content LEAD-ES10019AL.

16. LEADing Practice Process Architecture Reference Content LEAD-ES40004BP.

17. LEADing Practice Layered Enterprise Architecture Reference Content #LEAD-ES40001AL.

18. Lisa Dyer, Andrew Forget, Fahad Osmani, Jonas Zahn, *Creating a BPM Center of Excellence (CoE) IBM Redbooks* (2013), www.ibm/redbooks.

19. LEADing Practice Business Process Reference Content LEAD-ES20005BP.

20. LEADing Practice Process Architecture Reference Content LEAD-ES40004BP.

21. LEADing Practice Business Process Reference Content LEAD-ES20005BP.

22. LEADing Practice Business Process Reference Content #LEAD-ES20005BP.

23. LEADing Practice Business Process Reference Content #LEAD-ES20005BP.

24. LEADing Practice Portfolio Management Reference Content #LEAD-ES10019AL.

25. S.O.A. Governance Framework The Open Group (2009), http://www.opengroup.org/pubs/catalog/c093.htm.

26. A. Spanyi, BPM Governance. BPMInstitute.org (2008), http://www.bpminstitute.org.

27. LEADing Practice Maturity Reference Content #LEAD-ES60003AL.

28. LEADing Practice Enterprise Management Reference Content [#LEAD-ES10EMAS].

29. A part of the Change Management Reference Content (LEAD-ES60002AL).

30. H. Nauheimer, *The Change management tool book. Creative Commons Attribution-Share-Alike 3.0 Unported License.*

31. LEADing Practice Business Process Reference Content #LEAD-ES20005BP.

32. LEADing Practice Business Process Reference Content #LEAD-ES20005BP.

33. Asim Akram, *BPM Center of Excellence* (2013) www.bpminstitute.org.

34. LEADing Practice Business Process Reference Content #LEAD-ES20005BP.

［6］ K. Forbes, Driving Business Value Through Performance Management (2002) http://www.crmxchange.com/whitepapers/pdf/bluepumpkin-WP_PerformanceManagement.pdf.

［7］ Harold L. Sirkin, P. Keenan, and A. Jackson, "The hard side of change management." Harvard Business Review (2005).

［8］ Dyer L. and Ericksen J, "Complexity-based agile enterprises: putting self-organizing emergence to work", in The Sage Handbook of Human Resource Management, ed. A. Wilkinson et al (London, 2009).

［9］ Unlocking the Potential of Business Process Management", iGrafx (2013), www.igrafx.com.

［10］ LEADing Practice Business Process Reference Content #LEAD-ES20005BP.

［11］ Portfolio Management Reference Content LEAD-ES10019AL.

［12］ Portfolio Management Reference Content LEAD-ES10019AL.

［13］ Business Process Reference Content LEAD-ES20005BP.

［14］ M.O. George, The lean six sigma guide to doing more with less (John Wiley and Sons, 2010):261, ISBN: 9780470539576.

［15］ Taken from the Portfolio Management Reference Content LEAD-ES10019AL.

［16］ LEADing Practice Process Architecture Reference Content LEAD-ES40004BP.

［17］ LEADing Practice Layered Enterprise Architecture Reference Content #LEAD-ES40001AL.

［18］ Lisa Dyer, Andrew Forget, Fahad Osmani, Jonas Zahn, Creating a BPM Center of Excellence (CoE) IBM Redbooks (2013), www.ibm/redbooks.

［19］ LEADing Practice Business Process Reference Content LEAD-ES20005BP.

［20］ LEADing Practice Process Architecture Reference Content LEAD-ES40004BP.

［21］ LEADing Practice Business Process Reference Content LEAD-ES20005BP.

［22］ LEADing Practice Business Process Reference Content #LEAD-ES20005BP.

［23］ LEADing Practice Business Process Reference Content #LEAD-ES20005BP.

［24］ LEADing Practice Portfolio Management Reference Content #LEAD-ES10019AL.

［25］ S.O.A. Governance Framework The Open Group (2009), http://www.opengroup.org/pubs/catalog/c093.htm.

［26］ A. Spanyi, BPM Governance. BPMInstitute.org (2008), http://www.bpminstitute.org.

［27］ LEADing Practice Maturity Reference Content #LEAD-ES60003AL.

［28］ LEADing Practice Enterprise Management Reference Content［#LEAD-ES10EMAS］.

［29］ A part of the Change Management Reference Content (LEAD-ES60002AL).

［30］ H. Nauheimer, The Change management tool book. Creative Commons Attribution-ShareAlike 3.0 Unported License.

［31］ LEADing Practice Business Process Reference Content #LEAD-ES20005BP.

［32］ LEADing Practice Business Process Reference Content #LEAD-ES20005BP.

［33］ Asim Akram, BPM Center of Excellence (2013) www.bpminstitute.org.

［34］ LEADing Practice Business Process Reference Content #LEAD-ES20005BP.

Understanding Business Process Management Roles

Mark von Rosing, Neil Kemp, Maxim Arzumanyan

INTRODUCTION

In this chapter we will explore what a role is, discuss the features of a role, and seek to understand the motivation and challenges we face in identifying and using business process-centric roles today. This chapter also provides a summary of the profile of the factors or features needed to flesh out or give specific form to the roles in a business process management (BPM) center of excellence.

MOTIVATION FOR DEFINING YOUR BPM ROLES

Role modeling, role engineering, and various terms such as "rights," "privileges," and permissions"[1] may be used to communicate the idea of a part in which someone or something engages or does in a particular defined function, activity, or situation.

Successful BPM requires a structured and repeatable method of exploring, finding, specifying, and/organizing roles in both the organizational structure and its programs and projects. It also requires use of a complete set of roles that have stewardship, including, for example, the roles within a BPM center of excellence (CoE) (see chapter on BPM CoE).

Business processes and therefore BPM are concerned with three general types of things: process objects, consisting of the processes, gates, and events that are performed in a sequence; business objects that are consumed, transformed, and created by the process objects; and rules about participation in the work and ensuring that the work is done in accordance with its requirements (see chapter on BPM requirements management). Of the three, the controls on participation lead to the need to explore roles and is perhaps the least understood. It presents a great challenge to BPM today.

Within existing business process thought, there is a certain amount of confusion as to the nature of roles. Interestingly, for example, although the business process modeling notation (BPMN) standard, the reference standard for describing processes, refers to the term "role," it offers no specification or description as to what it means. It therefore is useful to examine the idea of "role" and its related concepts closely to tease apart the ideas and exploit the value and insight we can gain as a result. As we do this, it is important to keep some points in mind.

1. A repeatable process is needed, such that two analysts given the same problem are likely to come up with similar results: i.e., when determining the roles required to execute a particular design, they will be able to find a set of roles with similar properties or characteristics. This is a part of alignment and standardization.

The Complete Business Process Handbook. http://dx.doi.org/10.1016/B978-0-12-799959-3.00013-6

3.4　BPM相关角色认知

Mark von Rosing, Neil Kemp, Maxim Arzumanyan

3.4.1　介绍

在本节中,我们将探讨角色是什么、讨论角色的特征,并试图了解我们在识别和使用业务流程中心角色时所面临的动机和挑战。本节还概要介绍充实BPM CoE中的角色或为其提供特定形式所需的因素或特性。

3.4.2　定义BPM角色的动机

角色建模、角色工程和各种术语(如权利、特权和权限[1])可用于传达某人或某物参与或执行特定定义的功能、活动或情况的部分的概念。

成功的BPM需要一种结构化和可重复的方法来探索、查找、指定和安排组织结构中的及其程序和项目中的角色。它还需要使用一套完整的具有管理作用的角色,如在BPM CoE内的角色(请参阅有关BPM CoE的章节)。

因此,业务流程和BPM涉及三种一般类型的事物:流程对象,由按顺序执行的流程、接口和事件组成;流程对象消耗、转换和创建的业务对象;以及有关参与工作并确保按照其要求完成工作的规则(参见BPM需求管理章节)。在这三者中,对参与的控制导致了探索角色的需要,并且可能是最不被理解的。它对当今的BPM提出了巨大的挑战。

在现有的业务流程思想中,对于角色的性质存在一定程度的混淆。有趣的是,尽管描述流程的参考标准BPMN标准引用了术语(如角色),但它没有提供关于其含义的规范或描述。因此,仔细研究角色的概念及其相关概念有助于区分这些概念,并利用我们因此获得的价值和见解。当我们这样做的时候,记住一些要点是很重要的。

(1)需要一个可重复的流程,这样两个分析人员在同一个问题下可能会得出类似的结果:即当确定执行特定设计所需的角色时,他们将能够找到一组具有类似属性或特征的角色。这是校准和标准化的一部分。

2. We want to provide a method to separate the logical intent of the work being described from the constraints and realities of execution. Treating roles as logical constructs and position and actors as being purely physical in nature allows the subject matter expert to gather business requirements about the rules and constraints of behavior and therefore focus on the needs of the business, not the issues related to what the business has in terms of capabilities. This separation between logical and physical is critical because it give us the means to create forward-looking (to-be, should-be, and might-be) designs without being constrained by current restrictions, thinking, or organizational structure. Such structured thinking is part of role engineering and we believe that role modeling could greatly benefit from the insights gained through such an approach.

3. It is important to simplify the problems related to understanding work through the elimination of concerns about organizational aspects of where the work is done. This allows the analyst to concentrate on the nature of the work being explored without being distracted by whatever tradeoffs are part of the current organization design.

4. In the early stages of analysis, we do not want to spend time on capacity issues, i.e., the number or deployment of the actual resources that will be required to perform the work, when the specific nature of the work itself is unclear. Our requirement is to first address "what" and then address "how."

For these reasons, it is essential to develop a strong understanding of roles that reflects the value being sought by the strategy, which can in turn be communicated within an enterprise to address the above points.

RELEVANCE CONTEXT

Today's frequent practice for identifying the roles required to develop new processes designs or to otherwise transform or innovate the business into some future way of operations is to use the current roles, or worse, the current positions, or worse yet, the current people within the projected new processes. The result of designing new processes with incomplete insight into the logic of the work needed to perform the process really means that the results are literally compromised from the start.

In a Forrester Research survey of over 100 business process and application professionals, 86% of participants said that poor support for cross-functional processes is a significant or very significant problem. In addition, almost 70% reported that the lack of insight into process results is a significant or very significant problem for enterprise applications. The resolution of the methods to identify and allocate roles is one key to resolving this challenge.

WHAT IS A ROLE?

Roles and role relationships are the building blocks of any enterprise. The idea of a role is a business concept that can exist entirely without reference to technology, although it can be captured and enabled through automation.

（2）我们希望提供一种方法，将所描述的工作的逻辑意图与执行的约束和现实分开。将角色视为逻辑结构和位置，并将参与者视为纯粹的物理性质，这使得论题专家能够收集有关行为规则和约束的业务需求，从而专注于业务需求，而不是与业务能力相关的问题。逻辑和物理之间的这种分离是至关重要的，因为它给了我们创建前瞻性（to-be、should-be 和 might-be）设计的方法，而不受当前限制，以及思维或组织结构的约束。这种结构化思维是角色工程的一部分，我们相信角色建模可以从通过这种方法获得的领悟中受益匪浅。

（3）通过消除对组织完成工作方面的担忧，简化与理解工作相关的问题非常重要。这使得分析师可以专注于正在探索的工作的性质，而不会被当前组织设计的任何权衡分散注意力。

（4）在分析的早期阶段，当工作本身的具体性质不明确时，我们不希望将时间花在能力问题（即执行工作所需的实际资源的数量或部署）上，我们的要求是先解决 "what"，然后解决 "how"。

出于这些原因，为解决上述问题，有必要对反映战略所寻求的价值的角色进行深入了解，而战略又可以在企业内部进行沟通。

3.4.3 相关背景

今天，通常的做法是确定开发新流程设计所需的角色，或以其他方式将业务转换或创新为未来的运营方式，即使用当前的角色，或者更糟的是，使用计划新流程中的当前人员。设计新流程的结果对执行流程所需工作的逻辑不完全了解，这实际上意味着结果从一开始就受到严重损害。

在佛罗斯特研究公司（是一家独立的技术和市场调研公司）对100多名业务流程和应用程序专业人员进行的调查中，86%的参与者表示，对跨职能流程的支持不足是一个重要或非常重要的问题。此外，近70%的人报告说，缺乏对流程结果的洞察是企业应用程序的一个重要或非常重要的问题。识别和分配角色的方法的解决是破解这一挑战的关键之一。

3.4.4 什么是角色？

角色和角色关系是任何企业的构建基石。角色的概念是一个商业概念，完全可以不参考技术而存在，尽管可以通过自动化捕获和启用它。

The methods used to identify, define, and deploy these business components is of interest for executives or managers concerned with creating an organization capable of operating in the manner they desire. In particular, they are used to assigning process owners to each process across the enterprise and within the business process hierarchy.

Roles are often thought of as "assigned positions in the organization, each with specific accountabilities and authorities,"[2] or the set of activities that someone may perform to complete a process.[3]

Instead of treating the ideas of role, position (job title), and actor as equivalent or synonymous, we must, as The Open Group Architecture Framework (TOGAF) suggests, consider them to exist as distinct though related things,[4] or similarly, as LEAD states, consider that a role is "a part that someone or something has in a particular defined function, activity, or situation. A resource/actor may have a number of roles."[5]

We already know that good design practice tells us to separate the logical intent of a design from its physical description, whether designing software, a building, or an organization.[6] Because of this, these latter definitions are more appealing. Making a distinction between the concept of "role" versus that of "resource" or "actor" is therefore appealing. It meets our requirements and is therefore of more use to us than cases in which the differences are not as explicit.

If we limit the idea of a role to being **a set of rights granted in a business context and supported by the skill and knowledge needed to exercise those rights**, then within the design process we do not need to worry about anything more than what are the rights, skills, and knowledge. Later, in a separate and distinct design process, we can work out how to transition from the current or as-is rights to the target or to-be rights, which is an important aspect of business transformation. Alternatively, if the required skills and knowledge do not exist, gaps between what is required for the roles to be performed and what is required of available actors may be solved by training, or recruiting the resources, or through acquiring a going concern that has what is necessary.

A further problem and limitation in current thinking about roles is that many organizations do not appreciate that, as illustrated in Figure 1, business processes and their end-to-end flows cross multiple organizational business units, and therefore likely necessitate the involvement of multiple roles. These processes and the roles that interact with them must therefore be connected such that the various control types, properties, tasks, and objects in which a role is involved are each engaged in a sound manner.

Business processes not only run across multiple business units; when examined end-to-end, we see that as they flow they interact with multiple roles within the organization. For these reasons, the roles and how they are assembled in the organization to deliver the strategic intent change across each flow and are influenced by this context. Any exploration of role context must therefore include developing an understanding of the business units in which a role exists, but also the:

- Business areas: The highest-level meaningful grouping of the activities of the enterprise
- Business groups: An aggregation within an enterprise that is within a business area

用于识别、定义和部署这些业务组件的方法对于负责创建一个能够以他们希望的方式运行的组织的主管或经理来说是很重要的。特别是它们用于将流程所有人分配给整个企业和业务流程层次结构中的每个流程。

角色通常被认为是"在组织中分配的职位，每个职位都有特定的职责和权限"[2]，或者是某人完成一个流程可能执行的一组活动[3]。

我们不能将角色、职位（职务）和行动者的想法视为等价或同义的，而必须按照TOGAF的建议，将它们视为虽然相关或者类似但却截然不同的事物[4]，作为领导实践，将其视为"某人或某物在特定定义的功能、活动或情况中所具有的一部分。一个资源/参与者可能有许多角色[5]。"

我们已经知道，好的设计实践告诉我们将设计的逻辑意图与其物理描述分开，无论是设计软件、建筑还是组织[6]。正因为如此，后一种定义更具吸引力。因此，区分角色和资源或参与者的概念是很有吸引力的。它满足我们的要求，因此比差异不那么明显的情况更有用。

如果我们将角色的概念限制为在业务环境中授予的一组权利，并由行使这些权利所需的技能和知识支持，那么在设计流程中，我们不需要担心的不仅仅是什么是权利、技能和知识，而是任何事情都不需要担心。稍后，在一个单独的、独特的设计流程中，我们可以解决如何从当前或as-is的权利过渡到目标或to-be权利，这是业务转型的一个重要方面。或者，如果不存在所需的技能和知识，则可以通过培训、招募资源或通过获得具有所需内容的持续经营来解决执行角色所需的内容与现有行动者所需内容之间的差距。

当前对角色的另一个问题和限制是：如图1所示，许多组织不喜欢业务流程及其端到端流程跨越多个组织业务单元，因此可能需要多个角色的参与。因此，必须将这些流程及其相互作用的角色连接起来，以便将涉及角色的各种控制类型、属性、任务和对象都以良好的方式进行操作。

业务流程不仅跨多个业务单元运行。当检查端到端时，我们发现，当它们流动时，它们与组织中的多个角色交互。基于这些原因，组织中的角色及其组合方式将在每个流程中实现战略意图变更，并受此背景的影响。因此，对角色背景的任何探索都必须包括对存在角色的业务单位的理解，以及以下方面。

- 业务领域：企业活动的最高级别、有意义的分组；
- 业务组：业务范围内企业内的聚合；

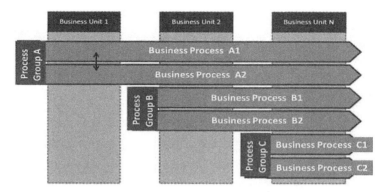

FIGURE 1

How business processes and end-to-end flows cross multiple organizational business units.

- Business functions: A cluster of tasks creating a specific class of jobs
- Service areas: A high-level, conceptual aggregation of provided services
- Service groups: An aggregation of services based on a common factor or domain that exist within a common service area.
- Business services: The externally visible ("logical") deed or effort performed to satisfy a need or to fulfill a demand, meaningful to the environment

The need to quantify the business process with all its role types and the related properties, tasks, etc., requires application of the principle of the separation of duties (The separation by sharing of more than one individual in one single task is an internal control intended to prevent fraud and error). This must be done without regard to the design or form of the organizational construct or the specific areas and groups in which the process activities take place. The result is that the analyst is better able to identify duplication of duties, business functions, tasks, services, and their associated resources and actors, and for employees, the actors in the enterprise that actually do the work, better see themselves in the creation of value. This is a significant point because when workers cannot see themselves in the work or do not understand how they add value, all sorts of problems may arise:

- If they are unable to see how their work directly contributes to a specific outcome that the customer values, they may be poorly motivated
- They do not have enough authority to do a good job
- A confusing process with multiple handoffs, many approvals, and rework caused by people not having complete and accurate information from an earlier step can lead to limited cooperation
- Limited cooperation can come from attitude again, or functional silos that are vying with one another

On the other hand, if they see a strong connection between the employee, and their work, the role they perform in carrying out this work makes sense to them.

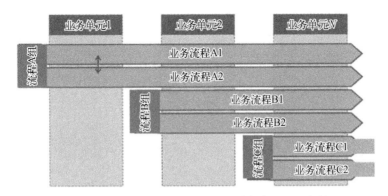

图1　业务流程和端到端流程如何跨越多个组织业务单元

- 业务功能：创建特定工作类别的一组任务；
- 服务领域：提供高层次、概念性的服务聚合；
- 服务组：基于公共服务区域内存在的公共因素或领域的服务聚合；
- 商业服务：为满足需求或满足对环境有意义的需求而进行的外部可见（逻辑）行为或努力。

出于对业务流程及其所有角色类型和相关属性、任务等进行量化的需要，需要应用职责分离原则（在一项任务中分享多个个人的做法是一种旨在防止欺诈和错误的内部控制）。这必须在不考虑组织结构的设计、形式或流程活动发生的特定区域和组别的情况下进行。结果是，分析师能够更好地识别重复的职责、业务功能、任务、服务及其相关资源和参与者，而对于员工、企业中实际完成工作的参与者，他们更能看到自己在创造价值。这一点很重要，因为当工人看不到自己在工作中的地位或不了解他们如何增加价值时，可能会出现各种各样的问题。

- 如果他们看不到自己的工作是如何直接为客户所重视的特定结果做出贡献的，那么他们可能缺乏动力；
- 他们没有足够的权力做好工作；
- 由于员工缺乏从早期步骤中获得的完整和准确的信息而导致多次移交、多个批准和返工的混乱流程，进而可能导致合作受到限制；
- 态度或相互竞争的职能性孤岛可能再次导致合作受限。

另一方面，如果他们看到员工和他们的工作之间有很强的联系，那么他们在执行这项工作中所扮演的角色对他们来说是有意义的。

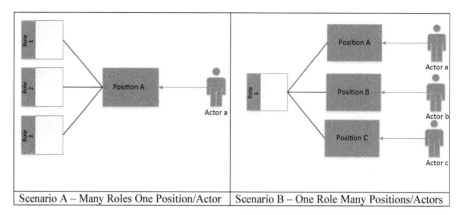

FIGURE 2

Scenarios relating roles to other components of the organization.

Similarly, during transformative change, it is always important that the employees be able to see themselves in the future plans of the enterprise; by providing strong links to the work, this is made easier.

As we see in Figure 2, many roles can be assigned to a single position or actor; similarly, one role can actually be performed, if it is deemed necessary, by many resources, who would then be granted all the same rights and have the personal competency to perform and engage in the same work. By separating the resource loading and work distribution questions that are inherently distinct from the role implementation and capacity issues, these can be ignored until the later stages of design as we focus on the parts that are played within the business process. In addition, hidden in this representation is resolution of conflict–separation of function and control: cashier versus daily depositor versus bookkeeper versus supervisor.

This should be consistent with personal experience; a person can act in many roles at the same time or at different times. We know in our personal lives that a person can, for example, be a daughter, mother, soccer coach, lawyer, and wife within the same person. Similarly, it would be reasonable for a large organization to have many accounts receivable clerks, many programmers, etc.

In the formal sense, a role is a set of rights and privileges (authorizations and accountabilities) together with human capability (skills, knowledge, and present and future decision-making ability) involved in the execution of a set of tasks:

1. The role of teacher involves specific instructional skills used to impart knowledge of a particular kind (e.g., mathematics, history, English literature) within an institutional framework (i.e., the authorizations).
2. The role of application security analyst involves the use of specific analysis skills that apply security principles (i.e., the knowledge) to the evaluation of computer programs (i.e., the accountability) for ensuring the protection of data and processes within an automated software product.[7]

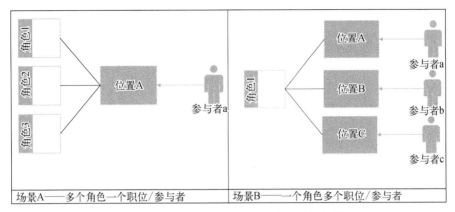

图2　与组织其他组成部分的角色相关的场景

同样,在革命性变更中,员工能够在企业未来的计划中看到自己,这一点始终很重要。通过提供与工作的强大链接,这一点变得更加容易。

如图2所示,可以将多个角色分配给一个职位或参与者。同样,如果认为有必要,一个角色也可以由许多资源执行,这些资源将被授予所有相同的权利,并具有执行和从事相同工作的个人能力。通过将资源加载和工作分配问题与角色实现和容量问题分开,可以忽略这些问题,直到设计的后期阶段,因为我们将重点放在业务流程中扮演的角色上。此外,这种表述中隐藏的是冲突的解决——职能和控制的分离,如出纳与日常储户、簿记员与主管。

这应该与个人经验相一致,一个人可以同时或在不同的时间扮演多个角色。例如,我们知道,在我们的个人生活中,同一个人可以是一个人的女儿、母亲、足球教练、律师和妻子。同样,对于一个大型组织来说,拥有许多应收账款职员、许多程序员等也是合理的。

在正式意义上,角色是一组权利和特权(授权和责任),以及执行一组任务所涉及的人类能力(技能、知识、当前和未来的决策能力)。

(1)教师的作用涉及用于在机构框架(即授权)内传授特定种类(如数学、历史、英国文学)知识的特定教学技能;

(2)应用程序安全分析人员的角色包括使用特定的分析技能,将安全原则(即知识)应用于计算机程序(即问责制)的评估,以确保自动化软件产品中数据和流程的保护[7]。

The context for the definition of role is applicable only within an organization, whether public, private, or not for profit. Although individuals acting on their own behalf have roles, they are not subject to the same methods of classification. For example, individuals acting on their own behalf in transformational or transactional work may be assigned the role of customer and act in this role throughout all views of the work.

STANDARDS THAT LINK TO ROLE CONCEPTS

Standard development bodies have categorized and classified how to think, work, and/or model roles within the context of their framework, method, and approaches. The standards developing organizations, standards bodies, or standards setting organization (such as governments) are, among others:

- The Object Management Group (OMG), a technology standards consortium. The OMG Task Forces develop enterprise integration standards for a wide range of technologies and an even wider range of industries. The OMG's modeling standards have specific aspects of roles including:
  - Model-driven architecture
  - BPMN
  - Business motivation model
  - Value Delivery Modeling Language
- Control Objectives for Information Technology (COBIT) - COBIT is an IT Governance Framework created by Information Systems Audit and Control Association (ISACA) as a centralized source of information and guidance in the field of auditing controls for computer systems. IT's Control Objective PO4.6 - Establishment of Roles and Responsibilities. is contained within define the IT Processes, Organisation, and Relationships.

- ITIL (formally known as Information Technology Infrastructure Library) is concerned with IT service management. The ITIL roles are employed to define responsibilities. In particular, they are used to assign process and service roles to the various ITIL processes and to illustrate responsibilities for the single activities within the detailed process descriptions used in all ITIL phases and steps
- The Open Group is a global consortium that enables the achievement of business objectives through IT standards. With more than 400 member organizations, it has a diverse membership that spans various sectors of the IT community. The Open Group, among others, has developed the framework and methods around TOGAF and TOGAF 9.X; although it has no artifacts to classify and/or categorize roles, it has the following meta objects and relationships related to a role:
  - Role is performed by an actor
  - Role accesses a function
  - Role decomposes role
- Although much of the body of work produced by these groups (OMG, COBIT, ITIL, TOGAF) is useful for understanding and applying roles in their specific context, it must generally be used with caution in another context, The

角色定义的背景仅适用于组织内,无论是公共的、私有的还是非营利的。虽然代表自己行事的个人有自己的角色,但他们不受相同的分类方法的约束。例如,在转型或事务性工作中以自己的名义行事的个人可以被分配为客户的角色,并在工作的所有视图中扮演此角色。

3.4.5　与角色概念相关的标准

标准开发机构已经对如何在其框架、方式和方法的背景中思考、工作或建模角色进行了分类和分级。标准制定组织、标准机构或标准制定组织(如政府)包括如下组织。

- OMG是一个技术标准联盟。OMG工作组为各种技术和更广泛的行业制定企业一体化标准。OMG的建模标准中与特定角色相关的,包括:
- 模型驱动架构;
- BPMN;
- 商业动机模型;
- 价值传递建模语言。
- 信息及相关技术控制目标(COBIT)——COBIT是信息系统审计与控制协会(Information Systems Audit and Control Association, ISACA)创建的一个IT治理框架,作为计算机系统审计控制领域的集中信息源和指导。IT的控制目标PO4.6——角色和职责的建立包含在流程中。定义IT流程、组织和关系中。
- ITIL涉及IT服务管理,ITIL角色用于定义职责。特别是它们用于为各种ITIL流程分配流程和服务角色,并在所有ITIL阶段和步骤中使用的详细流程描述中说明单个活动的责任。
- The Open Group(是一个全球联合体,是厂家中立、技术中立的联合会,旨在开放标准和全球互操作性的基础上,实现企业内部和企业之间的无边界的集成信息流)通过IT标准实现业务目标。它拥有400多个成员组织,成员多种多样,遍布IT界的各个部门。The Open Group等已经围绕TOGAF和TOGAF 9.X开发了框架和方法,尽管它没有用于分级和/或分类角色的构件,但它具有以下与角色相关的元对象和关系:
  - 角色由参与者执行;
  - 角色访问功能;
  - 角色分解角色。
- 尽管这些组织(OMG、COBIT、ITIL、TOGAF)的大部分工作对理解和应用其特定背景下的角色很有用,但通常必须谨慎地在另一个背景下使用:大

charters of most of the organizations and therefore their respective body of knowledge are focused almost exclusively on a specific way of thinking, working and modeling and are often very technology-centric based. The execution and enabling of roles and their function in the stewardship of technology is interesting in the context of this chapter, but so are the much larger questions about their nature, purpose, discovery, and use across all aspects of the business, not just with respect to the management of technology.

- The International Organization for Standardization is an international standard-setting body composed of representatives from various national standards organizations. The role modeling in ISO standards are not manged across the standards but rather within the specific standard.

- The Institute of Electrical and Electronics Engineers (IEEE) is a professional association dedicated to advancing technological innovation and excellence. The IEEE's constitution defines the purposes of the organization as "scientific and educational." In pursuing these goals, the IEEE serves as a major publisher of scientific journals and/organizer of conferences, workshops, and symposia (many of which have associated published proceedings). The role modeling in IEEE standards are not manged across their standards but rather within the specific standards.

- The LEADing Practice, with over 100 Enterprise Standards and its open source community, is one of the largest development organizations in the world. Its Enterprise Standards are the result of years of international industry research, mostly carried out by the Global University Alliance, which consist of over 400 universities that bring together professors, lecturers, and/or researchers from all over the world to improve enterprise architecture practices. Its Enterprise Standards are therefore de facto standards, capturing what has been shown to work well and expert consensus on repeatable patterns that can be reused and replicated. The various LEAD enterprise standards are packaged as reference content and are both tool agnostic and vendor neutral. The material is designed to be tailored and implemented by any organization, large or small, regardless of its various frameworks, methods, products/services, or activities. In terms of role modeling, LEADing practice has Role Modeling Reference Content LEAD ID ES20012BC; however roles are modeled in the following additional enterprise standards:
  - Competency Modeling Reference Content LEAD ID ES20013BC
  - Business Process Reference Content LEAD ID ES20005BP
  - Measurement Reference Content LEAD ID ES20014PG
  - Value Model Reference Content LEAD ID ES20007BCPG
  - Service Model Reference Content LEAD ID ES20008BCBS
  - Performance Modeling Reference Content LEAD ID ES20009BCPG
  - Operating Model Reference Content LEAD ID# ES20010BC

CURRENT METHODS

There is implied recognition in current techniques that one actor may actually have different roles, but there are no generally used techniques to tease the roles apart, understanding whether the roles are complete or appropriate.

多数组织的章程及其各自的知识体系几乎完全集中在特定的思考、工作和建模方式,通常以技术为中心。在本节中,角色的执行和启用及其在技术管理中的作用是很有趣的,但是关于它们的性质、目的、发现以及在业务的各个方面的使用的更大问题也不仅仅是关于技术管理。

- ISO是一个国际标准制定机构,由各国标准组织的代表组成。ISO标准中的角色建模不是跨标准管理的,而是在特定标准中管理的。
- IEEE是致力于推进技术创新和卓越发展的专业协会。IEEE章程将该组织的宗旨定义为"科学和教育"。在实现这些目标的过程中,IEEE是科学期刊的主要出版商,也是会议、研讨会和专题讨论会的组织者(其中许多都有相关的出版程序)。IEEE标准中的角色建模不是跨其标准进行管理,而是在特定标准中进行管理。
- 领导实践组织(The LEADing Practice, https: //www.leadingpractice.com/)拥有100多个企业标准及其开源社区,是世界上最大的开发组织之一。其企业标准是多年国际行业研究的成果,主要由全球大学联盟实施,该联盟由400多所大学组成,汇集了来自世界各地的教授、讲师和/或研究人员,以改进企业架构实践。因此,它的企业标准是事实上的标准,包含了已经证明工作良好的内容,并就可重用和复制的可重复模式达成了专家共识。各种LEAD企业标准被打包为参考内容,并且都是不可知论和供应商中立的工具。材料的设计是由任何组织(无论大小)定制和实施的,无论其各种框架、方法、产品/服务或活动如何。在角色建模方面,领导实践具有角色建模参考内容LEAD ID ES20012BC,但是,角色在以下附加企业标准中建模:
- 能力建模参考内容LEAD ID ES20013BC;
- 业务流程参考内容LEAD ID ES20005BP;
- 测量参考内容LEAD ID ES20014PG;
- 价值模型参考内容LEAD ID ES20007BCPG;
- 服务模型参考内容LEAD ID ES20008BCBS;
- 性能建模参考内容LEAD ID ES20009BCPG;
- 操作模型参考内容LEAD ID# ES20010BC。

3.4.6　现行方法

在当前的技术中,有一种隐含的认识,即一个参与者实际上可能具有不同的角色,但没有通常使用的技术来区分角色、了解角色是完整的还是适当的。

Traditional methods for identifying roles generally rely on interviews, observation, or questionnaires aimed at the current resources, performing the current processes, within the current organization to identify what the actor does.

The assumptions embedded in the current method are numerous:

1. That the existing organization is correct, complete, and reflective of the work to be done
2. That all processes exist and are at most subject to incremental change or other methods to find the current set of roles that exist in an organization
3. That the processes in which the subject is engaged are the correct processes, or that any variances can be addressed through minor adjustments
4. That the processes are positioned correctly with respect to their value proposition, i.e., that they exist inside a competency that reflects the business strategy and that the strategy has not and will not change, that the alignment of the organizational competency to the strategy is appropriate (core or non-core) and that both the process design and role profile are complete, leveled, and well-bounded (have consistent start and end points)

Even if these assumptions were false, those in the process cannot pinpoint problems with its current design and usually do not envision the best solutions.

As a further concern, the current practice has two fundamental limitations:

1. It provides no baseline with which to benchmark the completeness, veracity, or applicability of the role, its positioning, the work involved, its rights, or anything else
2. There is no way to determine from the list of roles created by this method whether the list itself is complete and/or whether the roles are well set

The net result is that suggestions or insights about changes are generally evolutionary and look at the problem only at its edges. That is to say, because the existing participants really only have experience with their existing work and their existing way of working, it is difficult for them to step out of this world and imagine a situation where the rules are different. Only someone who is outside the work in question has the context and tools, and may provide the insight to create truly innovative new processes, flows, and roles by challenging the assumptions embedded in these methods.

ROLE CONTEXT

When we think about anything, it is always good to place it in its proper context. The context of roles is critical to our understanding of what they are and to shape our thinking about how to use them.

Every role captures four essential lenses through which the world views the role and the role connects with the world. Generically, these are:

1. The way of thinking, which is the world view of the enterprise taken from the perspective of the purpose the role seeks to achieve and the goals it sets for itself.

识别角色的传统方法通常依赖于针对当前资源的访谈、观察或问卷调查,在当前组织内执行当前流程,以确定参与者所做的工作。

当前方法中嵌入的假设有很多:

(1)现有组织是正确的、完整的,并反映了要完成的工作;

(2)所有流程都存在,并且最多会受到增量更改或其他方法的影响,以查找组织中存在的当前角色集;

(3)主体参与的流程是正确的流程,或者任何差异都可以通过微小的调整来解决;

(4)就其价值主张而言,流程的定位是正确的,即它们存在于反映业务战略的能力中,并且战略没有也不会改变,组织能力与战略的一致性是适当的(核心或非核心),以及流程设计角色轮廓完整、层次分明、界限分明(起点和终点一致)。

即使这些假设是错误的,流程中的那些假设也不能精确地指出当前设计中的问题,并且通常无法预见最佳解决方案。

作为进一步关注,目前的做法有两个基本限制:

(1)它没有提供基准来衡量角色的完整性、准确性或适用性、其定位、涉及的工作、其权利或任何其他方面;

(2)无法从该方法创建的角色列表中确定列表本身是否完整或角色是否设置良好。

最终的结果是,有关变化的建议或见解通常是渐进的,只从问题的边缘看问题。也就是说,因为现有的参与者实际上只有他们现有的工作和他们现有的工作方式的经验,所以他们很难走出这个世界,想象一个规则不同的情况。只有工作之外的人才有背景和工具,并且可能通过挑战这些方法中嵌入的假设来提供见解,以创建真正创新的新流程、流和角色。

3.4.7 角色背景

当我们思考任何事情时,最好把它放在适当的环境中。角色的背景对于我们理解他们是什么以及如何使用他们的思路是至关重要的。

每个角色都包含四个"镜头",通过这些"镜头",世界可以看到角色与世界的联系。一般来说,这些描述如下。

(1)思维方式。即从角色所追求的目标和为自己设定的目标的角度来看待企业的世界观。

2. The way of working, which is the critical discipline applied to translate the identified drivers, expectations, and requirements into a form that guides the effort to expose the roles. It structures the arrangement of effort and work by translating the way of thinking into a structural approach for action.

3. The way of modeling, or the approach followed by the analyst so as to make an objective assessment of what is necessary and sufficient to complete the work. The way of modeling is the conceptual framework, language, and relationships.

4. The way of governing, which relates to decisions and guidance that define expectations and goals, grant power, or verify and ensure value identification and creation in the life cycle(s) of interest.

Figure 3 shows a role in the context of its way of thinking, way of working, way of modeling, and way of governing and how through the entire context the objects, the role interacts with, artifacts (templates), and life cycle steps are consistent. The figure further distinguishes between the properties of a role that are both inherent in the role and unchanging or static, and those that are dynamic and unique.

For every role described, there will be a set of properties or a profile that speak to each of these factors.

What many organizations do not realize is that there is something in common between the areas where roles need to be applied. The common things are the role objects. Global University Alliance research identifies the semantic relations of the

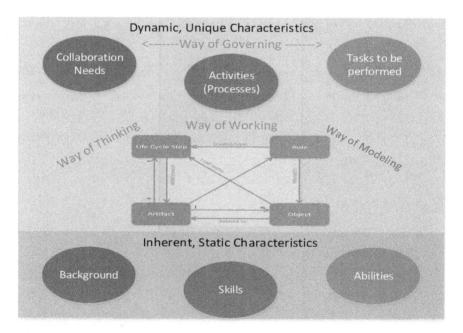

FIGURE 3

Context of role.

（2）工作方式。这是用于将已确定的驱动因素、期望和需求转换为指导识别角色的工作的形式的关键规程。它通过将思维方式转化为行动的结构方法来安排投入和工作。

（3）建模的方式，或分析员遵循的方法，以便对完成工作的必要性和充分性作出客观的评估。建模的方式是概念框架、语言和关系。

（4）管理方式。涉及定义期望和目标、授予权力或验证和确保价值识别和利益生命周期创造的决策和指导。

图3显示了一个角色在上下文中的思维方式、工作方式、建模方式和管理方式，以及角色如何在整个上下文中与对象/角色交互作用、制品（模板）和生命周期步骤是一致的。图3进一步区分了角色固有的、不变的或静态的属性，以及动态的和独一无二的属性。

对于所描述的每个角色，都将有一组与这些因素中的每一个相关的属性或简介。

许多组织没有意识到，在需要应用角色的领域之间有一些共同点。常见的是角色对象，全球大学联盟研究确定了各种角色对象的语义关系，以及它们如何在不同学科中应用[8]。角色对象的关系在其角色模板和制品中构建，如角色映射、角色

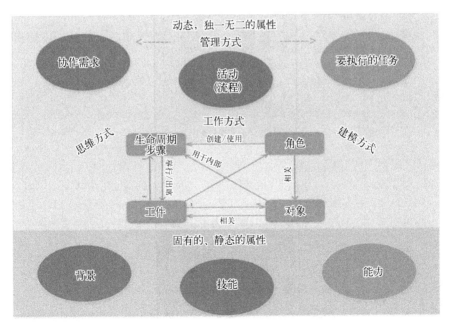

图3　角色背景

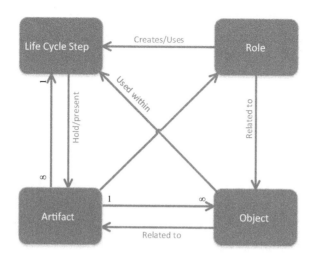

FIGURE 4

Key role relationships.

various role objects and how they can be applied within different disciplines.[8] The relations of the role objects are built into its role templates and artifacts, e.g., role maps, role matrices, and/or role models.

These properties influence the ability of the role to interact with other components, or objects within the enterprise. Figure 4 shows the logical relationships between the objects that have the most significant interaction with roles. In this figure, we see that a role captures or uses a life cycle step that has an object that is used within it and that holds or presents one or more artifacts. We also see that roles interact with artifacts and with objects.

The importance of these relationships and therefore this model is that it exposes and links the role in its complete context so that all of its parts may be fully understood.

ABILITIES TO ACT

There are many ways in which we describe the rights, privileges, powers, and immunities of which roles may be composed. Fundamentally, roles are described in terms of the permitted actions that can be performed on resources (including information objects). However, for the purpose of clarity it is useful to distinguish between those that are involved with business processes and those that are situationally dependent.

Rights versus Hierarchical Power

There is little agreement on what is meant by the rights to which a role has access. Although it is easy to say that rights are about how the role participates in the activity, it is not necessarily clear what we mean by this.

If we distinguish between the working relationships of roles and the relationship of a role with the work itself, we can see that two different sets of patterns emerge.

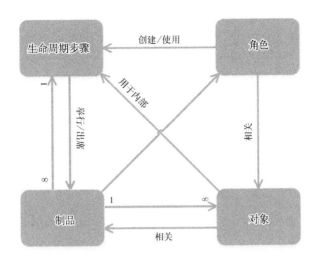

图4　关键角色关系

矩阵或角色模型。

这些属性影响角色与企业中其他组件或对象交互的能力。图4显示了与角色交互最重要的对象之间的逻辑关系。在这个图中，我们看到一个角色捕获或使用一个生命周期步骤，该步骤包含一个在其中使用的对象，并且包含或呈现一个或多个制品。我们还看到角色与制品和对象交互。

这些关系是至关重要的，这种模式揭示和链接了在完整上下文中的角色，因此可以充分理解角色的所有部分。

3.4.8　行动能力

我们有许多方法来描述权利、特权、权力和豁免权，这些权利、特权、权力和豁免权的角色可以组合。从根本上讲，角色是根据可以在资源（包括信息对象）上执行的允许操作来描述的。然而，为了清晰起见，区分涉及业务流程的流程和依赖于情境的流程是很有用的。

权利与等级权力

对于角色可以访问的权限的含义几乎没有一致意见。虽然很容易说权利是关于角色如何参与活动的，但我们不一定清楚这是什么意思。

如果我们区分角色的工作关系和角色与工作本身的关系，我们可以看到出现了两组不同的模式。

1. Role authority about either horizontal or vertical relationships between roles
2. Roles that are engaged in and have a direct relationship to work

In *Requisite Organization*, Elliot Jaques identified "task assignment role relationships" as "vertical" interactions and "task initiating role relationships" as "horizontal" in nature. Both these are made in the context of the organization structure or design and are relevant only in subordinate–superior relationship structures and are not core to BPM in a direct sense. What is important is the distinction between the nature of role relationships, which are about authority, and the work-centric roles, which are about responsibility. The difference between responsibility and authority is clear; responsibility is associated with execution, whereas accountability is about results and ultimate ownership. Actors possessing roles in a management hierarchy engage in task delegation, whereas within BPM, they exercise their rights and perform work. Whereas hierarchical structures and/organizational design work with roles, the rights associated with roles in this context are not directly part of BPM, which is the focus of this chapter.

ROLE PROFILE

To understand the specific set of rights granted to a role, its business context and the supported skill and knowledge needed to exercise those rights require an understanding of each of the areas identified in Table 1. These properties provide a profile of the role.

Categories of Working Roles

With business processes, the most common categorized role groups found in most organizations are business, process, service, and application. Table 2 gives an overview of the typical role categories within an organization (Table 2).

The purpose of having such a role categorization is to define how to organize and structure the viewpoints and role objects associated with the various disciplines, e.g., business role, process role, service role, and/or application role applying the concepts. This has proven to help companies with some of the most common and complex advanced role principles, dilemmas, and challenges that enterprises must confront today. This includes, but is not limited to:

- The role definitions used in the enterprise and their content; most large organizations have duplications and other redundancies in their roles and thereby their processes, leading to major inefficiencies and ineffectiveness
- The way specific roles link to the business model, and whether they are value creating is often unclear when viewing them just from the process perspective
- The link between roles and the competency type within which they work, e.g., core differentiating, core competitive, non-core. This is not possible to identify from process modeling alone; therefore, the link between roles and competency types is seen as vital
- The relationship of various roles with measurement and reporting for advanced decision making

（1）关于角色之间的水平或垂直关系的角色权限；

（2）从事并与工作有直接关系的角色。

在必需品组织中，Elliot Jaques 将"任务分配–角色关系"确定为"垂直"交互，同时"任务启动–角色关系"本质上是"水平"的。这两个方面都是在组织结构或设计的背景下进行的，并且只在下级–上级关系结构中相关，而不是直接意义上的 BPM 核心。重要的是，角色关系的本质（与权威有关）与工作中心角色（与责任有关）之间的区别。责任和权威之间的区别很明显，责任与执行有关，而责任与结果和最终所有权有关。在管理层级中拥有角色的参与者参与任务委托，而在 BPM 中，他们行使自己的权利并执行工作。虽然层次结构和组织设计与角色一起工作，但是与此上下文中的角色相关联的权限并不是 BPM 的直接部分，这是本章的重点。

3.4.9 角色简介

要了解授予某个角色的特定权利集、其业务背景以及行使这些权利所需的支持技能和知识，就需要了解表1中确定的每个领域，这些属性提供角色的简介。

工作角色类别

对于业务流程，大多数组织中最常见的分类角色组是业务、流程、服务和应用程序。表2概述了组织中的典型角色类别。

这样的角色分类的目的是定义如何组织和构造与各种规程相关联的观点和角色对象，如业务角色、流程角色、服务角色、应用概念的应用程序角色。事实证明，这有助于企业应对当今企业必须面对的一些最常见和最复杂的高级角色原则、困境和挑战。这包括但不限于以下几点。

- 企业中使用的角色定义及其内容。大多数大型组织在其角色及其流程中存在重复和其他冗余，从而导致严重的效率低下和效果不佳。
- 从流程的角度来看，特定角色与业务模型的链接方式，以及它们是否具有价值创造往往不清楚。
- 角色与他们工作的能力类型之间的联系，如核心差异化能力、核心竞争力、非核心能力。这不可能仅从流程建模中识别出来，因此，角色和能力类型之间的联系被认为是至关重要的。
- 各种角色与衡量和报告的关系，以便进行高级决策。

Table 1	Role Profile
Feature	Description
Organizational roles	The name of the role or roles included in the profile (multiple roles may be included to address different levels of experience)
Background and experience	Specialized accumulated knowledge or skill acquired by experience over a period of time and which is inherent in the work
Focus area	The class of output or functional area in which the role is involved
Relation to strategy	How the role connects to strategy and brings value to the enterprise
Business services	The business services in which the role participates and the nature of the participation
Task	Work that the role performs
Abstraction level	Identifies whether the context of the roles is strategic, tactical, and/or operational
Needed skill for abstraction level	Identifies the conceptual level of thinking of the role: concrete (thoughts and deeds tied to physical things), symbolic (descriptions connect to symbols that represent things which are treated as real or concrete), or conceptual (concepts that pull ideas together)
Products	The outputs of the work
Decisions	The choices the role is called upon to make
Scope	The boundary of decision making for the role
Way of governance	Objects of which the role participates in the life cycle
Life cycle phase	States within the life cycle of an object that are of concern to the role
Cross-competencies (roles)	Areas of similar background that overlap or work in conjunction with these roles

COMMON ROLES INVOLVED WITH ROLE MODELING

Now that the overall pattern or structure of a role has been given form, we can apply this knowledge to give general shape and direction to the roles needed. What is needed within BPM is the set of rights granted in a business context and supported by the skill and knowledge needed to exercise those rights, necessary to manage the business process life cycle

表 1 角色简介

特 征	描 述
组织角色	简介中包含的一个或多个角色的名称(可以包含多个角色,以解决不同级别的经验)
背景和经验	通过一段时间内的经验而获得的在工作中固有的专门知识或技能
聚焦区域	参与角色的输出或功能区的类别
与战略的关系	角色如何与战略联系起来并为企业带来价值
业务服务	角色参与的业务服务和参与的性质
任 务	角色执行的工作
抽象层次	确定角色的背景是战略的、战术的和/或操作的
抽象层次所需技能	确定角色的概念思维水平:具体的(与物理事物相关的思想和行为)、象征性的(描述与表示被视为真实或具体的事物的符号相连接)或概念性的(将思想结合在一起的概念)
产 品	工作成果
决 定	要求角色做出的选择
范 围	角色的决策边界
管理方式	角色参与生命周期的对象
生命周期阶段	与角色相关的对象生命周期内的状态
交叉能力(角色)	与这些角色重叠或共同工作的类似背景区域

3.4.10 角色建模涉及的常见角色

既然角色的整体模式或结构已经给出了形式,那么我们可以应用这些知识来为所需的角色提供一般的形状和方向。BPM中需要的是在业务环境中授予的一组权限,并由执行这些权限所需的技能和知识提供支持,这是管理业务流程生命周期所必需的。

Table 2 *Typical Role Categories within Enterprise Engineering, Enterprise Modeling, and Enterprise Architecture*

Business role	A part that someone or something has in a particular function, activity, or situation.
Owner role	A set of **responsibilities** assigned with the rights, competencies, and capabilities to take decisions about an object, its behavior, and properties Examples include the: • Process group owner, who is responsible for the production of value • Business process owner, who is responsible for the creation of a useable product • Process step owner, who is responsible for ensuring the process is in control • Process activity owner, who is responsible for creating value through the work • Application owner, who is responsible for decisions about an application
Process role	A prescribed set of expected **behavior** and rights (authority to act) that is meant to enable its holder to successfully carry out his or her work. Each role represents a set of allowable actions within the enterprise in terms of the rights that are required for the business to operate
Service role	A prescribed set of expected behavior and rights that is meant to enable its holder to successfully carry out his or her **accountability** in the delivery of value. Each role represents a set of allowable actions within the enterprise in terms of the rights that are required for the business to operate. Examples include the: • Service owner, who is accountable for the realization of value • Client service owner, who is accountable for obtaining value
Application role	A set of rights identified as part of the allowable behavior of a software application

ROLES WITHIN BPM

The transition to an enterprise that is permanently equipped to adopt change requires the ability to manage the targeted business processes. This necessitates an increased process orientation, leading to modifications in the organizational and operational structure of the modern enterprise.[9] To capture all the roles within BPM needed to achieve this objective would necessitate a complete specification of the strategic design of the enterprise in question. It would require identification of the situation or environmental variables relevant to the enterprise and strategic response to this

表2 企业工程、企业建模和企业体系结构中的典型角色类别

业务角色	某人或某物在特定功能、活动或情况下所具有的部分
所有者角色	一组职责,具有对对象、对象行为和属性进行决策的权限、能力和潜能。示例包括: • 流程组负责人,负责生产价值 • 业务流程所有人,负责创建可用产品 • 流程步骤负责人,负责确保流程处于受控状态 • 流程活动负责人,负责通过工作创造价值 • 应用程序所有者,负责有关应用程序的决策
流程角色	一套规定的预期行为和权利(行动授权),旨在使其持有人能够成功地执行其工作。每个角色代表企业内一组允许的行动,涉及业务运营所需的权限
服务角色	一套规定的预期行为和权利,旨在使其持有人在交付价值时成功履行其责任。每个角色代表企业内一组允许的行动,涉及业务运营所需的权限。示例包括: • 服务所有者,负责实现价值 • 客户服务所有者,负责获取价值
应用程序角色	作为软件应用程序允许行为的一部分标识的一组权限

3.4.11 BPM中的角色

向一个永久性地适应变化的企业过渡需要管理目标业务流程的能力。这就需要增加流程导向,从而导致现代企业的组织和运营结构发生变化[9]。要在BPM中获得实现这一目标所需的所有角色,就需要对所讨论的企业的战略设计进行完整的规范。它将要求以能力组的形式识别与企业相关的情况或环境变量,并作出战略响应,以保留生成业务能力的从属变量、战略和响应。一旦确定了业务活

situation in the form of the competency group as well as the subordinate variables, strategies, and responses that generate the business competencies. Once the context for the activity of the business has been established, it would be necessary to explore the process groups and the business services necessary to address these conditions to obtain the desired economies of scale, or to address the need for robustness and other variables affecting the strategic organizational design. Only then does the context exist to expose the full set of roles necessary to realize the process design and roles required.

This design would have to integrate a number of areas of concern that an enterprise may factor into, or connect with, its capacity to steward its business processes and therefore be part of its BPM strategy. The list of the knowledge areas might include:

- Strategic planning
- Business design
- Enterprise architect
- Change management
- Configuration management
- Process management
- Project management
- Value management
- Architecture alignment
- Performance measurement

Organizationally, these roles and their involvement may vary depending on the organizational style, strategy, and other factors; however following section provides an overview of the typical roles in a BPM (CoE).

TYPICAL BPM COE ROLES

One of the primary areas of responsibility for the delivery element of a BPM CoE is to provide the staffing, expertise, and experience required to execute the pipeline of programs and projects within the portfolio. This requires a staffing model and a resource pool, aligned with roles in the BPM CoE.

Any staffing model for BPM projects should take into account a need for scalability to keep pace with BPM adoption across the enterprise. This scalability can be enabled by understanding the specialization of various roles involved in successfully implementing a BPM project, and the recommended level of involvement by these roles in each project. These resources are likely to start being used for multiple projects as resource demands grow for the BPM CoE.[10]

Multiple additional roles can be found in a BPM CoE. This depends on many factors such as industry, BPM CoE size, the specific mission, and the focus of the portfolio, program, and/or project. The most common BPM roles are highlighted in Figure 5.[11,12]

BPM CoE Stakeholders

The BPM stakeholders could have many roles in the organization; however, by definition of being a stakeholder, they represent someone that has an interest in the

动的背景,就有必要探索处理这些条件所需的流程组和业务服务,以获得所需的规模经济,或解决对影响战略组织设计的稳健性和其他变量的需要。只有这样,背景才能公开实现流程设计所需的全部角色集和所需的角色。

这种设计必须整合企业可能会考虑或连接到的许多关注的领域,以管理其业务流程,从而成为其BPM战略的一部分。知识领域列表可能包括:

- 战略规划;
- 业务设计;
- 企业架构师;
- 变更管理;
- 配置管理;
- 流程管理;
- 项目管理;
- 价值管理;
- 架构调整;
- 绩效测量。

在组织上,这些角色及其参与可能因组织风格、策略和其他因素而有所不同,但是,以下部分概述了BPM(CoE)中的典型角色。

3.4.12　典型的BPM CoE角色

BPM CoE交付要素的主要职责之一是提供执行组合中项目和项目管道所需的人员、专业知识和经验。这需要一个人员配置模型和一个资源池,与BPM CoE中的角色保持一致。

为跟上企业对BPM的采用进程,针对BPM项目的任何人员配置模型都应考虑到可扩展性的需求。这种可测量性可以通过理解成功实现BPM项目所涉及的各种角色的专门化以及这些角色在每个项目中的推荐参与级别来实现。随着对BPM CoE资源需求的增长,这些资源可能开始用于多个项目[10]。

在一个 BPM CoE 中可以找到多个额外的角色。这取决于许多因素,如行业、BPM CoE大小、特定的任务以及产品组合、程序和/或项目的重点。最常见的BPM角色在图5中突出显示[11,12]。

1. BPM CoE利益相关者

BPM利益相关者可以在组织中扮演许多角色,但是,根据利益相关者的定义,他们代表了对BPM CoE感兴趣的人。这可能适用于整个产品组合或特定的程序

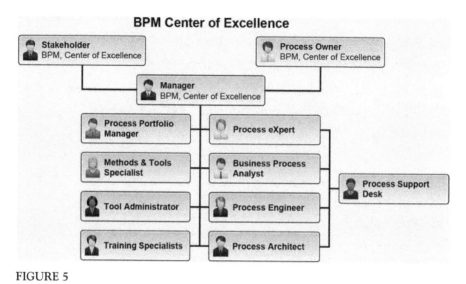

FIGURE 5

Typical business process management (BPM) CoE roles.[13]

BPM CoE. This could be for an entire portfolio or a specific program or project. No matter the position, this role is accountable for the successful initiation, implementation, and sustainability of the BPM CoE. Their responsibility is to:

- Deliver the necessary strategic guidance
- Provide direction
- Arrange alignment
- Enable funding for to the BPM CoE

Other stakeholders can, as mentioned, have an interest in a specific BPM portfolio, program, and/or project and therefore similarly advocate for these narrower areas of concern.

Process Owner

The process owner is responsible for the governance of process performance and process change and defines the process mission, vision, tactics, goals, and objectives, and selects the key performance indicators (KPIs) and measures that align with the organization's strategies. Process owners monitor and report process performance against these KPIs and on the health of execution versus plans. Furthermore, they are involved in synchronizing process improvement plans with other process owners within the value chain and other interfacing processes. Their process aim is to continuously increase the maturity of the process and sustain each level of maturity. The business process owner typically has the following role-specific responsibilities:

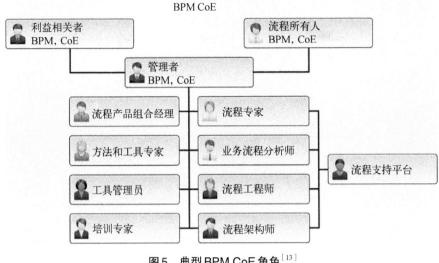

图5　典型BPM CoE角色[13]

或项目。无论职位如何,这个角色都要对BPM CoE的成功启动、实施和可持续性负责。他们的职责是:

- 提供必要的战略指导;
- 提供方向;
- 排列协调;
- 为BPM CoE提供资金。

如前所述,其他利益相关者可以对特定的BPM组合、程序或项目感兴趣,因此也同样主张缩小这些关注领域。

2. 流程所有人

流程所有人负责管理流程绩效和流程变更,并定义流程任务、愿景、策略、目的和目标,并选择与组织战略一致的KPI和度量。流程所有人根据这些KPI监控和报告流程绩效,并根据计划监控执行状况。此外,他们还参与将流程改进计划与价值链和其他接口流程中的其他流程所有人同步。他们的流程目标是不断地提高流程的成熟度并保持每一个成熟度级别。业务流程所有人通常具有以下特定于角色的职责:

- 是团队的头号支持者;
- 为流程的目的与业务部门协调;
- 对当前和新开发的流程具有权威性和责任感;

- Is number one cheerleader for team
- Coordinates with the business on purpose of the processes
- Has authority and accountability for current and newly developed processes
- Has end-to-end focus and responsibility
- Creates and charters team based on organization's business goals
- Appoints process team leader
- In conjunction with the process portfolio manager sets schedule for team
- Provides guidance, information, and support to team
- Meets periodically with BPM project manager for team status
- Informs other parts of organization about the BPM team's work
- Stays informed of activities happening elsewhere that may impact team's activities and coordinates between team and those activities

BPM CoE Manager

The BPM CoE manager supervises and manages all process management activities at a group or business unit level, and manages a team of business process experts that report into the BPM CoE and whose mandate is organization-wide. The BPM CoE manager should have experience with BPM methods and should be comfortable with a leadership role. The BPM CoE manager typically has the following role-specific responsibilities:

- Develop the BPM CoE strategy
- Set the BPM CoE objectives
- Business process management CoE representation to the business
- Brings BPM awareness to the business
- Work with the chief process officer
- Ensure BPM CoE funding
- Work with stakeholders to define BPM needs and wants
- Work with the BPM portfolio managers
- Provides BPM governance coordination between the process owner
- Coaches BPM CoE portfolio and project managers

Business Process Portfolio Manager

The business portfolio manager coordinates the organization-wide portfolio of projects, along with the associated alignment to organization strategy, the development of the BPM pipeline of projects, and their approval and subsequent execution in a program or project. The business process portfolio manager typically has the following role-specific responsibilities:

- Cross-functional representation
- Provides portfolio objectives
- Work with stakeholders to define BPM portfolio
- Provides consultation on process portfolio to process owner

- 具有端到端的关注点和责任感；
- 根据组织的业务目标创建和组建团队；
- 任命流程团队负责人；
- 与流程组合经理一起为团队制定时间表；
- 为团队提供指导、信息和支持；
- 定期与BPM项目经理会面，了解团队状态；
- 通知组织的其他部门有关BPM团队的工作；
- 随时了解其他可能影响团队活动的活动，以及团队与这些活动之间的协调。

3. BPM CoE 经理

BPM CoE经理监督和管理集团或业务部门级别的所有流程管理活动，并管理一个业务流程专家团队，该团队向BPM CoE报告，其任务是整个组织范围的。BPM CoE经理应具有BPM方法的经验，并应熟悉领导角色。BPM CoE经理通常具有以下特定于角色的职责：

- 制定BPM CoE战略；
- 设定BPM CoE目标；
- 表达BPM CoE业务；
- 为企业带来BPM意识；
- 与CPO合作；
- 确保BPM CoE融资；
- 与利益相关者合作，定义BPM需要和需求；
- 与BPM组合经理合作；
- 在流程所有人之间提供BPM治理协调；
- 指导BPM CoE组合和项目经理。

4. 业务流程组合经理

业务组合经理协调整个组织范围内的项目组合，以及与组织战略的关联一致性、项目的BPM路径的开发及程序或项目中的批准和后续执行。业务流程组合经理通常具有以下特定于角色的职责：

- 跨职能代表；
- 提供组合目标；
- 与利益相关者一起定义BPM产品组合；
- 向流程所有人提供流程组合咨询；

- Provides process alignment methodology to program and project team
- Coordinates training for team
- Coaches portfolio team leader on process agenda and strategic direction
- Provides link to portfolio stakeholders
- Assists BPM team in portfolio methodology execution so that portfolio goals and milestones are met
- Brings portfolio awareness of group dynamics uses to BPM team

Process Project Manager

This role is typically a senior project manager who has experience with large projects, establishing methods, and providing governance, and quality control across multiple projects. Typically, this role is a full-time commitment. Responsibilities of the process project manager include:

- Estimates, plans, and manages the overall BPM project
- Creates, manages, and drives the BPM project methodology within the BPM CoE
- Enables the delivery team with project methods and a business value focus
- Leads the BPM Program sub-team and is a member of this sub-team
- Work with project stakeholders
- Cross-functional project representation
- Brings project awareness to BPM team

Process Expert

The process expert combines business analysis and various process subject matter expertise with enhanced process modeling skills. The role is to interpret the business requirements, and based on specific process knowledge and expert modeling skills, to develop the process models with both business and application context. The process expert typically has the following role-specific responsibilities:

- Be a subject matter expert in process-related project
- Develop input to business case for BPM project
- Analyze processes, including process landscape
- Capture process requirements
- Design processes, including process transformation and process innovation
- Build process models, e.g., BPMN
- Process deployment and rollout
- Process monitoring
- Continuous process improvement

Business Process Analyst

Business process experts work as internal process management consultants on projects and are in charge of executing BPM excellence and in rolling out BPM

- 为项目和项目团队提供流程协调方法；
- 协调团队培训；
- 指导组合团队负责人处理流程议程和战略方向；
- 提供与组合利益相关者的链接；
- 协助BPM团队执行项目组合方法，以实现项目组合目标和里程碑；
- 为BPM团队带来合作意识。

5. 流程项目经理

该角色通常是高级项目经理，具有大型项目、建立方法、提供治理和跨多个项目的质量控制的经验。通常，这个角色是一个全职的委任。流程项目经理的职责包括：

- 评估、计划和管理整个BPM项目；
- 在BPM CoE中创建、管理和驱动BPM项目方法论；
- 使交付团队能够以项目方法和业务价值为中心；
- 领导BPM项目子团队，并且是该子团队的成员；
- 与项目利益相关者合作；
- 跨职能项目代表；
- 为BPM团队带来项目意识。

6. 流程专家

流程专家将业务分析和各种流程主题专业知识与增强的流程建模技能相结合。此角色是为了解释业务需求，并基于特定的流程知识和专家建模技能，开发具有业务和应用程序上下文的流程模型。流程专家通常具有以下特定于角色的职责：

- 成为流程相关项目的主题专家；
- 为BPM项目开发对业务案例的输入；
- 分析流程，包括流程布局；
- 捕获流程要求；
- 设计流程，包括流程转换和流程创新；
- 构建流程模型，如BPMN；
- 流程部署和推广；
- 流程监控；
- 持续流程改进。

7. 业务流程分析师

业务流程分析师是项目的内部流程管理顾问，负责执行卓越的BPM，并在所有项目中推广BPM知识。他们在BPM方面拥有所需的专业知识和技能，并且对

knowledge into all projects. They have the required expertise and skills in BPM as well as good understanding of project management. The business process analyst typically has the following role-specific responsibilities:

- Be a subject matter expert in business-related matters
- Work with the business stakeholders
- Develop the business case for BPM projects
- Participate in process analysis
- Capture business requirements
- Agree and work on process transformation and process innovation
- Work with process owners
- Business process management alignment
- Process deployment and rollout
- Continuous process improvement

Process Engineer

Process engineers focus on the design, operation, control, and optimization of business, application, and technology processes. They use specific engineering principles to enable better enterprise related processes. Process engineers typically have the following role-specific responsibilities:

- Construct and maintain the endeavor-specific process from the process landscape
- Evaluate process tools for consistency with the organizational process landscape and process life cycle and/or endeavor-specific
- Ensures that the endeavour-specific process is constructed based on endeavour-specific needs prior to process tool selection, rather than being driven by the early selection of a potentially inappropriate process tool
- Provide input to the environment team regarding required process tool support
- Provide local guidance and mentoring in the proper adoption and use of the endeavor process.
- Identify, document the enterprise's own leading practices, disseminate, and evangelize industry best practices and common best
- Work to support strategic process initiatives including recommending improvements to the organizational process framework
- Support multiple endeavors within a local region
- Staff regional process help desks
- Present local training on process-related topics
- Research advances in process engineering (e.g., new software development methods) practices

Process Architect

The process architect works with senior business stakeholders as a process linkage, structure, and change agent in shaping and fostering continuous improvement, business transformation, and business innovation initiatives related to processes. Process architects

项目管理有良好的理解。业务流程分析师通常具有以下特定于角色的职责：

- 成为业务相关事务的主题专家；
- 与业务利益相关者合作；
- 为BPM项目开发业务案例；
- 参与流程分析；
- 捕获业务需求；
- 同意并致力于流程变革和流程创新；
- 与流程所有人合作；
- BPM协调；
- 流程部署和推广；
- 持续流程改进。

8. 流程工程师

流程工程师专注于业务、应用和技术流程的设计、操作、控制和优化。他们使用特定的工程原理来实现更好的企业相关流程。流程工程师通常具有以下特定于角色的职责：

- 从流程布局构建和维护努力特定流程；
- 评估流程工具，以确保与组织流程布局和流程生命周期或努力特定的协调性；
- 确保在选择流程工具之前，根据流程工具的具体需求，而不是由可能不适当的流程工具的早期选择驱动，构建流程工具；
- 向环境团队提供有关所需流程工具支持的信息；
- 在适当采用和使用努力流程中提供本地指导和监控；
- 识别、记录企业自己的领导实践，传播和宣传行业最佳实践和共同最佳实践；
- 支持战略流程举措，包括建议改进组织流程框架；
- 支持一个地区内的多种尝试；
- 员工区域流程帮助问讯台；
- 就流程相关主题提供本地培训；
- 流程工程（如新软件开发方法）实践方面的研究进展。

9. 流程架构师

为实现流程链接、结构和变更代理，塑造和促进与流程相关的持续改进、业务转型和业务创新计划，流程架构师应该与高级业务相关人紧密合作。流程架构师

build an effective description of the structural design of the general process system for the BPM team/project that makes up the business process programs or the change programs. Process architects typically have the following role-specific responsibilities:

- Provide advice on the structure of the link between business models and process models
- Link between strategy, objectives, and process purpose and goals
- Relationship between process measurements, business reporting, and analytical decision making
- Association between business rules and process rules
- Increase the level of process automation
- Link between process flow, information flow, and service flow

Enterprise Architect

An enterprise architect is a person responsible for the organization, administration of conceptual, logical, and physical relationships and connectivity of specific objects and artifacts within and across the enterprise to each other (and the environment) to understand to enable transformation, performance, or value.

Process Support Desk

The task of the process support desk is to assist in maintaining and supporting ongoing BPM initiatives as a first line of contact to resolve BPM issues and risks that are raised against BPM operational and project-related queries. The process support desk can also be responsible for BPM issue management support, BPM governance support, and even BPM change management support.

Methods and Tools Specialist

Methods and tools specialists' role is to ensure that the correct BPM tool set is being used consistently across all projects. They share knowledge about BPM tool usage with project teams and BPM CoE members, and research and look for improvements in the current BPM tool sets and set out methods for their adoption.

Business Process Management Tool Administrator

The BPM tool administrator controls and administers management of the BPM tool set. This includes managing the authorization profiles and ensuring that business process owners maintain governance over their area of processes through the available BPM tool set.

Training Specialist

The BPM training specialists are individuals who evangelize the business with BPM understanding and knowledge transfer. They maintain the BPM training curricula

为组成业务流程程序或变更程序的BPM团队或项目构建通用流程系统结构设计的有效描述。流程架构师通常具有以下特定于角色的职责：

- 就业务模型和流程模型之间的结构链接提供建议；
- 构建战略、目标、流程目的和目标之间的联系；
- 构建流程度量、业务报告和分析决策之间的关系；
- 在业务规则和流程规则之间实现关联；
- 提高流程自动化水平；
- 在流程、信息流和服务流之间构建链接。

10. 企业架构师

企业架构师是负责组织、管理概念上的、逻辑上的和物理上的关系，以及企业内部和跨企业的特定对象和制品之间（以及环境之间）的连接性的人员，以便理解转换、性能或价值。

11. 流程支持平台

流程支持平台的任务是协助维护和支持正在进行的BPM计划，作为解决针对BPM操作和项目相关查询提出的BPM问题和风险的第一面平台。流程支持平台还可以负责支持BPM问题管理、BPM治理，甚至BPM变更管理。

12. 方法和工具专家

方法和工具专家的角色是确保在所有项目中始终使用正确的BPM工具集。他们与项目团队和BPM CoE成员分享有关BPM工具使用的知识，并研究和寻找当前BPM工具集的使用和改进方法。

13. BPM工具管理员

BPM工具管理员控制和管理BPM工具集。这包括管理授权文件，并确保业务流程所有人通过可用的BPM工具集维护其流程领域的治理。

14. 培训专家

BPM培训专家是通过BPM理解和知识转移来宣传业务的人员。他们维护

and ensure that project teams have the latest information and/or updates on the BPM tools and methods.

ADDITIONAL ROLES

In the context of the enterprise, ancillary roles are not secondary roles although the actors involved may well have roles that require them to be involved in processes owing to the nature of those roles.

Overseeing the process centric roles will be the competency group owner, who is accountable for ensuring the enterprise maintains and exercises the wherewithal to respond to external forces, drivers, or situations that are within the purview of the competency group. The competency group owner is answerable for achieving strategic business objectives set by the enterprise owner. As such, the competency group owner is a source of oversight on the collective process groups within the competency group, but also an escalation and exception point to ensure exception conditions for these processes are addressed in a manner consistent with the strategy. Similarly, the competency owner has the same accountability and process involvement as the competency group owner, but only for their competency.

These roles become involved in process management either directly through their oversight and stewardship functions or within escalation and exception management. As with the process-centric roles, these roles each may be supported by a role that administers the flow of information.

At the apex of the accountability chain is the role with the greatest scope. This, of course, is the enterprise owner, who is ultimately answerable for the strategy and alignment of the processes, for investment in their capabilities, and as the final escalation and oversight point.

Each of these roles will be supported by an administrative role whose purpose is to exercise the duties needed to support the affairs of the principal. It should be remembered that this generic set of roles does not imply a specific number of actors, nor does it imply anything about the organizational hierarchy; a design decision is a function of the size of the organization, its strategy, and other factors.

Such roles within the BPM CoE ensure that there is an accountability framework for creating decisions and determining the services, architecture, standards, and policies for continuous management of business processes. Over all, they ensure the sustainability, maturity, governance, and alignment of processes horizontally from end-to-end across the enterprise, but in a manner that ensures their connection and contribution to strategy.

Specific roles actually work with the definition, categorization, and classification of roles in an organization. The role that creates and defines other roles is often found in the form of the human resource officer, whereas the roles that categorize and classify other roles in the organization are also found in the disciplines of enterprise modeling, engineering, and architecture. Table 3 gives an overview of the typical roles involved with the categorization and classification of roles in an organization.

BPM培训课程,并确保项目团队拥有关于BPM工具和方法的最新信息或更新。

3.4.13 其他角色

在企业环境中辅助角色不是次要角色,尽管由于这些角色的性质很可能让所涉及的参与者要求他们参与流程的角色。

能力部门(人力资源部门)负责人负责监督以流程为中心的各个角色,负责确保企业维护和行使应对能力组职责范围内的外部力量、驱动因素或情况所需的资金。能力部门负责人负责实现企业所有者设定的战略业务目标。因此,能力部门负责人是对能力组内的集体流程组进行监督的来源,也是一个升级和例外点,以确保这些流程的例外情况以符合策略的方式得到解决。同样,能力所有者与能力组所有者具有相同的责任和流程参与,但仅限于其能力。

这些角色直接通过其监督和管理功能或在升级和异常管理中参与流程管理。这与以流程为中心的角色一样,这些角色都可以得到管理信息流中的角色的支持。

问责制度链的顶端是最大范围的角色。当然,这是企业所有者,他们最终要对流程的战略和协调、对其能力的投资负责,并作为最终的升级和监督点。

每一个角色都将由一个负责管理的角色来支持,其目的是行使支持负责人的职责。应该记住,这种通用的角色集并不意味着特定数量的参与者,也不意味着关于组织的层次结构,设计决策是对组织规模、策略和其他因素的影响。

此类角色在BPM CoE中负责确保存在用于创建决策和确定业务流程持续管理的服务、体系结构、标准和策略的问责框架。总而言之,它们确保了整个企业内端到端的流程的可持续性、成熟度、治理和一致性,并保证它们与战略的联系和贡献。

特定角色实际上与组织中角色的定义、分类和分级一起使用。创建和定义其他角色的角色通常以人力资源主管的形式出现,而在组织中对其他角色进行分类和分级的角色也可以在企业建模、工程和体系结构的学科中找到。表3概述了组织中角色的分类和分级所涉及的典型角色。

Table 3 *Typical Roles Involved with the Definition, Categorization, and Classification of Roles in an Organization*

Enterprise Modelers	Enterprise Engineers	Enterprise Architects
Business analyst	System engineer	Business architect
Process expert	Process engineer	Process architect
Information expert	Value engineer	Service-oriented architect
Service expert	Quality engineer	Information architect
Change/transformation expert	Software engineer	Solution architect
	Technology engineer	Data architect
		Technology architect
		Enterprise architect

ROLE PROFILE WITHIN BPM CoE

Table 4 applies the role profile to record sets of rights granted the roles within the BPM CoE, the business context, and the supported skill and knowledge needed to exercise those rights required to perform the work in each of the areas.

Table 4 *Role Profile for Process Management Roles*

Feature	Description
Organizational roles	• Senior process expert and consultant • Senior process method specialist • Process engineer • Quality/production/manufacturing engineer • Junior and senior process architect
Background and experience	• Business management and business engineering
Focus area	• Identify architectural process requirements • Focus on architectural pain points and bottlenecks • Define the process architecture standards • Ensure cross-flows to process and information • Continuous architectural process optimization
Relation to strategy	• Align business processes to architectural goals • Architectural alignment based on process objectives • Link process architecture to business model transformation
Task and services	• Work with process owner • Identify cross process requirements • Categorize processes • Benchmark process architecture maturity • Define process architecture standardization and integration • Align process measures and monitoring to information

Continued

表3 与组织中角色的定义、分类和分级相关的典型角色

企业建模人员	企业工程师	企业架构师
业务分析师	系统工程师	业务架构师
流程专家	流程工程师	流程架构师
信息专家	价值工程师	面向服务的架构师
服务专家	质量工程师	资讯架构师
变更/转型专家	软件工程师	解决方案架构师
	技术工程师	数据架构师
		技术架构师
		企业架构师

3.4.14 BPM CoE 中的角色简介

表4将角色简介应用于记录在BPM CoE中授予角色的权限集、业务背景以及在每个领域执行工作所需的支持技能和知识。

表4 流程管理角色的简介

特　　征	描　　　　述
组织角色	• 高级流程专家和顾问 • 高级流程方法专家 • 流程工程师 • 质量/生产/制造工程师 • 初级和高级流程架构师
背景和经验	• 业务管理和业务工程
聚焦区域	• 确定体系结构流程要求 • 关注架构难点和瓶颈 • 定义流程架构标准 • 确保流程和信息的交叉流动 • 持续的架构流程优化
与战略的关系	• 使业务流程与架构目标保持一致 • 基于流程目标的架构调整 • 将流程架构与业务模型转换联系起来
任务和服务	• 与流程所有人合作 • 确定跨流程需求 • 对流程进行分类 • 基准流程架构成熟度 • 定义流程架构标准化和一体化 • 根据信息调整流程措施和监控

Table 4 *Role Profile for Process Management Roles—cont'd*	
Feature	**Description**
Abstraction level	• Tactical • Operational (solution specific processes)
Needed skill for abstraction level	• Concrete • Descriptive • Design
Map, matrix, and models	• Stakeholder matrix • Process owner matrix • Process requirement map and business case • Process map/matrix, and models (BPMN) • Objects matrix • Performance map, matrix, and model • Reporting matrix and model • Process roles matrix • Process rules matrix • Operating model • Business process modeling notation model • Process media and channel • Process maturity model • Process innovation and transformation matrix and model
Decisions	• Architectural requirement decisions; process, task, event, and gateways • Architecture scenario decisions; rules, flow, and measurements
Scope	• Enterprise, area, • Project • Process flow
Way of governance	• Process life cycle • Value life cycle (performance) • Transformation life cycle (optimization)
Life cycle phase	• Process analysis • Process design • Process implementation • Continuous process improvement
Cross-competencies (roles)	• Value expert • Service expert • Process architect

ROLE PROFILE WITHIN BPM PROJECTS

In BPM projects, these role profiles can be applied in various forms depending on their business context in terms of title, roles, and responsibility. Thus, in addition to the role categorizations and classifications specified in Table 4, we suggest specifying the various forms the BPM roles could have in the various projects. This includes, as illustrated in Table 5 specification of the project title, roles and responsibility in the project.

（续表）

特　征	描　述
抽象层次	● 战术 ● 运营（解决方案特定流程）
抽象级别所需的技能	● 具体 ● 描述性 ● 设计
流程图、矩阵和模型	● 利益相关者矩阵 ● 流程所有人矩阵 ● 流程需求图和业务案例 ● 流程图/矩阵和模型（BPMN） ● 对象矩阵 ● 绩效图、矩阵和模型 ● 报告矩阵和模型 ● 流程角色矩阵 ● 流程规则矩阵 ● 操作模式 ● 业务流程建模符号模型 ● 流程媒体和渠道 ● 流程成熟度模型 ● 流程创新和转型矩阵和模型
决定	● 体系结构需求决策；流程、任务、事件和网关 ● 架构场景决策；规则、流程和度量
范围	● 企业、地区 ● 项目 ● 工艺流程
治理方式	● 流程生命周期 ● 价值生命周期（绩效） ● 转换生命周期（优化）
生命周期阶段	● 流程分析 ● 流程设计 ● 流程实施 ● 持续流程改进
交叉能力（角色）	● 价值专家 ● 服务专家 ● 流程架构师

3.4.15　BPM项目中的角色简介

在BPM项目中，这些角色简介可以以各种形式应用，具体取决于它们在标题、角色和职责方面的业务上下文。因此，除表4中指定的角色分类和分级之外，我们建议指定BPM角色在不同项目中可能具有的各种形式。如表5所示，这包括项目标题的规范、项目中的角色和责任。

Table 5 *Example of Project Role Specification*

Project Title	Project Role	Project Responsibilities
<Title>	<Role>	• <Responsibility>
<Title>	<Role>	• <Responsibility>
<Title>	<Role>	• <Responsibility>
<Title>	<Role>	• <Responsibility>
<Title>	<Role>	• <Responsibility>

CONCLUSION

We have shown that the concept of "role" is separate and distinct from the persons or thing that access the rights and perform its responsibilities. We have shown that as a set of rights granted in a business context and supported by the skill and knowledge needed to exercise those rights, a role can be treated and managed as a conceptual thing of significance in the design of enterprises and organizations.

We have applied a standard way of thinking, working, modeling, and governing to the exploration of the nature of roles to create a standard and repeatable method for identifying, characterizing, and documenting roles, and then applied this approach to finding and describing the roles needed within the BPM CoE.

End Notes

1. Newman J., Rhona Newman, "Role, Right and Rationality in the Business Process," in *ElnfUhrung yon CSCW-Systemen in Organlsationen*, ed. Ulrich Hasenkamp (Braunschweig/Wiesbaden: © Friedr.Vieweg & Sohn Verlagsgesellschaft mbH, 1994).
2. Jaques E., Requisite Organization.
3. Body of Knowledge (BABOK)™.
4. Version 9.1 "Enterprise Edition".
5. LEAD Object Definitions.
6. John A. Zachman, *Conceptual, Logical, Physical: It Is Simple.*
7. http://www.modernanalyst.com/Careers/InterviewQuestions/tabid/128/ID/1197/How-do-you-define-a-role.aspx.
8. http://www.globaluniversityalliance.net/research-areas/industry-standards/.
9. Stefan Eicker, Jessica Kochbeck and Peter M. Schuler, Employee Competencies for Business Process Management (University of Duisburg-Essen).
10. Lisa Dyer, Andrew Forget, Fahad Osmani, Jonas Zahn, *Creating a BPM Center of Excellence (CoE) IBM Redbooks* (2013), www.ibm/redbooks.
11. Taken from the Business Process Reference Content LEAD-ES20005BP.
12. Taken from the Process Architecture Reference Content LEAD-ES40004BP.
13. LEADing Practice Business Process Reference Content #LEAD-ES20005BP.

表5 项目角色规范示例

项 目 名 称	项 目 角 色	项 目 职 责
<名称>	<角色>	● <职责>
<名称>	<角色>	● <职责>
<名称>	<角色>	● <职责>
<名称>	<角色>	● <职责>
<名称>	<角色>	● <职责>

3.4.16 结论

我们已经表明,角色的概念是独立的,并且与获得权利并履行其责任的人或事物不同。我们已经表明,作为在业务环境中授予的一组权利,并由行使这些权利所需的技能和知识支持,角色可以被视为在企业和组织设计中具有重要意义的概念性事物来对待和管理。

我们采用了一种标准的思考、工作、建模和管理方式来探索角色的本质,从而创建了一种标准的、可重复的方法来识别、描述和记录角色,然后将此方法应用于查找和描述 BPM CoE 中所需的角色。

参考文献

[1] Newman J., Rhona Newman, "Role, Right and Rationality in the Business Process," in ElnfUhrung yon CSCW-Systemen in Organlsationen, ed. Ulrich Hasenkamp (Braunschweig/Wiesbaden: © Friedr.Vieweg & Sohn Verlagsgesellschaft mbH, 1994).

[2] Jaques E., Requisite Organization.

[3] Body of Knowledge (BABOK)TM.

[4] Version 9.1 "Enterprise Edition".

[5] LEAD Object Definitions.

[6] John A. Zachman, Conceptual, Logical, Physical: It Is Simple.

[7] http://www.modernanalyst.com/Careers/InterviewQuestions/tabid/128/ID/1197/How-do-you-define-a-role. aspx.

[8] http://www.globaluniversityalliance.net/research-areas/industry-standards/.

[9] Stefan Eicker, Jessica Kochbeck and Peter M. Schuler, Employee Competencies for Business Process Management (University of Duisburg-Essen).

[10] Lisa Dyer, Andrew Forget, Fahad Osmani, Jonas Zahn, Creating a BPM Center of Excellence (CoE) IBM Redbooks (2013), www.ibm/redbooks.

[11] Taken from the Business Process Reference Content LEAD-ES20005BP.

[12] Taken from the Process Architecture Reference Content LEAD-ES40004BP.

[13] LEADing Practice Business Process Reference Content #LEAD-ES20005BP.

Working with the Business Process Management (BPM) Life Cycle

Mark von Rosing, Ulrik Foldager, Maria Hove, Joshua von Scheel, Anette Falk Bøgebjerg

INTRODUCTION

Business processes are collections of one or more linked activities that realize a business objective or policy goal, such as fulfilling a business contract and/or satisfying a specific customer need. The life cycle of a business process involves everything from setting up process goals and requirements, capturing the process in a computerized representation, as well as automating the process. This typically includes specific steps for measuring, evaluating, and improving the process. Currently, commercially available workflow management systems (WFMSs) and business process modeling tools (BPMTs) provide for complementary aspects of business process life-cycle management.

Furthermore, new concepts and interoperating tools in these categories are emerging to provide comprehensive support for managing the entire business process life cycle. In this chapter, we provide an overview and an evaluation of the Process Life Cycle phases, as well as details around process modeling, analysis, automation, and coordination capabilities. The life cycle represents the course of developmental changes through which the process evolves in terms of transformation and/or innovation as it passes through six different phases during its lifetime. From process analysis, design, construction, deployment, implementation, as well as governance and continuous improvement. The life cycle helps guide the practitioner to complete categorizations of process areas and groups, mapping of processes, their steps, activities, operations, improvements, and planned changes for the future by using change management as the driving force in the project.

The Process Life Cycle consists of a set of steps and phases in which each step and phase uses the results of the previous one. It provides a highly useful sequence of actions that any Business Analyst, Process Expert, Process Engineer, Process Architect, Business Architect, and/or Enterprise Architect can follow during any process-oriented projects. This can be used in combination of various process methods and approaches such as Business Process Reengineering [BPR], Business Process Management (BPM), Lean, and Six Sigma exist today, but no end-to-end BPM Life-cycle models have been developed in the market thus far. However, parts of the BPM Life Cycle can be found within Control Objectives for Information and related Technology (COBIT) and Information Technology Infrastructure Library (ITIL) v2 and v3, which are both Application and Service Life-cycle concepts, but they concentrate only very little on process maturity and the architectural aspects of processes.

The Complete Business Process Handbook. http://dx.doi.org/10.1016/B978-0-12-799959-3.00014-8

3.5　使用BPM生命周期

Mark von Rosing, Ulrik Foldager, Maria Hove, Joshua von Scheel, Anette Falk Bøgebjerg

3.5.1　介绍

业务流程是实现业务目标或策略目标（如履行业务合同或满足特定客户需求）的一个或多个链接活动的集合。业务流程的生命周期涉及建立流程目标和需求、流程捕获展示方式、流程自动化等所有方面。这通常包括：测量、评估和流程改进的具体步骤。目前，商用工作流管理系统（workflow management systems，WFMs）和BPMTs提供了业务流程生命周期管理的补充方面。

此外，这些类别中的新概念和交互操作工具正在出现，为管理整个业务流程生命周期提供全面的支持。在本节中，我们提供流程生命周期阶段的概述和评估，以及有关流程建模、分析、自动化和协调能力的详细信息。生命周期代表了发展变化的过程，通过这一过程，流程在其生命周期中经历了六个不同的阶段，从流程分析、设计、构建、部署、实施，以及治理和持续改进开始，该过程在转型和创新方面进行了演变。生命周期通过将变更管理作为项目中的驱动力，帮助指导从业者完成流程领域和组别的分类、流程的制作、步骤、活动、操作、改进和未来的计划变更。

流程生命周期由一套步骤和阶段组成，其中每个步骤和阶段使用前一个步骤和阶段的结果。它提供了一系列非常有用的操作，任何业务分析师、流程专家、流程工程师、流程架构师、业务架构师和企业架构师都可以在任何面向流程的项目中遵循这些操作。这可以用于各种流程方法的组合，如BPR、BPM、精益和六西格玛，但目前市场上尚未开发端到端的BPM生命周期模型。然而，BPM生命周期的某些部分可以在COBIT和ITIL第二版和第三版中找到，这些都是应用程序和服务生命周期概念，但他们只关注流程成熟度和流程的体系结构方面。

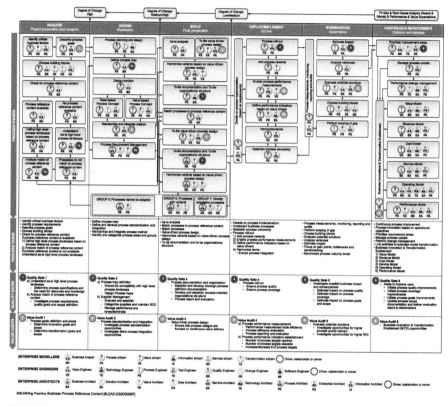

FIGURE 1

The BPM Life Cycle at a glance.

Ref. 1.

The proposed BPM Life Cycle concept interlinks with, and can be integrated with, the previously mentioned life cycles and helps practitioners place focus on all process-relevant aspects from business and application requirements to process modeling, engineering and architecture (see Figure 1).

PHASE 1: ANALYZE—PROJECT PREPARATION AND BLUEPRINT

The ultimate goal when we model business processes is to describe what the business does in a hierarchy of detail from the top level down to the level at which documents and other types of specific information components become visible. When we analyze processes, the information we discover will come from many sources and at many levels of abstraction and granularity. This information helps ensure consistency and completeness if we try to answer the same questions for each process that we encounter throughout the process landscape. In this context,

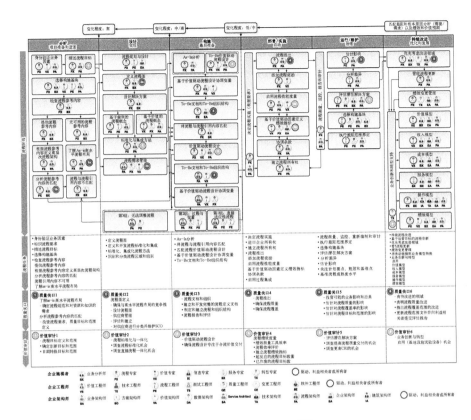

图1　BPM生命周期一览[1]

本文所提到的BPM生命周期概念与前面提到的生命周期相互关联,并且可以与之结合,并帮助从业者将重点放在从业务和应用程序需求到流程建模、工程和体系结构的所有流程相关方面(图1)。

3.5.2　第一阶段:分析(项目准备和蓝图)

当我们对业务流程建模时,最终的目标是以详细的层次结构描述业务所做的工作,在这个层次上,文档和其他类型的特定信息组件变得可见。当我们分析流程时,在许多抽象和粒度级别上,我们发现的信息来自许多来源。如果我们尝试回答我们在整个流程布局中遇到的每个流程的同样的问题,那么这些信息有助于确保一致性和完整性。在此背景下,流程分析是对所有相关方面的逐步细分,包括输

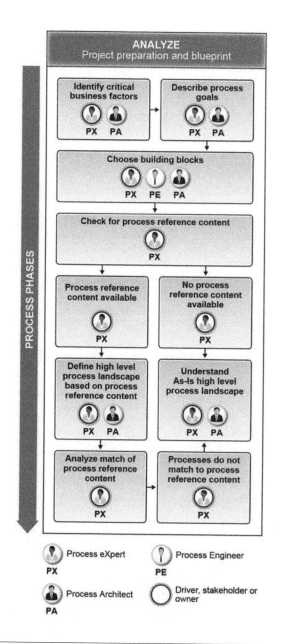

FIGURE 2

The Analysis phase of the BPM Life Cycle.

Ref. 2.

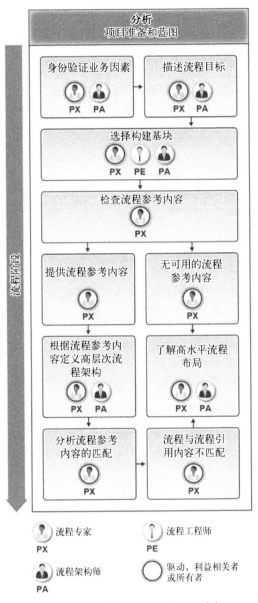

图2　BPM生命周期的分析阶段[2]

process analysis is a step-by-step breakdown of all the relevant aspects, including inputs, outputs, and the BPM Center of Excellence (CoE) operations that take place during the phase.

As an example, if our goals are strategic, we would be taking a top-down approach and interviewing senior executives or managers with a holistic and big picture view of an organization to identify the critical business factors and the process goals. Process recognition on this level tends to yield processes that are very abstract or very generic, partitioning activity into large, goal-oriented chunks. Among the questions, the answers for which describe processes at this level, are:

- What are the critical business factors?
- Which processes exist?
- What are the names of the processes?
- What are the goals and/or purposes of the processes?
- What industries, functional areas, or organizations are involved with the processes?
- Who are the stakeholders, owners, and/or participants in the processes?
- What is the process landscape?
- Does process reference content exist that could be used?
- Do the current processes present problems?

The process analysis phase can, therefore, be used to improve understanding of how the process operates, determine potential targets for process alignment with business goals, and identify increasing efficiency. Asking questions and recording their answers in a disciplined way rapidly creates a web of related information about interconnected processes from which we can develop models. The various BPM roles will get more information that is useful if they both ask questions and record the answers, using:

1. a standard vocabulary and definitions for the various process concepts described (see the BPM Ontology chapter or take a look at Figure 3).
2. process reference content that already exists within the domain in which the organization is working [3].

In this chapter, we will show the kinds of reference content that already exist. The 1st phase, the Process Analysis Phase (see Figure 2), is the phase in which the organization's processes are analyzed, captured, and defined based on the business goals and specific process requirements (e.g., business needs and wants), as well as on any interlinked business and process demands. Process goals and detailed process requirements are defined, choices are clarified through blueprinting, and the initial process maps are populated with the identified processes. Traditional output of the analyze phase would be problem analysis, As-Is analysis, measurement analysis, as well as the establishment of business goals. This phase includes a link to change management and the continuous improvement loop through the change management of the BPM Life Cycle. The degree of changes made during this phase is considered high.

入、输出和在阶段中发生的BPM CoE操作。

例如，如果我们的目标是战略性的，我们将采取自上而下的方法，用一个组织的整体和全局视图观面见高级管理人员或经理，以确定关键的业务因素和流程目标。在这一级别上的流程识别倾向于生成非常抽象或非常通用的流程，将活动划分为大型的、面向目标的模块。在这些问题中，描述这一级别流程的答案是：

- 关键业务因素是什么？
- 存在哪些流程？
- 流程名称是什么？
- 流程的目标和目的是什么？
- 流程涉及哪些行业、职能领域或组织？
- 谁是流程中的利益相关者、所有者和参与者？
- 流程布局是什么？
- 是否存在可使用的流程参考内容？
- 当前流程是否存在问题？

因此，流程分析阶段可用于提高对流程如何运行的理解，确定流程与业务目标一致的潜在目标，并确定提高的效率。以一种有纪律的方式提出问题并记录他们的答案，很快就形成了一个关于相互连接的流程的相关信息网络，从中我们可以开发模型。如果他们都提出问题并记录答案和应用，不同的BPM角色将获得更多有用的信息。

（1）各种流程概念的标准词汇和定义描述（请参阅BPM本体论一节或查看图3）；

（2）存在于组织运行领域中的流程参考内容[3]。

在本节中，我们将展示已经存在的各种参考内容。第一阶段是流程分析阶段（图2），是根据业务目标和特定流程需求（如业务需要和需求）以及任何相互关联的业务和流程需求对组织的流程进行分析、捕获和定义的阶段。该阶段定义了流程目标和详细的流程需求，通过蓝图阐明了选择，并且初始流程图中填充了已识别的流程。分析阶段的传统输出将是问题分析、as-is分析、度量分析，以及建立业务目标。此阶段包括到变更管理的链接，以及通过BPM生命周期的变更管理实现的持续改进循环。在这一阶段所做的改变程度被认为是很高的。

Process Lifecycle Verb Taxonomy

ANALYZE	DESIGN	BUILD	DEPLOY/IMPLEMENT	RUN/MAINTAIN	CONT. IMPROVEMENT
Analyze	Aim	Accept	Accomplish	Administer	Adjust
Appraise	Align	Adapt	Achieve	Assign	Alter
Approximate	Arrange	Assemble	Activate	Audit	Amend
Ascertain	Begin	Assure	Apply	Calculate	Boost
Assess	Blueprint	Build	Assimilate	Chronicle	Change
Capture	Categorize	Chart	Carry out	Communicate	Condense
Clarify	Characterize	Check	Cause	Conserve	Convert
Collate	Classify	Codify	Close	Control	Coordinate
Collect	Cluster	Combine	Complete	Engage	Correct
Consider	Commence	Compile	Conclude	Exchange	Decrease
Count	Compare	Compose	Conduct	Fix	Diminish
Demand	Convene	Configure	Conform	Govern	Eliminate
Detain	Describe	Confirm	Deliver	Handle	Enhance
Detect	Design	Constitute	Deploy	Keep	Escalate
Diagnose	Determine	Construct	Do	Maintain	Improve
Discover	Devise	Craft	Educate	Manage	Incorporate
Estimate	Display	Create	Employ	Measure	Moderate
Evaluate	Draft	Customize	Evolve	Monitor	Modernize
Examine	Draw	Define	Execute	Operate	Modify
Explore	Drive	Develop	Finish	Oversee	Optimize
Find out	Enter	Enact	Generate	Preserve	Realign
Forecast	Enumerate	Enlarge	Get done	Process	Reassess
Formulate	Establish	Erect	Implement	Oversee	Reconsider
Gage	Form	Expand	Include	Promote	Redevelop
Gather	Format	Extend	Initiate	Protect	Redirect
Gauge	Found	Fabricate	Instigate	Reconcile	Redraft
Identify	Idea	Increase	Integrate	Record	Reduce
Inspect	List	Itemize	Interlink	Recover	Reevaluate
Investigate	Negotiate	Make	Launch	Register	Reexamine
Judge	Obtain	Manufacture	Migrate	Reintroduce	Reform
Learn	Organize	Match	Perform	Report	Refresh
Observe	Outline	Pilot	Present	Respond	Regulate
Recognize	Plan	Procure	Progression	Retain	Renew
Reflect on	Plot	Provide	Put into action	Retire	Renovate
Research	Prepare	Purchase	Put into operation	Run	Reorganize
Review	Prioritize	Raise	Put into service	Save	Reprioritize
Revise	Propose	Rank	Realize	Service	Restore
Search	Quantify	Scan	Reallocate	Set up	Restructure
See	Recommend	Secure	Set off	Supervise	Revert
Seek out	Select	Shape	Shift	Support	Revolutionize
Study	Sketch	Systemize	Teach	Turn on	Rework
Survey	Start	Test	Train	Update	Standardize
Think about	Suggest	Translate	Transfer	Uphold	Transfigure
Understand	Verify	Unify	Transition	Withdraw	Transform

FIGURE 3

The process life-cycle verb taxonomy model can be used to help identify the terminology associated with the organization's BPM life cycle around analyzing, designing, building, implementing, running, and improving business processes.

Ref. 4.

流程生命周期动词分类

分析	设计	构建	部署/实施	运行/维护	继续改进
分析	旨在	接受	完成	管理	调整
评价	排列	适应	实现	分配	改变
接近	安排	集合	激活	审计	修正
查明	启动	保证	应用	计算	促进
评估	蓝图	建造	同化	记录	变化
捕获	分类	记录	开展	沟通	压缩
澄清	表征	检查	造成	保存	转换
校对	分类	编纂	关闭	控制	使协调
收集	集群	联合	完成	雇用	修正
考虑	着手	编译	断定	交换	减少
计算	比较	撰写	断定	使固定	减少
需求	召开	安置	顺应	治理	消除
留置	描述	确认	交付	处理	增强
查明	设计	构成	部署	保持	逐步升级
诊断	确定	构建	执行	维护	改进
发现	发明	精心制作	教育	管理	包含
估计	显示	创造	雇用	测量	缓和
评价	草案	订制	进化	监视	现代化
检查	描绘	定义	执行	操作	修改
探索	驱动	发展	完成	监督	优化
发现	进入	制定	生成	监督	重新排列
预报	列举	扩大	完成	加工	重新评估
制定	建立	建立	实施	监督	重新考虑
量规	形式	展开	包括	促进	再开发
聚集	格式化	延伸	启动	保护	重定向
量规	发现	制作	使开始	调和	重新起草
识别	构思	增加	整合	记录	减少
检查	列清单	逐项列举	互连	恢复	重新评价
调查	谈判	制作	发射	注册	复查
判断	获得	生产	迁移	重新引入	改革
学习	组织	相配	执行	报告	刷新
观察	概述	试行	显示	回应	调节
辨别	计划	设法获得	进展	保持	重新开始
反思	标出	提供	付诸行动	退休	整修
研究	准备	购买	投入运行	运作	重组
回顾	优先化	提高	投入使用	保存	重新排序
修订	提议	分级	实现	服务	恢复
搜索	量化	扫描	重新分配	设置	重组
看见	推荐	保卫	出发	监督	还原
寻找	选择	形成	转移	支撑	革命化
学习	简述	系统化	讲授	打开	返工
调查	开始	试验	训练	更新	标准化
考虑	建议	翻译	转移	支持	易形
理解	验证	统一	过渡	撤回	转换

图3　流程生命周期动词分类模型可用于帮助识别与组织的BPM生命周期相关的术语,包括分析、设计、构建、实现、运行和改进业务流程[4]

Step 1: Identify Critical Business Factors

To identify and define process goals effectively, an organization must first and foremost identify the existence and possible impact of any internal and external value and performance drivers and how they relate to the critical business factors. This is primarily done to identify the drivers for change, plan for the requirements, identify the resources needed, and relate to the process landscape the most critical business factors that are expected to impact the organization. For the most part, this step typically involves identifying both the internal and external value and performance drivers that may impact the design and creation of new business processes and/or to create an effective and reliable environment in which to support an eventual reengineering of existing processes. The output of step 1 is consumed by step 2.

Typical tasks that are done within this step:

- Identify which critical value and performance drivers are impacting and/or influencing the process landscape
- Prioritize value and performance drivers based on the level of impact, severity, and/or urgency
- Associate the value and performance drivers to the critical success factors (CSFs) of the various business areas/groups (based on the level of opportunity, priority, and/or importance)

Typical templates that are used:

- Forces and Drivers Map
- Vision, Mission and Goals Map
- Stakeholder Map

Typical BPM CoE roles involved:

- Process Experts/Business Analysts
- Value Experts
- Process Architects

Step 2: Describe Process Goals

Process goals have to be clearly defined, documented, and agreed upon by stakeholders, process owners, and all involved process roles of the organization. Defining the process goals of the organization is the natural continuation from the identification of critical business factors, as those are the main tools—and critical for the goal descriptions—to compare against process goals and plan and prepare for the upcoming design and construction phases of the life cycle. Many BPM CoE organizations are in the habit of working with process goals and, in many cases, it is business as usual to define and document them. Among typical process goals are reducing complexity, aligning processes, identifying duplication, creating new processes by which new business goals are supported, or removing business bottlenecks through automation. Process goals can, however, also link to very multifaceted

步骤1：确定关键业务因素

为了有效地确定和定义流程目标，组织必须首先确定任何内部、外部价值和绩效驱动因素的存在和可能的影响，以及它们与关键业务因素的关系。这主要是为了确定变更的驱动因素、需求计划，确定所需的资源，以及与流程布局相关的最关键的业务因素，这些因素预计会影响组织。在大多数情况下，此步骤通常涉及确定可能影响新业务流程设计和创建的内部和外部价值及绩效驱动因素或创建有效和可靠的环境，以支持现有流程的最终重新设计。步骤1的输出被步骤2使用。

在此步骤中完成的典型任务包括：

- 确定哪些关键价值和绩效驱动因素正在影响或将影响流程布局；
- 根据影响程度、严重程度和紧急程度确定价值和绩效驱动因素的优先级；
- 将价值和绩效驱动因素与各个业务领域／集团的关键成功因素（CSF）（基于机会、优先级和重要性的等级）关联起来。

使用的典型模板：

- 力量和驱动程序图；
- 愿景、任务和目标图；
- 利益相关者图。

涉及的典型BPM CoE角色：

- 流程专家／业务分析师；
- 价值专家；
- 流程架构师。

步骤2：描述流程目标

流程目标必须由利益相关者、流程所有人和组织所有涉及的流程角色清晰地定义、记录和商定。定义组织的流程目标是识别关键业务因素的自然延续，因为这些因素是目标描述的主要工具和关键，可以与流程目标进行比较，并为生命周期的未来设计和构建阶段进行计划和准备。许多BPM CoE组织都习惯于处理流程目标，在许多情况下，定义和记录这些目标是正常的业务。典型的流程目标包括降低复杂性、调整流程、识别重复、创建支持新业务目标的新流程，或通过自动化消除业务瓶颈。然而，流程目标也可以链接到非常多方面的业务创新和转型计划，因此

business innovation and transformation initiatives and therefore be difficult and complex to relate to individual processes. More on how this is done is found in the value-driven process design.

The output of step 2 is consumed by multiple other steps, such as:

- Step 6 to lay the foundation of goals and requirements for the upcoming process planning and design steps, and then moves back again to step 3.
- Step 12 for the purpose of identifying and defining both the high-level requirements as well as the detailed requirements of the entire process landscape of the organization.

Typical tasks that are done within this step:

- Develop the process goals based on the specific critical business factors
- Develop process goals based on the business innovation and transformation initiatives
- Split process goals to specific main, supporting, and management processes
- Define and document process goals
- Relate and connect all of the defined process goals to established business, application, and technology goals.

Typical templates that are used:

- Vision, Mission, and Goals Map and/or Matrix
- Stakeholder Map and/or Matrix
- Process Map

Typical BPM CoE roles involved:

- Process eXperts/Business Analysts
- Process Architects

Value Audit 1:

- Process goals definition and scope
- Determine innovation goals and scope
- Determine transformation goals and scope

Step 3: Choose Building

Similar to software development, for example, building blocks are reusable pieces of content. In BPM, the most common building blocks would be strategic, organizational, process, and technology contexts. The following is an example of typical BPM Building Blocks and thereby reusable aspects that need to be considered again and again (Figure 4).

Choosing the right building blocks for the creation of new business processes or relating to strategy, and linking to organizational context or technology automation are daunting tasks for any BPM CoE organization. These choices should always be based upon review, identification, and documentation of the current baseline

与个别流程联系起来是困难和复杂的。关于如何实现这一点的更多信息,请参见价值驱动流程设计。

步骤2的输出由多个其他步骤使用,例如:

- 步骤6为即将到来的流程规划和设计步骤奠定目标和要求的基础,然后再次返回到步骤3;
- 步骤12的目的是识别和定义组织整个流程布局的高级需求和详细需求。

在此步骤中完成的典型任务:

- 根据特定的关键业务因素制定流程目标;
- 根据业务创新和转型计划制定流程目标;
- 将流程目标分解为特定的主要、支持和管理流程;
- 定义和记录流程目标;
- 将所有已定义的流程目标与已建立的业务、应用程序和技术目标联系起来。

使用的典型模板:

- 愿景、任务和目标图或矩阵;
- 利益相关者图或矩阵;
- 流程图。

涉及的典型BPM CoE角色:

- 流程专家/业务分析师;
- 流程架构师。

价值审计1:

- 流程目标定义和范围;
- 确定创新目标和范围;
- 确定转型目标和范围。

步骤3:选择性构建

例如,构建与软件开发类似可重用的内容模块。在BPM中,最常见的构建模块是战略、组织、流程和技术语义。下面是典型的BPM构建模块的一个例子,因此需要反复考虑可重用方面(图4)。

对于任何BPM CoE组织来说,选择正确的构建模块去创建新的业务流程或与策略相关联,并链接到组织环境或技术自动化,都是一项艰巨的任务。这些选择应始终基于对组织的当前基线的审查、识别和记录(包括已经涉及流程布局的所有

STRATEGIC CONTEXT		ORGANIZATIONAL CONTEXT		PROCESS CONTEXT		TECHNOLOGY CONTEXT	
Strategy	Principles & Rules	Organization	Guidelines & Standards	Business Processes	Process Development	IT Operations	IT Enablement
Vision & Mission	Stakeholder Management	Organizational Structure	Organizational Interaction	Process Models	Process Owner	IT Management	IT Enablement
Market Approach	Business Issues / Problem Chain	Critical Success Factors	Business Governance	Pain Chain	Process Governance	IT Operations	IT Business Model
Strategic Business Objectives	Operating Model	Roles & Tasks	Reward & Motivation	Business Process Execution	Process Performance Indicators	End-user Focus IT Support	IT Documentation
Business Model	Business Value Management	Key Performance Indicators	Program Management	Process Drivers	Process-based Rewards	Software Capabilities / Competencies	IT Standardization
Strategy Map	Value Audits	Change Management	Change Policy	Process Audits	Process Domains	Application Lifecycle Management	IT Integration
Value Map	Value Clusters	Information Need	Business Competency Modelling	Process Policies	IT Process Parameters	Service Orientation Competencies	IT Process Flow
Scorecards	Business Performance Management	Goal Chain	Portfolio Management	Process Measurements	Process Architecture	IT Governance	Process Means / Tools
Enterprise Value Architecture	Business Maturity Models	Training & Education	Organizational Maturity Models	Continuous Improvement	Process Maturity Models	Service Level Agreement	Technology Maturity Models

FIGURE 4

Example of BPM building blocks.

Ref. 5.

(including all of the systems that are already involved with the process landscape) of the organization, and then executing the choices from a perspective of the needed functionality that is required to create new or to re-engineer any existing processes.

The output of step 3 is by step 4.

Typical tasks that are done within this step:

- Identification of relevant building blocks
- Alignment and unification of building blocks across areas
- Review and document current baseline and compare viewpoints of possible solutions

Typical templates that are used:

- Process Map and/or Matrix
- Object Map and/or Matrix
- Service Map and/or Matrix
- Application Service Map and/or Matrix
- Data Service Map and/or Matrix

Typical BPM CoE roles involved:

- Process eXperts
- Business Analysts
- Value eXperts

战略环境		组织背景		流程上下文		技术背景	
策略	原则与规则	组织	指南和标准	业务流程	流程开发	信息技术操作	IT启用
愿景与使命	利益相关者管理	组织结构	组织互动	流程模型	流程所有者	IT管理	IT启用
市场途径	业务问题/问题链	关键成功因素	企业治理	疼痛链	流程治理	IT操作	IT业务模型
战略业务目标	运行模型	角色与任务	奖励与动力	业务流程执行	流程绩效指标	最终用户关注IT支持	IT文档
业务模型	企业价值管理	关键绩效指标	项目管理	流程驱动程序	基于流程的奖励	软件能力/能力	IT标准
策略图	价值审计	变更管理	变更政策	流程审计	流程领域	应用生命周期管理	信息一体化
价值图	价值集群	信息需求	业务能力建模	流程策略	IT流程参数	服务导向能力	IT流程流
记分卡	业务绩效管理	目标链	产品组合管理	流程测量	流程体系结构	IT治理	工艺手段/工具
企业价值架构	业务成熟度模型	培训与教育	组织成熟度模型	持续改进	程成熟度模型	服务水平协议	技术成熟型

图4　BPM构建模块示例[5]

系统),然后从创建新流程或重新设计任何现有流程所需的功能的角度执行选择。

步骤3的输出由步骤4完成。

在此步骤中完成的典型任务:

- 相关构建模块的识别;
- 跨区域的构建模块调整和统一;
- 审查和记录当前基线并比较可能解决方案的观点。

使用的典型模板:

- 流程图及矩阵;
- 目标图及矩阵;
- 服务图及矩阵;
- 应用服务图及矩阵;
- 数据服务图及矩阵。

涉及的典型BPM CoE角色:

- 流程专家;
- 业务分析师;
- 价值专家;

- Process Engineers
- Process Architects

Step 4: Check for Process Reference Content

Today, most BPM CoE organizations already have a process landscape. The process landscape might not be very well defined or documented, however, and to define the expectations and purpose of both new and existing processes, it is essential to clearly define and document the existing process landscape of the organization. This is achieved through extensive process mapping across all business units, and through meticulous and detailed documentation work. The documentation and mapping task is likely going to require a substantial number of man-hours to allow both process owners and the involved process roles to get an established overview of the "As-Is" situation. It also enables decision makers to make their decisions based on which processes to create and/or reuse (Figure 5).

The output of step 4 is consumed by step 4a (if any process reference content is available) and step 5 (if no currently available process reference content).

Typical tasks that are done within this step:

- Review the existing process portfolio for any available process reference content

Typical templates that are used:

- Information Map and/or Matrix
- Process Map and/or Matrix
- Object Map and/or Matrix
- Service Map and/or Matrix

Typical BPM CoE roles involved:

- Process eXperts
- Business Analysts

Step 4a: Process Reference Content Available

If process content is available for the BPM CoE organization to re-use within the existing process landscape, the process reference content should be clearly identified, reviewed, documented, and prepared for evaluation at the time of analysis and comparison to the existing process landscape. The output of step 4a is consumed by step 4b.

Typical tasks that are done within this step:

- Identify and review the current process reference content
- Document (as needed) and prepare the current process reference content for analysis and comparison

Typical templates that are used:

- Information Map and/or Matrix
- Process Map and/or Matrix

- 流程工程师;
- 流程架构师。

步骤4：检查流程参考内容

今天，大多数BPM CoE组织已经有了一个流程布局。然而，流程布局可能没有很好地被定义或记录，而要定义新的和现有的流程预期和目的，就必须清楚地定义和记录组织的现有流程布局。这是通过对所有业务部门进行广泛的流程映射，并通过细致详细的文档工作实现的。文档和映射任务可能需要大量的工时，以允许流程所有人和相关流程角色对as-is情况进行既定的概述。它还使决策者能够根据创建或重用哪些流程来做出决策（图5）。

步骤4的输出由步骤4a（如果有任何流程参考内容可用）和步骤5（如果没有当前可用的流程参考内容）使用。

在此步骤中完成的典型任务：
- 审查现有流程组合中任何可用的流程参考内容。

使用的典型模板：
- 信息图及矩阵;
- 流程图及矩阵;
- 目标图及矩阵;
- 服务图及矩阵。

涉及的典型BPM CoE角色：
- 流程专家;
- 业务分析师。

步骤4a：可用流程参考内容

如果流程内容可供BPM CoE组织在现有流程布局中重新使用，则应在分析和比较现有流程布局时，清楚地识别、审查、记录流程参考内容，并为评估做好准备。步骤4a的输出被步骤4b使用。

在此步骤中完成的典型任务：
- 识别和审查当前流程参考内容;
- 记录（根据需要）并准备当前流程参考内容，以便进行分析和比较。

使用的典型模板：
- 信息图及矩阵;
- 流程图及矩阵;

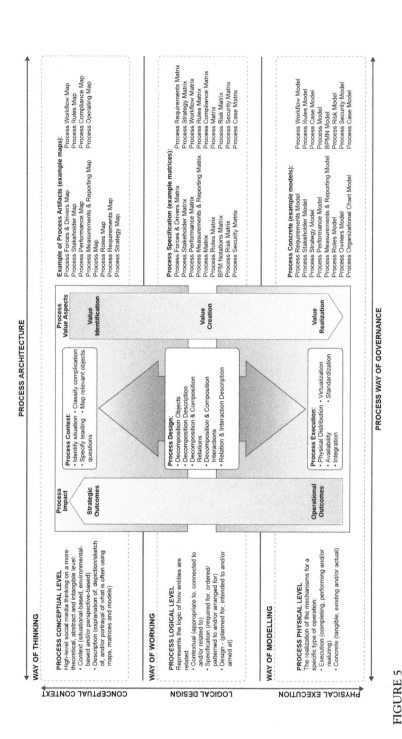

FIGURE 5

Example of a model showing how and when the different parts (maps, matrices, and models) of process reference content is used across the Layered Process Architecture of an organization.

Ref. 6.

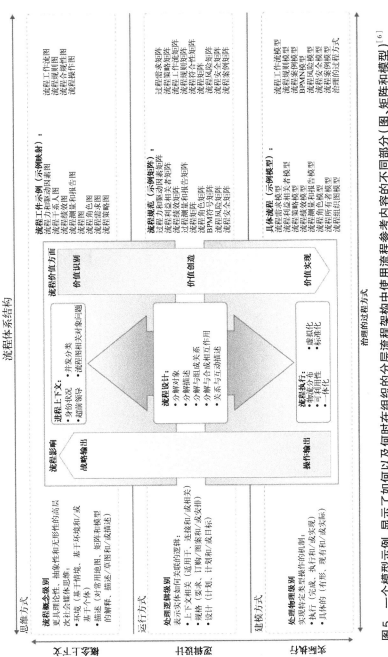

图 5 一个模型示例，显示了如何以及何时在组织的分层流程架构中使用流程参考内容的不同部分（图、矩阵和模型）[6]

- Object Map and/or Matrix
- Service Map and/or Matrix

Typical BPM CoE roles involved:

- Process eXperts
- Business Analysts

Step 4b: Define High-Level Process Landscape Based on Process Reference Content

The definition of a high-level process landscape includes using architecture standards and creating relationships between the different process levels used by the organization. The high-level process landscape is often referred to as the Value Chain View, and is a visualization of a process that is used to illustrate how an organization's business units (both primary and supporting) work together to facilitate and execute the organization's business model. The high-level process landscape of an organization is typically used as a tool for setting a standard for how the rest (when diving deeper into the detailed view of the process landscape) of the organization is expected to (and should) use processes. The output of step 4b is consumed by step 4c.

Typical tasks that are done within this step:

- Define the high-level process landscape by defining the objectives and intended content of process areas and process groups (process levels 1–2)
- Define and document the main process areas (process level 1) and process groups (process level 2)
- Define and document the management process areas (process level 1) and process groups (process level 2)
- Define and document the supporting process areas (process level 1) and process groups (process level 2)

Typical templates that are used:

- Information Map and/or Matrix
- Requirement Map and/or Matrix
- Role Map and/or Matrix
- Owner Map and/or Matrix
- Process Map and/or Matrix
- Object Map and/or Matrix
- Service Map and/or Matrix
- Application Service Map and/or Matrix
- Data Service Map and/or Matrix

Typical BPM CoE roles involved:

- Process eXperts
- Process Architects

- 目标图及矩阵；
- 服务图及矩阵。

涉及的典型BPM CoE角色：

- 流程专家；
- 业务分析师。

步骤4b：根据流程参考内容定义高级流程布局

高级流程布局的定义包括使用架构标准和在组织使用的不同流程级别之间创建关系。高级流程布局通常被称为"价值链视图"，它是一个流程的可视化，用于说明组织的业务单元(首要和支撑)如何协作以促进和执行组织的业务模型。一个组织的高级流程布局通常被用作一个工具，用于为组织的其余部分(当深入到流程布局的详细视图中时)预期(并且应该)如何使用流程设定标准。步骤4b的输出被步骤4c使用。

在此步骤中完成的典型任务：

- 通过定义流程区域和流程组的目标和预期内容(流程级别1～2)来定义高级流程布局；
- 定义并记录主要流程区域(流程1级)和流程组(流程2级)；
- 定义和记录管理流程区域(流程1级)和流程组(流程2级)；
- 定义和记录支持流程区域(流程1级)和流程组(流程2级)。

使用的典型模板：

- 信息图及矩阵；
- 需求图及矩阵；
- 角色图及矩阵；
- 所有者图及矩阵；
- 流程图及矩阵；
- 目标图及矩阵；
- 服务图及矩阵；
- 应用服务图及矩阵；
- 数据服务图及矩阵。

涉及的典型BPM CoE角色：

- 流程专家；
- 流程架构师。

Step 4c: Analyze Match of Process Reference Content

When moving towards the analysis of the process reference content that is available, and when comparing it to the existing process landscape, it is important to draw upon the documentation work done during steps 4a and 4b. Also note that the analysis of all of the processes of the organization is likely to become a very time-consuming process in itself, although much of the work has already been done during the previous review and documentation steps. This analysis and comparison focuses almost exclusively on extracting knowledge, know-how, efficiencies, advantages, and other benefits and nuggets of wisdom that might somehow aid the BPM CoE organization in creating new processes in the future, or to prepare them for process optimization through process reengineering. The output of step 4c is consumed by step 4d. It is important to note, however, that the output of step 4c may (if wanted or deemed required) be consumed directly by step 18 to immediately begin matching existing processes to the available process reference content.

Typical tasks that are done within this step:

- Select the process reference content that *cannot* be used within the currently established process landscape
- Categorize the main, supporting, and management business processes from the process reference content that *can* be used within the currently established process landscape

Typical templates that are used:

- Information Map and/or Matrix
- Process Map and/or Matrix
- Object Map and/or Matrix
- BPM Notations Map and/or Matrix
- Application Service Map and/or Matrix
- Data Service Map and/or Matrix

Typical BPM CoE roles involved:

- Process eXperts
- Business Analysts

Quality Gate 1b:

- Analyze match of process reference content
- Undergo detailed comparison of process reference content to the existing process landscape
- Investigate process requirements, quality goals, and scope definition

Step 4d: Processes That Do Not Match Process Reference Content

Based on the analysis of process during step 4c, there will be a need to collect and gather the processes that does not match the current process reference content. In

步骤4c：分析流程参考内容的匹配

在分析可用的流程参考内容时，以及在将其与现有流程布局进行比较时，重要的是利用步骤4a和步骤4b中完成的文档工作。还要注意：尽管在之前的审查和文件编制步骤中已经完成了许多工作，但对组织所有流程的分析本身很可能成为一个非常耗时的过程。这种分析和比较几乎完全集中在提取知识、技术诀窍、效率、优势以及其他可能在某种程度上帮助BPM CoE组织在未来创建新流程或通过流程再造为流程优化做好准备的益处和智慧。步骤4d使用步骤4c的输出。但是，重要的是要注意，步骤4c的输出可以（如果需要或认为需要）在步骤18直接使用，以立即开始将现有流程与可用的流程引用内容匹配。

在此步骤中完成的典型任务：

- 选择当前已建立的流程布局中无法使用的流程参考内容；
- 根据流程参考内容对主要、支持和管理业务流程进行分类，这些内容可在当前已建立的流程布局中使用。

使用的典型模板：

- 信息图及矩阵；
- 流程图及矩阵；
- 目标图及矩阵；
- BPM符号图及矩阵；
- 应用服务图及矩阵；
- 数据服务图及矩阵。

涉及的典型BPM CoE角色：

- 流程专家；
- 业务分析师。

质量门1b：

- 分析流程参考内容的匹配；
- 详细比较流程参考内容与现有流程全景；
- 调查流程要求、质量目标和范围定义。

步骤4d：与流程引用内容不匹配的流程

根据步骤4c中的流程分析，需要收集和聚集与当前流程参考内容不匹配的

this regard, it is necessary to thoroughly examine these processes for the purpose of discarding them entirely based on the process goals and the newly defined high-level process landscape, or to prepare them for any reengineering purposes. The output of step 4d is consumed by step 4e.

Typical tasks that are done within this step:

- Collect and gather existing processes that do not match the available process reference content
- Examine and document the processes that do not match the available process reference content

Typical templates that are used:

- Information Map and/or Matrix
- Process Map and/or Matrix
- Object Map and/or Matrix
- BPM Notations Map and/or Matrix
- Application Service Map and/or Matrix
- Data Service Map and/or Matrix

Typical BPM CoE roles involved:

- Process eXperts
- Business Analysts

Step 4e: Understand As-Is High-Level Process Landscape

If no current process reference content is available or if the current process landscape does not match the existing process reference content, it is essential to generate a mutual understanding throughout the organization of the current As-Is high-level process landscape. The output of step 4e is consumed by step 7, yet step 6 is preceded by step 5 that focuses entirely on process planning and design.

Typical tasks that are done within this step:

- High-level process landscape identification and documentation
- Teams must collaborate and spread awareness of the current As-Is process situation
- Examine process requirements and resources needed
- Identify and categorize the main process areas and process groups
- Identify and categorize the management process areas and process groups
- Identify and categorize the supporting process areas and process groups

Typical templates that are used:

- Information Map and/or Matrix
- Process Map and/or Matrix
- Object Map and/or Matrix
- Service Map and/or Matrix

流程。在这方面,有必要彻底检查这些流程,以便完全基于流程目标和新定义的高级别流程全景抛弃这些流程,或为任何重新设计目的做好准备。步骤4d的输出被步骤4e使用。

在此步骤中完成的典型任务:

- 收集和聚集与可用流程参考内容不匹配的现有流程;
- 检查和记录与可用流程参考内容不匹配的流程。

使用的典型模板:

- 信息图及矩阵;
- 流程图及矩阵;
- 目标图及矩阵;
- BPM符号图及矩阵;
- 应用服务图及矩阵;
- 数据服务图及矩阵。

涉及的典型BPM CoE角色:

- 流程专家;
- 业务分析师。

步骤4e:了解高级流程布局

如果当前流程参考内容不可用,或者当前流程布局与现有流程参考内容不匹配,则必须在当前组织中形成相互理解,就像高级流程布局一样。步骤4e的输出由步骤7使用,而步骤6前面是步骤5,它完全集中于流程规划和设计。

在此步骤中完成的典型任务:

- 高级流程布局识别和记录;
- 团队必须协作并传播对当前流程状况的认识;
- 检查流程要求和所需资源;
- 识别和分类主要流程区域和流程组;
- 识别和分类管理流程区域和流程组;
- 识别和分类支持流程区域和流程组。

使用的典型模板:

- 信息图及矩阵;
- 流程图及矩阵;
- 目标图及矩阵;
- 服务图及矩阵。

Typical BPM CoE roles involved:

- Process eXperts
- Process Architects

Quality Gate 1a:

- Understand As-Is high-level process landscape
- Determine process specifications and the need for resources and knowledge

Step 5: No Process Reference Content Available

If no current process reference content is available for the organization to further build upon, the output of step 5 is consumed by step 4e.

Typical tasks that are done within this step:

- Examine availability of process reference content
- Research for resources of knowledge, know-how, and firmly established best and leading practices to support the creation and/or reengineering of processes

Typical templates that are used:

- Information Map and/or Matrix
- Process Map and/or Matrix
- Object Map and/or Matrix
- BPM Notations Map and/or Matrix
- Application Service Map and/or Matrix
- Data Service Map and/or Matrix

Typical BPM CoE roles involved:

- Process eXperts
- Business Analysts

In the first phase of the BPM Life Cycle, we focused on the identification of critical business factors, description of process goals, and choosing the correct building blocks for the upcoming design phase. We also put much of our effort on process analysis and checking whether any process reference content was currently available. In the next phase of the BPM Life Cycle, the Design phase, we will begin designing our process solutions through planning, definitions, requirements, and standardization (Figure 6).

PHASE 2: DESIGN—PROJECT REALIZATION AND DESIGN

Business process design is the method by which an organization understands and defines the business activities that enable it to function. Process design is concerned with designing business processes to ensure that they are optimized, effective, meet customer requirements and demands, and support and sustain organizational development and growth. A well-designed process will improve efficiency, deliver greater

涉及的典型BPM CoE角色：

- 流程专家；
- 流程架构师。

质量门1a：

- 了解高级流程布局；
- 确定流程规范和对资源和知识的需求。

步骤5：无可用流程参考内容

如果没有当前的流程参考内容可供组织进一步构建，则步骤4e使用步骤5的输出。

在此步骤中完成的典型任务：

- 检查流程参考内容的可用性；
- 研究知识资源、技术诀窍以及稳固的最佳和领导实践，以支持流程的创建及重新设计。

使用的典型模板：

- 信息图及矩阵；
- 流程图及矩阵；
- 目标图及矩阵；
- BPM符号图及矩阵；
- 应用服务图及矩阵；
- 数据服务图及矩阵。

涉及的典型BPM CoE角色：

- 流程专家；
- 业务分析师。

在BPM生命周期的第一个阶段，我们重点关注关键业务因素的识别、流程目标的描述，以及为即将到来的设计阶段选择正确的构建模块。我们还将大量精力放在流程分析和检查当前是否有任何流程参考内容上。在BPM生命周期的下一个阶段，即设计阶段，我们将开始通过计划、定义、需求和标准化来设计流程解决方案（图6）。

3.5.3 第二阶段：设计（项目实现与设计）

业务流程设计是组织理解和定义使其能够运行的业务活动的方法。流程设计涉及设计业务流程，以确保它们得到优化、有效、满足客户要求和需求，以及支持和维持组织的发展和成长。一个设计良好的流程将提高效率，提高生产力，创造更多

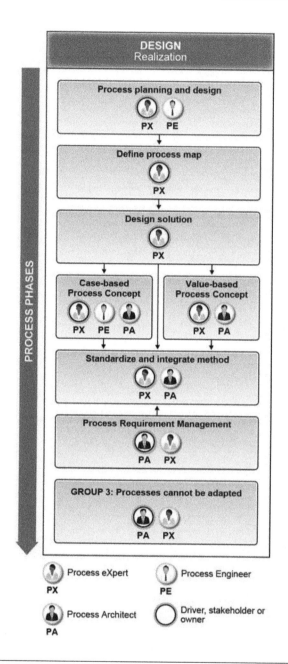

FIGURE 6

The Design Phase of the BPM Life Cycle.

Ref. 7.

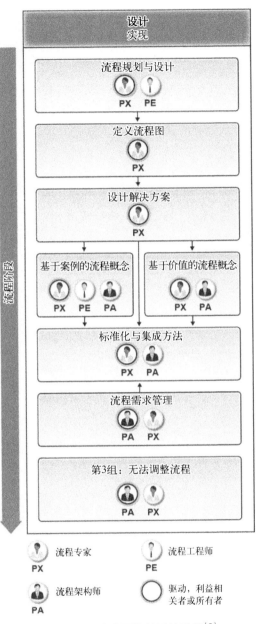

图6　BPM生命周期的设计阶段[7]

productivity, and create more business value.[8] The most common initiatives behind business process design projects are:

- customer and supply chain management;
- operational performance improvement;
- business process integration, standardization, and automation;
- cost reduction; and
- creating new business opportunities.

Business process design typically occurs as an early, critical phase in BPM projects, rather than as an end in itself. The goal of the overall project is to implement business change, whether that change is primarily organizational (improving business operating processes), technical (implementing or integrating software systems), or a combination of the two.

In a process improvement project, the focus of the business process design phase is to streamline the process: to understand and measure the requirements, and to eliminate the risk of losing value through inefficient or inappropriate activities. In a technology implementation project, the focus is on understanding the processes that are being automated, and ensuring that the appropriate technology is selected, configured, and implemented to support them. In both cases, the process design activities can range from modest (e.g., tweak existing processes and look for some quick wins) to aggressive (e.g., identify major opportunities to increase value or drive down costs through radical process improvement or outsourcing). In short, business process design is a tool that can serve many different kinds of projects.

The 2nd phase, the Process Design Phase (see Figure 6), is the phase in which the BPM CoE organization initiates, aligns, arranges, categorizes, defines, determines, as well as quantifies, drafts, outlines, and designs the processes and the process structures. The process design phase considers the identified business requirements and the specific process design considerations for the processes, steps, and activities, as well as events and gateways. Relating requirements and goals to the identified processes applies composition principles and, therefore, process matrices are created to assist project teams in relating the relevant aspects. This phase also includes change management aspects of the defined process innovation and/or transformation. The continuous improvement feedback loop through change management of the BPM Life Cycle, and the likelihood of changes made during this phase, is considered to be medium/high.

Step 6: Process Planning and Design

As a direct continuation from defining and describing the overall process goals, planning and design steps are initiated with the purpose of designing new processes from scratch and/or plan the redesign and reengineering requirements of existing processes. The level of detail in this area is also increased dramatically, as the design process slowly moves away from the high-level process landscape and into a much more detailed process landscape (Figure 7).

的商业价值[8]。业务流程设计项目背后最常见的计划是：

- 客户和供应链管理；
- 运营绩效改进；
- 业务流程一体化、标准化和自动化；
- 降低成本；
- 创造新的商业机会。

业务流程设计通常在BPM项目中作为一个早期的、关键的阶段出现，而不是作为其本身的一个结尾。整个项目的目标是实现业务变更，无论该变更主要是组织的（改进业务操作流程）、技术的（实现或整合软件系统）还是两者的结合。

在流程改进项目中，业务流程设计阶段的重点是简化流程：了解和测量需求，并消除因效率低下或不当活动而失去价值的风险。在技术实施项目中，重点是了解正在被自动化的流程，并确保选择、配置和实施适当的技术来支撑它们。在这两种情况下，流程设计活动的范围都可以是适度的（例如，调整现有流程并寻找一些快速的收益），也可以是积极的（例如，通过彻底的流程改进或外包，确定增加价值或降低成本的主要机会）。简而言之，业务流程设计是一种可以服务于许多不同类型项目的工具。

第二阶段（流程设计阶段，图6）是BPM CoE组织发起、协调、安排、分类、定义、确定以及量化、起草、概述和设计流程和流程结构的阶段。流程设计阶段考虑已确定的业务需求以及流程、步骤和活动及事件和网关的特定流程设计注意事项。应用构成原理将实现需求和目标与已确定的流程相关联，因此，创建流程矩阵以帮助项目团队关联相关方面。此阶段还包括定义的流程创新和转换的变更管理方面，贯穿整个BPM生命周期的变更管理的持续改进反馈循环，以及在此阶段进行变更的可能性被认为是具有中等或高等的重要性。

步骤6：流程规划和设计

作为定义和描述整个流程目标的直接延续，开始规划和设计步骤的目的是从头开始设计新流程或对现有流程的重新设计。随着设计流程慢慢从高层次的流程转移到更为详细的流程，这一领域的细节水平也显著提高（图7）。

The output of step 6 is consumed by step 7.
Typical tasks that are done within this step:

- Determine the need for new main, management, and/or supporting (classification of) processes
- Organize and structure process hierarchy
- Determine and define each required process level
- Gather and categorize process steps, process activities, and events and gateways
- Collect information around process meta objects

Process Workflow Connection Diagram

	STRATEGIC					TACTICAL												OPERATIONAL											
---	1	2	3	4	5	6	7	8	9	10	11	12	13	14	15	16	17	18	19	20	21	22	23	24	25	26	27	28	29
1. Strategy (S)	x	x	x	x	x					x				x															
2. Plan (S)		x	x		x					x		x	x	x						x									
3. Forecast (S)		x	x		x						x																		
4. Value Management (S)	x	x	x		x			x			x					x													
5. Budget (S)		x	x		x			x			x							x		x									
6. Strategic Advice (T)	x	x	x	x						x																			
7. Strategic Guidance & Compliance (T)																			x					x			x	x	
8. Monitoring (T)								x	x	x						x	x												
9. Reporting (T)	x									x					x	x	x												
10. Evaluation and/or Audit (T)																x	x												
11. Policy (T)	x	x				x	x					x	x		x	x	x												
12. Procedures (T)															x	x	x				x								
13. Measurements (T)							x		x							x	x												
14. Administration (T)	x																												
15. Communication (T)																													
16. Performance Management (T)	x			x				x	x	x			x	x							x								
17. Risk Management (T)				x				x																					
18. Administration (O)																		x											
19. Issue Management (O)									x													x	x			x			
20. Operational Planning (O)																		x									x		x
21. Process Management (O)																		x											x
22. Monitoring (O)									x							x						x		x	x	x	x		x
23. Reporting (O)									x											x			x		x		x		x
24. Evaluation and/or Audit (O)									x											x			x		x				x
25. Measurements (O)							x		x							x				x			x	x		x			x
26. Procedures (O)									x									x			x							x	x
27. Operational Advice and/or Support (O)																				x	x								x
28. Operational Guidance & Compliance (O)																				x	x			x	x	x			x
29. Processing (O)																		x					x			x			

FIGURE 7

The process workflow connection diagram is a process matrix that shows the connectivity between the services delivered by business processes in the process landscape. This is a very powerful and important tool to use when designing an organization's business processes as it shows how strategic, tactical, and operational service deliverables relate to one another.

Ref. 9.

步骤6的输出被步骤7使用。

在此步骤中完成的典型任务：

- 确定对新的主要流程、管理流程及支持（分类的）流程的需求；
- 组织和构建流程层次结构；
- 确定和定义每个所需的流程级别；
- 收集和分类流程步骤、流程活动、事件和网关；
- 收集和流程元对象有关的信息。

使用的典型模板：

- 信息图及矩阵；
- 流程图及矩阵；

流程工作流连接图

	战略性					战术性												操作性											
	1.策略(S)	2.计划(S)	3.预测(S)	4.价值度量(S)	5.预算(S)	6.战略性建议(T)	7.战略指导与合规(T)	8.监测(T)	9.报告(T)	10.评估和/或审计(T)	11.政策(T)	12.程序(T)	13.测量(T)	14.管理(T)	15.交流(T)	16.绩效管理(T)	17.风险管理(T)	18.管理(O)	19.问题管理(O)	20.运营计划(O)	21.流程管理(O)	22.检测(O)	23.报告(O)	24.评估和/或审计(O)	25.测量(O)	26.程序(O)	27.操作建议和/或支持(O)	28.操作指导与合规(O)	29.处理(O)
战略性 1.策略(S)	×	×	×	×					×					×															
2.计划(S)		×	×	×					×		×	×	×							×									
3.预测(S)			×	×					×																				
4.价值度量(S)	×	×	×	×				×					×																
5.预算(S)			×	×	×													×	×										
战术性 6.战略性建议(T)	×	×	×	×																									
7.战略指导与合规(T)																			×					×			×	×	
8.监测(T)								×	×	×						×	×												
9.报告(T)	×								×							×	×	×											
10.评估和/或审计(T)																×	×												
11.政策(T)	×	×				×	×				×	×																	
12.程序(T)											×	×	×																
13.测量(T)								×	×							×	×												
14.管理(T)	×																												
15.交流(T)																													
16.绩效管理(T)		×		×				×	×	×											×								
17.风险管理(T)			×						×																				
操作性 18.管理(O)																		×											
19.问题管理(O)									×										×	×			×						
20.运营计划(O)																		×									×		×
21.流程管理(O)																		×											×
22.检测(O)									×							×						×		×	×	×	×		×
23.报告(O)									×														×						×
24.评估和/或审计(O)									×															×		×			×
25.测量(O)							×		×							×						×			×	×			
26.程序(O)									×															×				×	×
27.操作建议和/或支持(O)																			×	×						×	×		
28.操作指导与合规(O)																			×	×		×				×			
29.处理(O)																		×					×						

图7　流程工作流连接图是一个流程矩阵，显示流程布局中业务流程提供的服务之间的连接。这是在设计组织的业务流程时使用的一个非常强大和重要的工具，因为它显示了战略、战术和运营服务交付之间的关系[9]

Typical templates that are used:

- Information Map and/or Matrix
- Process Map and/or Matrix
- Object Map and/or Matrix
- BPM Notations Map and/or Matrix
- Role Map and/or Matrix
- Owner Map and/or Matrix
- Requirement Map and/or Matrix
- Workflow Map and/or Matrix
- Application Service Map and/or Matrix
- Data Service Map and/or Matrix
- Application Rule Map and/or Matrix
- Data Rule Map and/or Matrix

Typical BPM CoE roles involved:

- Process eXperts
- Process Engineers
- Enterprise Architects

Step 7: Define Process Map

When preparing for the design of the process solution, definition of the organization's process maps, matrices, and models is needed to create a foundation upon which to build the organization's future processes. The design foundation also defines criteria for the redesign and/or reengineering of existing processes. The output of step seven is consumed by step 8.

Typical tasks that are done within this step:

- Define process content and the process maps, matrices, and models to be used
- Define relationships between process levels 1–5 and business goals and objectives
- Identify and define business process areas and groups
- Identify and define business processes, steps, and activities
- Identify and define stakeholders, process owners, managers, and roles
- Identify and define the required resources

Typical templates that are used:

- Process Map
- Service Map
- Application Service Map
- Data Service Map

Typical BPM CoE roles involved:

- Process eXperts
- Business Analysts

- 目标图及矩阵；
- BPM 符号图及矩阵；
- 角色图及矩阵；
- 所有者图及矩阵；
- 需求图及矩阵；
- 工作流程图及矩阵；
- 应用服务图及矩阵；
- 数据服务图及矩阵；
- 应用程序规则图及矩阵；
- 数据规则图及矩阵。

涉及的典型 BPM CoE 角色：

- 流程专家；
- 流程工程师；
- 企业架构师。

步骤7：定义流程图

在准备流程方案设计时，需要定义组织的流程图、矩阵和模型，以创建组织未来流程的基础。设计基础还定义了现有流程的重新设计及业务流程再设计的标准。步骤7的输出被步骤8使用。

在此步骤中完成的典型任务：

- 定义流程内容和要使用的流程图、矩阵和模型；
- 定义流程级别1 ~ 5与业务目的和目标之间的关系；
- 识别和定义业务流程领域和组别；
- 识别和定义业务流程、步骤和活动；
- 识别和定义利益相关者、流程所有人、经理和角色；
- 确定和定义所需资源。

使用的典型模板：

- 流程图；
- 服务图；
- 应用服务图；
- 数据服务图。

涉及的典型 BPM CoE 角色：

- 流程专家；
- 业务分析师。

Quality Gate 2a:

- Process map definition
- Ensure full compatibility with high-level process landscape
- Design process maps

Step 8: Design Solution

Designing the organization's process solution is a huge task in itself, but is fully and extensively supported by the detailed process analysis and documentation done during phase 1. The process solution is designed through an extensive use of detailed process maps, matrices, and models. The process maps, matrices, and models represent information of how the processes relate to the business layer (purpose and goals, value aspects, business competencies, and business services), the application layer (software and data), as well as the underlying technology layer (platform and infrastructure) of the organization. The output of step 8 is consumed by step 9; however, if the need for business cases is present, it would be consumed by step 10, and then consumed by step 11.

Typical tasks that are done within this step:

- Develop and design process maps, matrices, and models
- Develop and design the process meta model to illustrate the connections and relationships between process meta objects and the identified business, application, and technology meta objects

Typical templates that are used:

- Information Map and/or Matrix
- Process Map and/or Matrix
- Object Map and/or Matrix
- BPM Notations Map and/or Matrix
- Role Map and/or Matrix
- Owner Map and/or Matrix
- Requirement Map and/or Matrix
- Workflow Map and/or Matrix
- Application Service Map and/or Matrix
- Data Service Map and/or Matrix
- Application Rule Map and/or Matrix
- Data Rule Map and/or Matrix

Typical BPM CoE roles involved:

- Process eXperts
- Business Analysts

Step 9: Case-Based Process Concept

While designing the process solution, it is of great importance to simultaneously develop a case-based process concept that will serve to illustrate how new

质量门 2a:

- 流程图定义;
- 确保与高级流程布局完全兼容;
- 设计流程图。

步骤 8: 设计方案

设计组织的流程方案本身就是一项艰巨的任务,但在第一阶段完成的详细流程分析和文档为其提供了全面和广泛的支持。流程方案是通过广泛使用详细的流程图、矩阵和模型来设计的。流程图、矩阵和模型表示流程如何与组织的业务层(目的和目标、价值方面、业务能力和业务服务)、应用层(软件和数据)以及基础技术层(平台和基础设施)相关的信息。步骤 8 的输出由步骤 9 使用,但是,如果存在对业务案例的需求,那么它将由步骤 10 使用,然后再由步骤 11 使用。

在此步骤中完成的典型任务:

- 开发和设计流程图、矩阵和模型;
- 开发和设计流程元模型,以说明流程元对象与已识别的业务、应用程序和技术元对象之间的连接和关系。

使用的典型模板:

- 信息图及矩阵;
- 流程图及矩阵;
- 目标图及矩阵;
- BPM 符号图及矩阵;
- 角色图及矩阵;
- 所有者图及矩阵;
- 需求图及矩阵;
- 工作流程图及矩阵;
- 应用服务图及矩阵;
- 数据服务图及矩阵;
- 应用程序规则图及矩阵;
- 数据规则图及矩阵。

涉及的典型 BPM CoE 角色:

- 流程专家;
- 业务分析师。

步骤 9: 基于案例的流程概念

在设计流程方案时,同时开发一个基于案例的概念流程非常重要,该概念流程

processes will function in practice, and how re-engineered processes will support the objectives and goals of the organization. In reality, one may consider a case-based process concept to be constructed much like a business case, although this particular case evolves predominantly around the technical requirements, capabilities, and functionality of both the high level and the detailed process structures that are to be built for the organization. The output of step 9 is consumed by step 11, unless a business case is needed, then the output of step 9 would be consumed by step 10.

Typical tasks that are done within this step:

- Include and document process meta objects
- Include and relate process meta objects to application and technology meta objects
- Create a storyline with illustrations of graphical models that show functionality, principles, and behavior

Typical templates that are used:

- Value Map, Matrix, and/or Model
- Cost Map, Matrix, and/or Model
- Revenue Map, Matrix, and/or Model
- Competency/Business Model Map, Matrix, and/or Model
- Requirement Map, Matrix, and/or Model
- Vision, Mission and Goals Map, Matrix, and/or Model
- Stakeholder Map, Matrix, and/or Model
- Strategy Map, Matrix, and/or Model
- Case Map, Matrix, and/or Model

Typical BPM CoE roles involved:

- Process eXperts
- Process Engineers
- Process Architects

Step 10: Value-Based Process Concept

Much like a case-based process concept, the value-based process concept focuses almost exclusively on the relationship between business processes and the value concepts of the organization. It is typically created to demonstrate how processes relate specifically to each business goal and objective as well as the previously documented value and performance drivers of the organization. The output of step 10 is consumed by step 11.

Typical tasks that are done within this step:

- Relate the value objects, i.e., strategic business objectives (SBOs), critical success factors (CSFs), and performance indicators (KPIs) to the processes
- Sort processes according to the SBOs and CSFs
- Document the connection between process objects and value objects

将有助于说明新流程在实践中的作用,以及重新设计的流程如何支持组织的目的和目标。实际上,我们可以考虑构建一个与业务案例非常相似的基于案例的概念流程,尽管这个特定的案例主要围绕以下需求不断演化:适用于组织的高级和详细流程结构的技术需求、能力和功能进行开发。步骤9的输出由步骤11使用,除非需要业务案例,否则步骤9的输出将由步骤10使用。

在此步骤中完成的典型任务:

- 包含和记录流程元对象;
- 将流程元对象包括并关联到应用程序和技术元对象;
- 创建一个带有图形模型插图的故事情节,显示功能、原则和行为。

使用的典型模板:

- 价值图、矩阵及模型;
- 成本图、矩阵及模型;
- 收入图、矩阵及模型;
- 能力/业务模型图、矩阵及模型;
- 需求图、矩阵及模型;
- 愿景、任务和目标图、矩阵及模型;
- 利益相关者图、矩阵及模型;
- 战略图、矩阵及模型;
- 案例图、矩阵及模型。

涉及的典型BPM CoE角色:

- 流程专家;
- 流程工程师;
- 流程架构师。

步骤10:基于价值的流程概念

与基于案例的流程概念非常相似,基于价值的流程概念几乎只关注业务流程和组织的价值概念之间的关系。通常创建它是为了演示流程如何与每个业务目的和目标以及以前记录的组织价值和绩效驱动因素相关。步骤10的输出被步骤11使用。

在此步骤中完成的典型任务:

- 将价值对象,即战略业务目标(SBO)、CSF和KPI与流程相关联;
- 根据SBO和CSF对流程进行排序;
- 记录流程对象和价值对象之间的联系。

Typical templates that are used:

- Value Map, Matrix, and/or Model
- Cost Map, Matrix, and/or Model
- Revenue Map, Matrix, and/or Model
- Competency/Business Model Map, Matrix, and/or Model
- Requirement Map, Matrix, and/or Model
- Vision, Mission and Goals Map, Matrix, and/or Model
- Stakeholder Map, Matrix, and/or Model
- Strategy Map, Matrix, and/or Model
- Case Map, Matrix, and/or Model

Typical BPM CoE roles involved:

- Value eXperts
- Process eXperts
- Process Architects

Step 11: Standardize and Integrate

Standardization and integration of the previously designed solution is a necessary step of any BPM project. The definitions have to be clearly documented to categorize which processes can, or should be, standardized, and, in continuation thereof, how they will be integrated. For clear BPM definitions, see the chapter on BPM Ontology. The standardization and integration of the processes are done in the following sequence: output of step 11 is consumed by step 14 and step 15 simultaneously to document the As-Is analysis to prepare for the creation of the To-Be value-driven process design. The output of step 11 is also consumed by step 18 for matching processes to existing reference content.

Typical tasks that are done within this step:

- Identify which processes need to be standardized and/or integrated
- Investigate and thoroughly examine available methods for standardization
- Investigate and thoroughly examine available methods for integration
- Agree on and choose best possible methods for process standardization and integration
- Standardize and integrated chosen processes
- Document changed processes

Typical templates that are used:

- Process Map, Matrix, and/or Model
- Service Map, Matrix, and/or Model
- Operating Map, Matrix, and/or Model
- Information Map, Matrix, and/or Model

Typical BPM CoE roles involved:

- Process eXperts
- Process Architects

使用的典型模板：

- 价值图、矩阵及模型；
- 成本图、矩阵及模型；
- 收入图、矩阵及模型；
- 能力/业务模型图、矩阵及模型；
- 需求图、矩阵及模型；
- 愿景、任务和目标图、矩阵及模型；
- 利益相关者图、矩阵及模型；
- 战略图、矩阵及模型；
- 案例图、矩阵及模型。

涉及的典型BPM CoE角色：

- 价值专家；
- 流程专家；
- 流程架构师。

步骤11：标准化和整合

以前设计方案的标准化和一体化是任何BPM项目的必要步骤。必须清楚地记录定义，以便对哪些流程可以或应该标准化进行分类，并在后续的流程中，对这些流程进行整合。有关清晰的BPM定义，请参阅有关BPM本体的章节。流程的标准化和整合按照以下顺序进行：步骤11的输出被步骤14和步骤15同时使用，以记录As-Is分析，去准备创建To-Be价值驱动的流程设计。步骤11的输出也被步骤18使用，用于将流程与现有引用内容匹配。

在此步骤中完成的典型任务：

- 确定哪些流程需要标准化及整合；
- 调查和彻底检查可用的标准化方法；
- 调查和彻底检查可用的整合方法；
- 商定并选择流程标准化和整合的最佳方法；
- 标准化和整合所选流程；
- 记录变更流程。

使用的典型模板：

- 流程图、矩阵及模型；
- 服务图、矩阵及模型；
- 操作图、矩阵及模型；
- 信息图、矩阵及模型。

涉及的典型BPM CoE角色：

Value Gate 2:

- Investigate process standardization opportunities
- Investigate direct process integration opportunities
- Document changed processes

Step 12: Process Requirement Management

Process requirements need to be clearly defined and documented. They are essential and critical information for the purpose of the process design solution, and will contain useful information, such as requirements around:

- Business resources/actors
- Organizational requirements (i.e., does the organization contain process-educated personnel)
- Knowledge of process architecture and process modeling
- Application-layer-specific requirements (i.e., has the process design software been decided upon, does the data support thereof exist, etc.)
- Technology-layer-specific requirements (i.e., does the platform and infrastructure inventory support the automation of processes, etc.)

For the purpose of defining and delivering process requirements for each of these steps, the output of step 12 can be simultaneously consumed by steps 15, 16, 17, and 18.

Typical tasks that are done within this step:

- Identify and categorize process requirements based on resources, business needs and wants, as well as application- and technology-layer aspects

Typical templates that are used:

- Process Map, Matrix, and/or Model
- Requirement Map, Matrix, and/or Model
- Service Map, Matrix, and/or Model
- Application Service Map and/or Matrix
- Data Service Map and/or Matrix

Typical BPM CoE roles involved:

- Process Architects
- Process eXperts
- Value eXperts

Quality Gate 2b:

- Supplier management
- Evaluate and establish
- Categorize suppliers and maintain supply chain design (SCD)
- Manage performance and renew/terminate

- 流程专家；
- 流程架构师。

价值门2：

- 调查流程标准化机会；
- 调查直接流程整合机会；
- 记录变更流程。

步骤12：流程需求管理

流程需求需要明确定义和记录。它们是流程设计方案的基本和关键信息，将包含有用的信息，如以下方面的要求：

- 业务资源/参与者；
- 组织要求（即组织是否包含流程培训人员）；
- 流程架构和流程建模知识；
- 应用层特定要求（即是否已确定流程设计软件，是否存在其数据支持等）；
- 技术层特定需求（即平台和基础设施库存是否支持流程自动化等）。

为了定义和交付每个步骤的流程要求，步骤12的输出可以同时被步骤15、16、17和18使用。

在此步骤中完成的典型任务：

- 根据资源、业务需要和需求以及应用程序和技术层方面识别和分类流程要求。

使用的典型模板：

- 流程图、矩阵及模型；
- 需求图、矩阵及模型；
- 服务图、矩阵及模型；
- 应用服务图及矩阵；
- 数据服务图及矩阵。

涉及的典型BPM CoE角色：

- 流程架构师；
- 流程专家；
- 价值专家。

质量门2b：

- 供应商管理；
- 评估和建立；
- 对供应商进行分类并维护供应链设计（supply chain design，SCD）；
- 管理绩效和更新/终止。

Step 13: Processes Cannot Be Adapted

Unadaptable processes that are encountered during the Design phase often require back stepping to step 4d to establish a new or, at the very least, an updated As-Is high-level process landscape. This proactive move enables process owners and decision makers to continuously analyze and plan for the support and requirement of substitute processes that are to be designed and built during the Design and Build phases. The output of step 13 is consumed by step 4d in the Analyze phase because these processes are evaluated as being unadaptable.

Typical tasks that are done within this step:

- Document and compile a list of processes that are expected to be re-engineered and modified to fit with the established process landscape
- Document and compile a list of unadaptable processes
- Eliminate processes that are evaluated as unusable or irrelevant to the established process landscape

Typical templates that are used:

- Process Map, Matrix, and/or Model
- Requirement Map, Matrix, and/or Model
- Information Map, Matrix, and/or Model
- Object Map, Matrix, and/or Model
- Service Map, Matrix, and/or Model

Typical BPM CoE roles involved:

- Process Architects
- Process eXperts

In the second phase of the BPM Life Cycle, we have developed process design solutions to create a strong foundation for business process development to prepare for the upcoming build phase. In the third phase of the BPM Life Cycle, the Build phase (see Figure 8), we turn our attention toward executing upon the previously defined process solution plans and framework based upon the As-Is situation, aiming toward a To-Be value-driven process design. We will be matching existing processes to any available process reference content, and harmonize variants, while documenting changes prior to process release and going live.

PHASE 3: BUILD—FINAL PROJECT PREPARATION

Process models (on all process levels 1–5) are created as flow charts to give a clear, graphical indication of *what* happens *when*.

High-level process models (process levels 1–2)—usually referred to as Value Chain Diagrams (VCDs)—are used to illustrate how primary and supporting (secondary) business units work together to fulfill one or more specific goals. A value chain is then a chain of activities that an organization performs to deliver a product or a service to the customer (whether internal or external).

步骤13：无法适应的流程

在设计阶段遇到的不可适应的流程通常需要后退到步骤4d,以建立新的或至少是更新的As-Is高级流程布局。这一积极主动的行动将帮助流程所有人和决策者能够持续分析和策划在设计和构建阶段被设计和构建的替代流程的支持和需求。步骤13的输出在分析阶段由步骤4d使用,因为这些流程被评估为不可适应。

在此步骤中完成的典型任务：

- 记录并编制一份流程清单,该清单预计将重新设计和修改,以适应既定的流程布局;
- 记录和编制不可适应流程列表;
- 消除被评估为不可用或与既定流程布局无关的流程。

使用的典型模板：

- 流程图、矩阵及模型;
- 需求图、矩阵及模型;
- 信息图、矩阵及模型;
- 目标图、矩阵及模型;
- 服务图、矩阵及模型。

涉及的典型BPM CoE角色：

- 流程架构师;
- 流程专家。

在BPM生命周期的第二阶段,我们开发了流程设计方案,为业务流程开发创造了坚实的基础,为即将到来的构建阶段做准备。在BPM生命周期的第三个阶段,即构建阶段(图8),我们将注意力转向根据现状执行先前定义的流程方案计划和框架,以实现价值驱动的流程设计。我们将使现有流程与任何可用的流程参考内容相匹配,并协调变体,同时在流程发布和上线之前记录更改。

3.5.4　第三阶段：构建（最终项目准备）

流程模型(在所有流程级别1～5上)都创建为流程图,来给出一个清晰的图形化的什么时候要发生什么的标示。

高级流程模型(流程级别1～2)通常被称为价值链图(VCDs),用于说明首要和支撑(次要)业务单位如何协作以实现一个或多个特定目标。价值链是组织为向客户交付产品或服务(无论是内部的还是外部的)而执行的一系列活动。

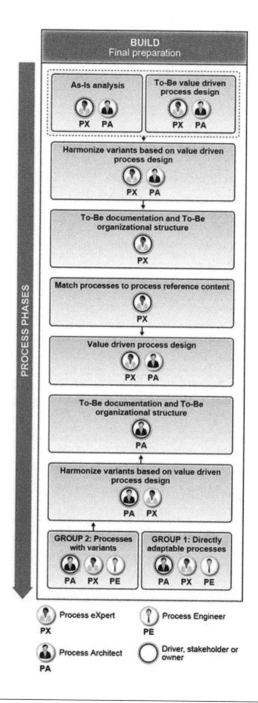

FIGURE 8

The Build phase of the BPM Life Cycle.

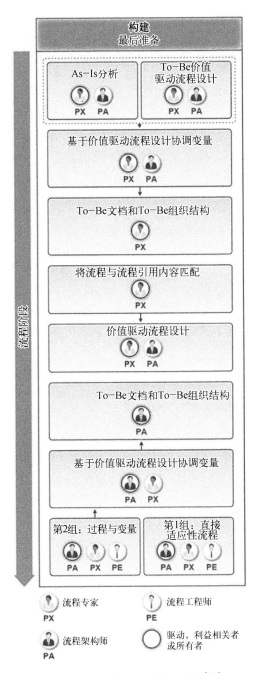

图8 BPM生命周期的构建阶段[10]

Business process models (process level 3) are used to demonstrate the activities of an organization or within or around a specific project team (they can also move across multiple business units), such as planning activities, people's actions (what they do), and reactions (internal and/or external outputs) necessary to carry out given tasks within the organization. Creating a business model can, for an example, give the employees of an organization an easy-to-use reference guide that outlines the tasks that they are expected of them to carry out, including their responsibilities and the steps necessary to complete each task correctly and proficiently.

Business Process Modeling Notation (BPMN) models (process levels 4 and 5) is a standard for business process modeling that provides a graphical notation for specifying business processes in a Business Process Diagram (BPD), based on a flowcharting technique very similar to activity diagrams from Unified Modeling Language (UML). The objective of BPMN is to support business process management for both technical and business users, by providing notation that is intuitive to business users, yet able to represent complex process semantics. The BPMN specification also provides a mapping between the graphics of the notation and the underlying constructs of execution languages, particularly Business Process Execution Language (BPEL).

The primary goal of BPMN is to provide a standard notation readily understandable by all stakeholders. These include the process experts, process engineers, and process architects who create and refine the processes, the technical developers (process engineers and process architects) responsible for implementing them, and the business managers (process owners, process experts, process architects, and business analysts) who monitor and manage them. Consequently, BPMN serves as a common language, bridging the communication gap that frequently occurs between business process design, development, execution, monitoring and optimization.

Regardless of whether a business process model's level (levels 1–5, the higher the more detailed, i.e., levels 4 and 5 are considered BPMN diagrams), the same rules apply to all process models.

A business process, therefore:

1. Has a goal.
2. Has specific inputs.
3. Has specific outputs.
4. Uses resources.
5. Has a number of activities that are performed in some order.
6. May affect more than one organizational unit.
7. Creates value of some kind for the customer (internal or external).

The 3rd phase, the Process Build Phase (see Figure 8), is the phase in which the BPM CoE organization builds, creates, develops, and crafts the processes and the process structures of the organization (see Figure 9). The process build phase takes into account the identified business requirements and the different process design solutions that have been generated for the purpose of process construction.

业务流程模型(流程级别3)用于演示组织或特定项目团队内部或周围的活动(它们也可以跨多个业务单元移动),如计划活动、人员的操作(它们所做的)和执行在组织中给定的任务所需的反应(内部及外部输出)。例如,创建一个业务模型可以为组织的员工提供一个易于使用的参考指南,该指南概述了他们期望执行的任务,包括他们的职责和正确、熟练地完成每个任务所需的步骤。

业务流程建模标记法(BPMN)模型(流程级别4和5)是业务流程建模的标准,它基于流程图技术,与UML中的活动图非常相似,为在业务流程图(BPD)中指定业务流程提供了图形符号。BPMN的目标是通过提供对业务用户直观的符号为技术用户和业务用户支持BPM,同时能够表示复杂的流程语义。BPMN规范还提供了符号图形与执行语言,特别是业务流程执行语言(BPEL)的底层结构之间的映射。

BPMN的主要目标是提供一个所有利益相关者都容易理解的标准符号。其中包括创建和完善流程的流程专家、流程工程师和流程架构师,负责实施流程的技术开发人员(流程工程师和流程架构师)以及监控和管理流程的业务经理(流程所有人、流程专家、流程架构师和业务分析师)。因此,BPMN作为一种通用语言,弥合了业务流程设计、开发、执行、监控和优化之间经常出现的通信鸿沟。

不管业务流程模型的级别(级别1～5,越高越详细,即级别4和5被视为BPMN图),相同的规则适用于所有流程模型。

因此,一个业务流程:

(1)有一个目标;

(2)具有特定的输入;

(3)有特定的输出;

(4)使用资源;

(5)具有按某种顺序执行的多个活动;

(6)可能影响多个组织单位;

(7)为客户创造某种价值(内部或外部)。

第三阶段的流程构建阶段(图8)是BPM CoE组织构建、创建、开发和制作流程和组织的流程结构的阶段(图9)。流程构建阶段考虑到已确定的业务需求和为流程构建而生成的不同流程设计解决方案。在这个阶段中创建流程模型。此阶段

This phase is where the process models are created. This phase also includes change management aspects of the To-Be process innovation and/or transformation enabled in the value-driven process design. The continuous improvement feedback loop through change management of the BPM Life Cycle, and the degree of changes made during this phase, is considered low/medium.

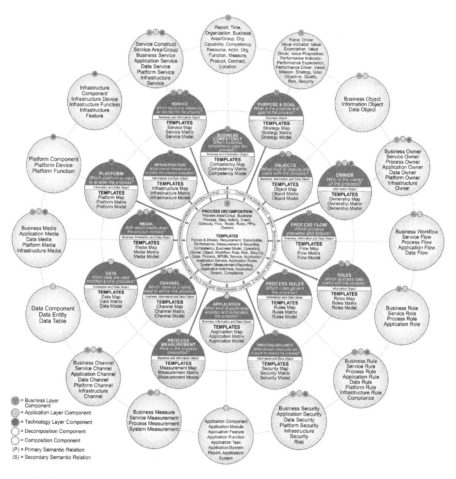

FIGURE 9

The process decomposition and composition model is used to show which meta-objects are combined when building (composing) a business process and which meta-objects are extracted by disassembling (decomposing) a business process. This particular illustration shows intricate detail, as it also connects each specific process-related area with which type of objects (business, information, and data) are used in the creation process. Not only that, it also shows whether the semantic relation between the related meta-objects is primary (required) or secondary (optional).

Ref. 11.

还包括价值驱动流程设计中实现的待处理创新及转换的变更管理方面。通过对BPM生命周期的变更管理持续改进反馈循环，在此阶段所做的变更程度被认为是低/中的。

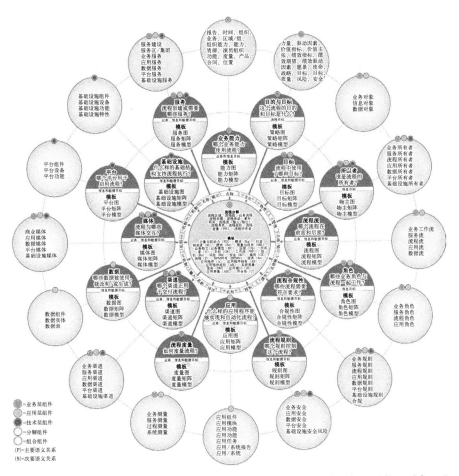

图9　流程分解和组合模型用于显示在构建（组合）业务流程时组合了哪些元对象，以及通过拆解（分解）业务流程提取了哪些元对象。这个特殊的插图显示了复杂的细节，因为它还连接了在创建过程中使用的对象类型（业务、信息和数据）的每个特定流程相关区域。不仅如此，它还显示了相关元对象之间的语义关系是首要的（必需的）还是次要的（可选的）[11]

Step 14: As-Is Analysis

The As-Is analysis of the currently developed process design serves as a staging point for the creating the To-Be value-driven process design. In today's organizations, the focus on getting the most value out of the process requires understanding the as-is situation and thereby understanding the different capabilities. The output of step 14 is consumed by step 16.

Typical tasks that are done within this step:

- Conduct thorough process analysis across the current process landscape
- Specify which processes should be blueprinted (as-is blueprint) for the build phase
- Put the As-Is situation into BPM Notations
- Identify which challenges correspond to the processes

Typical templates that are used:

- Process Map, Matrix, and/or Model
- Information Map, Matrix, and/or Model
- Object Map, Matrix, and/or Model
- Service Map, Matrix, and/or Model
- BPM Notations Map, Matrix, and/or Model
- Application Service Map and/or Matrix
- Data Service Map and/or Matrix

Typical BPM CoE roles involved:

- Process eXperts
- Process Architects

Step 15: To-Be Value-Driven Process Design

Value-driven process design has but a single focus point; to create and build processes that generate and deliver previously unacquired value to the company and its customers or to enhance already established processes that have been re-engineered to produce even more value than before. The output of step 15 is consumed by step 16.

Typical tasks that are done within this step:

- Align value objects (based on step 10) with the defined process goals
- Sort processes according to the SBOs and CSFs
- Focus on the creation of process designs that put process objects into relationships with value objects across the organization
- Focus on value designs shaped by process structures
- Model the process based on relationships and connectivity between process and value objects
- Identify duplication of processes, business functions, and services
- Review and document current baseline and compare viewpoints of possible solutions

步骤14：As-Is（现状）分析

对当前开发的流程设计的As-Is（现状）分析是创建价值驱动的流程设计的一个阶段。在当今的组织中，要想从流程中获得最大价值就需要了解As-Is情况，从而了解不同的能力。步骤14的输出被步骤16使用。

在此步骤中完成的典型任务：

- 对当前流程布局进行全面的流程分析；
- 为构建阶段指定应蓝图化的流程（如As-is蓝图）；
- 将As-Is情况放入BPM符号中；
- 确定哪些挑战对应于流程。

使用的典型模板：

- 流程图、矩阵及模型；
- 信息图、矩阵及模型；
- 目标图、矩阵及模型；
- 服务图、矩阵及模型；
- BPM符号图、矩阵及模型；
- 应用服务图及矩阵；
- 数据服务图及矩阵。

涉及的典型BPM CoE角色：

- 流程专家；
- 流程架构师。

步骤15：To-Be（未来）价值驱动流程设计

价值驱动的流程设计只有一个重点：创建和构建流程，为公司及其客户创造并提供以前未获得的价值，或者增强已经重新设计的已建立的流程，以产生比以前更多的价值。步骤15的输出被步骤16使用。

在此步骤中完成的典型任务：

- 使价值对象（基于步骤10）与定义的流程目标保持一致；
- 根据SBOs和CSFs对流程进行排序；
- 专注于创建流程设计，将流程对象与整个组织的价值对象建立关系；
- 关注流程结构形成的价值设计；
- 根据流程和价值对象之间的关系和连接建立流程模型；
- 识别流程、业务功能和服务的重复；
- 审查和记录当前基线并比较可能方案的观点。

Typical templates that are used:

- Process Map, Matrix, and/or Model
- Information Map, Matrix, and/or Model
- Object Map, Matrix, and/or Model
- Service Map, Matrix, and/or Model
- BPM Notations Map, Matrix, and/or Model
- Value Map, Matrix, and/or Model
- Application Service Map and/or Matrix
- Data Service Map and/or Matrix

Typical BPM CoE roles involved:

- Process eXperts
- Value eXperts
- Process Architects

Value Gate 3:

- Value-driven process design
- Ensure that process designs are focused on continuous value delivery

Step 16: Harmonize Variants Based on Value-Driven Process Design

In the context of Business Process Management, harmonization defines the extent of standards and how they fit together, but it does not attempt to make different standards uniform. Harmonization avoids a one-size-fits-all approach. It makes the trade-off between too many and too few process standards and avoids inconsistencies between standards.[12] The output of step 16 is consumed by step 17.

Typical tasks that are done within this step:

- Identify, assess, and establish process level commonality across the organization
- Identify, assess, and establish process harmonization opportunities across the organization
- Identify, assess, and establish process standardization opportunities across the organization

Typical templates that are used:

- Process Map, Matrix, and/or Model
- Information Map, Matrix, and/or Model
- Object Map, Matrix, and/or Model
- Service Map, Matrix, and/or Model
- BPM Notations Map, Matrix, and/or Model
- Value Map, Matrix, and/or Model
- Application Service Map and/or Matrix
- Data Service Map and/or Matrix

使用的典型模板：

- 流程图、矩阵及模型；
- 信息图、矩阵及模型；
- 目标图、矩阵及模型；
- 服务图、矩阵及模型；
- BPM符号图、矩阵及模型；
- 价值图、矩阵及模型；
- 应用服务图及矩阵；
- 数据服务图及矩阵。

涉及的典型BPM CoE角色：

- 流程专家；
- 价值专家；
- 流程架构师。

价值门3：

- 价值驱动的流程设计；
- 确保流程设计专注于持续价值交付。

步骤16：基于价值驱动的流程设计的协调变量

在BPM的阐述中，协调定义了标准的范围以及它们如何组合在一起，但它并不试图统一标准。协调避免了一刀切的做法，它使得过多和过少的流程标准之间实现了平衡，避免了标准之间的不一致[12]。步骤16的输出被步骤17使用。

在此步骤中完成的典型任务：

- 识别、评估和建立整个组织的流程级共性；
- 识别、评估和建立整个组织的流程协调机会；
- 识别、评估和建立整个组织的流程标准化机会。

使用的典型模板：

- 流程图、矩阵及模型；
- 信息图、矩阵及模型；
- 目标图、矩阵及模型；
- 服务图、矩阵及模型；
- BPM符号图、矩阵及模型；
- 价值图、矩阵及模型；
- 应用服务图及矩阵；
- 数据服务图及矩阵。

Typical BPM CoE roles involved:

- Process eXperts
- Process Architects
- Value eXperts

Step 17: To-Be Documentation and To-Be Organizational Structure

The To-Be definitions of the process, its structure, and how it relates to the organizational structure (the As-Is as well as the To-Be) is relevant to both the process designs as well as the innovation and transformation potential of the organization. The way this is done is to:

1. Relate it through the process maps and process matrices to processes on all 5 levels:
   a. Process Level 1: Process Areas
   b. Process Level 2: Process Groups
   c. Process Level 3: Business Processes
   d. Process Level 4: Process Steps
   e. Process Level 5: Process Activities (including sub-processes)
2. Adapt the BPM Notations based on the To-Be changes

   The output of step 17 is consumed by step 21.
   Typical tasks that are done within this step:

- Document process direction and organizational structure
- Document high-level To-Be process landscape
- Document detailed To-Be process landscape
- Document process descriptions and technicalities

   Typical templates that are used:

- Process Map and/or Matrix
- Organizational Chart Map and/or Matrix
- Service Map and/or Matrix

   Typical BPM CoE roles involved:

- Process eXperts
- Enterprise Architects

   Quality Gate 3:

- Process documentation and organization
- Establish and develop thorough process definition documentation
- Develop and establish process-oriented organizational structure
- Process report and evaluation

涉及的典型BPM CoE角色：

- 流程专家；
- 流程架构师；
- 价值专家。

步骤17：To-Be（未来）文档和To-Be（未来）组织结构

流程的To-Be定义、其结构及其与组织结构（As-Is和To-Be）的关系既与流程设计有关，也与组织的创新和转型潜力有关。这样做的目的如下。

（1）通过流程图和流程矩阵将其与所有5个级别的流程关联起来。

a. 流程级别1：流程区域；

b. 流程级别2：流程组；

c. 流程级别3：业务流程；

d. 流程级别4：流程步骤；

e. 流程级别5：流程活动（包括子流程）。

（2）根据要更改的内容调整BPM标记。

步骤17的输出被步骤21使用。

在此步骤中完成的典型任务：

- 记录流程方向和组织结构；
- 记录高水平的流程布局；
- 详细记录流程布局；
- 记录流程说明和技术指标。

使用的典型模板：

- 流程图及矩阵；
- 组织结构图及矩阵；
- 服务图及矩阵。

涉及的典型BPM CoE角色：

- 流程专家；
- 企业架构师。

质量门3：

- 流程文档和组织；
- 建立和开发完整的流程定义文档；
- 开发和建立面向流程的组织结构；
- 流程报告和评估。

Step 18: Match Processes to Process Reference Content

Matching of existing processes to any available process reference content of the organization is required. This enables process owners and other organizational roles to reduce the number of processes that are used by the organization, and to lessen the chance of overly complex process portfolios in the process landscape. The output of step 17 is consumed by step 15.

Typical tasks that are done within this step:

- Compare existing processes to the process reference content (if any)
- Chose relevant process reference content
- Tailor process reference content
- Agree and align tailor process reference content with own process content

Typical templates that are used:

- Process Map and/or Matrix
- Service Map and/or Matrix
- Object Map and/or Matrix
- Application Service Map and/or Matrix
- Data Service Map and/or Matrix

Typical BPM CoE roles involved:

- Process eXperts
- Business Analysts

Step 19: Processes with Variants

If the initiation of a BPM Life-cycle project has been decided upon and is to be carried out with the intent of updating, reengineering, and/or re-evaluating an already established existing process landscape, processes with different variants will likely already exist in multiple places within the organization. Therefore, it is essential that such processes are identified, accounted for, and re-engineered through process variant harmonization to support the goals of the value-driven process design that has previously been decided upon by the management. In this regard, see step 16. Step 19 is grouped with steps 13 and 20, and relate directly to the process reference content in Phase 1: Analyze. The output of step 19 is consumed by step 16.

Typical tasks that are done within this step:

- Identify and document processes with variants
- Outline their behavior compared to existing processes of the same type
- Catalog a process portfolio of all identified process variants
- Identify duplication (see harmonization and standardization)

Typical templates that are used:

- Process Map and/or Matrix
- Service Map and/or Matrix
- Object Map and/or Matrix

步骤18：将流程与流程参考内容匹配

需要将现有流程与组织的任何可用流程引用内容进行匹配。这使流程所有人和其他组织角色能够减少组织使用的流程数量，并减少流程环境中过于复杂的流程组合的机会。步骤17的输出被步骤15使用。

在此步骤中完成的典型任务：

- 将现有流程与流程参考内容（如果有）进行比较；
- 选择相关的流程参考内容；
- 定制流程参考内容；
- 同意并协调裁剪流程参考内容与自己的流程内容。

使用的典型模板：

- 流程图及矩阵；
- 服务图及矩阵；
- 目标图及矩阵；
- 应用服务图及矩阵；
- 数据服务图及矩阵。

涉及的典型 BPM CoE 角色：

- 流程专家；
- 业务分析师。

步骤19：流程变更

如果已经决定启动一个 BPM 生命周期项目，并打算以更新、重新设计及重新评估一个已经建立的现有流程布局的目的来执行，则组织内的多个地方可能已经存在具有不同变体的流程。因此，必须通过流程变量协调来识别、说明和重新设计这些流程，以支持先前由管理层决定的价值驱动过程设计的目标。在这方面，请参见步骤16。步骤19与步骤13和步骤20分组，并直接与阶段1：分析中的流程参考内容相关。步骤19的输出被步骤16使用。

在此步骤中完成的典型任务：

- 识别和记录具有变体的流程；
- 与同一类型的现有流程相比，概述其行为；
- 列出所有已识别流程变量的流程组合；
- 识别重复（参考协调和标准化）。

使用的典型模板：

- 流程图及矩阵；
- 服务图及矩阵；
- 目标图及矩阵；

- Application Service Map and/or Matrix
- Data Service Map and/or Matrix

Typical BPM CoE roles involved:

- Process Architects
- Process eXperts
- Process Engineers

Step 20: Directly Adaptable Processes

If the initiation of a BPM Life-cycle project has been decided upon and is to be carried out with the intent of updating, reengineering, and/or re-evaluating an already established existing process landscape, directly adaptable processes simply need to be either customized or re-aligned to support the plans for the value-driven process design. For a value-driven process concept design, see steps 10 and 17. This step is grouped with steps 13 and 19, and relates directly to the process reference content in phase 1. Step 20 output is consumed by step 17.

Typical tasks that are done within this step:

- Identify process variants that can be directly adapted
- Document the process variants that can be directly adapted to fit the defined process landscape
- Catalog a process portfolio of all identified and directly adaptable processes
- Publish directly adaptable processes

Typical templates that are used:

- Process Map and/or Matrix
- Service Map and/or Matrix
- Object Map and/or Matrix

Typical BPM CoE roles involved:

- Process Architects
- Process eXperts
- Process Engineers

In the 3rd phase of the BPM Life Cycle, we went through a series of steps intent on crafting the business processes of the organization. In the 4th phase of the BPM Life Cycle however, the Deploy/Implement phase (see Figure 10), it is time to prepare for Release and Deployment Management of the business processes and going live. We do this by creating a process-rollout plan, adding rewards and incentives as well as prepare and enable performance measurements.

PHASE 4: DEPLOY/IMPLEMENT—GO LIVE

The 4th phase, the Process Deployment and Implementation phase (see Figure 10), is the phase in which the organization launches, implements, executes, deploys, activates, completes, concludes, and transitions the processes to execution (go live).

- 应用服务图及矩阵；
- 数据服务图及矩阵。

涉及的典型BPM CoE角色：

- 流程架构师；
- 流程专家；
- 流程工程师。

步骤20：流程的直接应用

如果已经确定了BPM生命周期项目的启动，并且要对流程潜力进行更新、重新设计或重新评估，那么直接应用流程只需进行定制或重新调整，以支持价值驱动的流程设计计划有关价值驱动的流程概念设计，请参阅步骤10和步骤17。此步骤与步骤13和步骤19分组，并直接与阶段1中的流程参考内容相关。步骤20的输出由步骤17使用。

在此步骤中完成的典型任务：

- 识别可直接适应的流程变量；
- 记录可直接适应所定义流程布局的流程变体；
- 列出所有已识别和直接适应性流程的流程组合；
- 直接发布适应性强的流程。

使用的典型模板：

- 流程图及矩阵；
- 服务图及矩阵；
- 目标图及矩阵。

涉及的典型BPM CoE角色：

- 流程架构师；
- 流程专家；
- 流程工程师。

在BPM生命周期的第三个阶段，我们经历了一系列旨在制定组织业务流程的步骤。然而，在BPM生命周期的第四个阶段，即部署/实施阶段（图10），是时候为业务流程的发布和部署管理以及上线做准备了。我们通过创建流程展示计划、添加奖赏和奖励以及准备和启用绩效度量来实现这一点。

3.5.5　第四阶段：部署/实施（上线）

第四阶段是流程部署和实施阶段（图10），是组织启动、实施、执行、部署、激活、完成、结束和将流程转换为执行（上线）的阶段。BPM生命周期中的流程发布

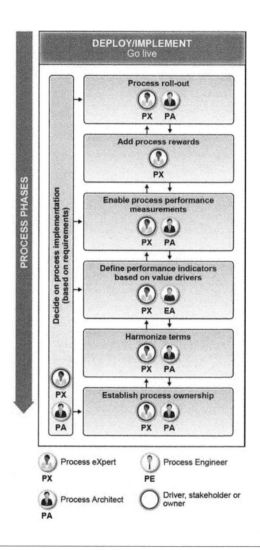

FIGURE 10

The Deploy/Implement phase of the BPM Life Cycle.

Ref. 13.

The Process Release and Deployment Management in the BPM Life Cycle aims to plan, schedule, and control the movement of releases to test in live environments. The primary goal of Release and Deployment Management is to ensure that the integrity of the live environment is protected and that the correct components are released on time and without errors.

Release and Deployment Management aims to build, test, and deliver services to the customers specified by process design by deploying releases into operation, and establishing effective use of the service to deliver value to the customer. As illustrated in Figure 11, process implementation involves multiple aspects from coordination with process owners, change management, to process training.

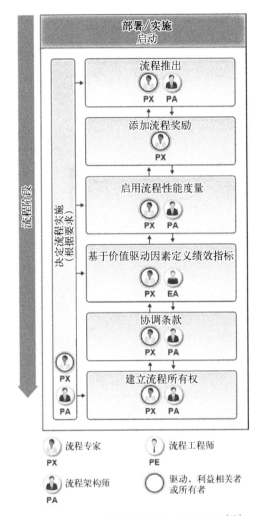

图10 BPM生命周期的部署/实施阶段[13]

和部署管理旨在计划、安排和控制发布在活动环境中进行测试的活动。发布和部署管理的主要目标是确保活动环境的完整性得到保护,并确保正确的组件按时发布且没有错误。

发布和部署管理的目的是通过将已发行的流程设计部署到运行中,并建立有效的服务使用来为客户提供价值,从而为指定的客户构建、测试和交付服务。如图11所示,流程实施涉及多个方面,从与流程所有人的协调、变更管理到流程培训。

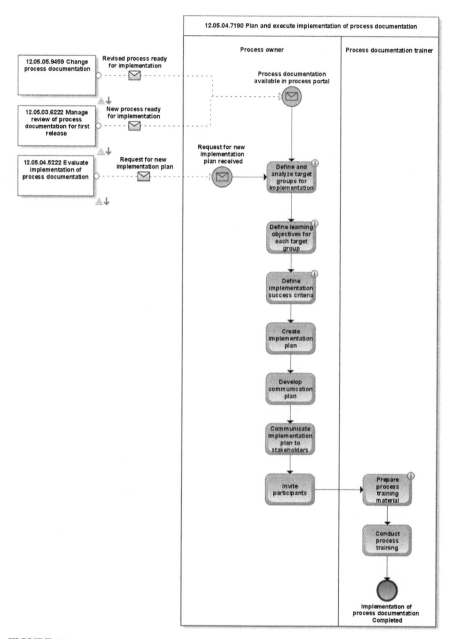

FIGURE 11

Example of a process rollout diagram (Lego Group, Anette Falk Bøgebjerg, Director).

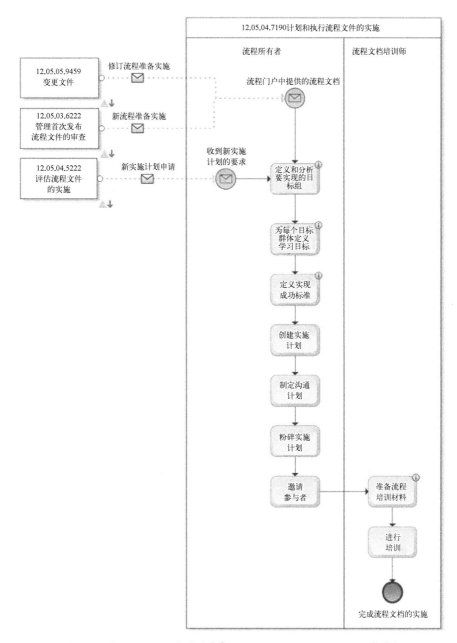

图11　流程展示图示例（乐高集团，Anette Falk Bogebjerg，董事）

The purpose of Release and Deployment Management is to:

- Define and agree release and deployment plans with customers/stakeholders
- Ensure that each release package consists of a set of related assets and service components that are compatible with each other
- Ensure that integrity of a release package and its constituent components is maintained throughout the transition activities and recorded accurately in the configuration management system
- Ensure that all release and deployment packages can be tracked, installed, tested, verified, and/or uninstalled or backed out, if appropriate
- Ensure that change is managed during the release and deployment activities
- Record and manage deviations, risks, and issues related to the new or changed service, and take necessary corrective action
- Ensure knowledge transfer to enable the customers and users to optimize their use of the service to support their business activities
- Ensure that skills and knowledge are transferred to operations and support staff to enable them to effectively and efficiently deliver, support, and maintain the service, according to required warranties and service levels

Plans for release and deployment will be linked into the overall service transition plan. The approach is to ensure an acceptable set of guidelines is in place for the release into production/operation. Release and deployment plans should be authorized as part of the change management process.

The plan should define the:

- Scope and content of the release
- Risk assessment and risk profile for the release
- Customers/users affected by the release
- Change advisory board (CAB) members that approved the change request for the release and/or deployment
- Team who will be responsible for the release
- Delivery and deployment strategy
- Resources for the release and deployment

Build and test planning establishes the approach to building, testing, and maintaining the controlled environments prior to production. The activities include:

- Developing build plans from the service design package, design specifications, and environment configuration requirements
- Establishing the logistics, lead times, and build times to set up the environments
- Testing the build and related procedures
- Scheduling the build and test activities
- Assigning resources, roles, and responsibilities to perform key activities
- Preparing build and test environments
- Managing test databases and test data

发布和部署管理的目的是：

- 与客户/利益相关者定义并同意发布和部署计划；
- 确保每个发行包包含一组相互兼容的相关资产和服务组件；
- 确保在整个过渡活动中保持发行包及其组成组件的完整性，并在配置管理系统中准确记录；
- 确保所有发行和部署包都可以被跟踪、安装、测试、验证及卸载，或者在适当的情况下退出；
- 确保在发布和部署活动期间对变更进行管理；
- 记录和管理与新服务或变更服务相关的偏差、风险和问题，并采取必要的纠正措施；
- 确保知识传授，使客户和用户能够优化服务的使用，以支持其业务活动；
- 确保将技能和知识传授给运营和支持人员，使他们能够根据所需的保证和服务级别有效、高效地提供、支持和维护服务。

发布和部署计划将链接到整体服务过渡计划中。该方法是为了确保为发布到生产/操作中制定一套可接受的指导方针。发布和部署计划应作为变更管理流程的一部分进行授权。

该计划应规定：

- 发布的范围和内容；
- 发布的风险评估和风险概况；
- 受发布影响的客户/用户；
- 批准发布及部署变更请求的变更咨询委员会（change advisory board，CAB）成员；
- 负责发布的团队；
- 交付和部署策略；
- 发布和部署的资源。

构建和测试计划建立了在生产之前构建、测试和维护受控环境的方法。这些活动包括：

- 根据服务设计包、设计规范和环境配置要求制定构建计划；
- 建立物流、交付周期和构建时间以建立环境；
- 测试构建和相关程序；
- 安排构建和测试活动；
- 分配资源、角色和职责以执行关键活动；
- 准备构建和测试环境；
- 管理测试数据库和测试数据；

- Software license management

Procedures, templates, and guidance should be used to enable the release team to build an integrated release package efficiently and effectively. Procedures and documents will be required for purchasing, distributing, installing, moving, and controlling assets and components that are relevant to acquiring, building, and testing a release.[14]

Step 21: Decide on Process Implementation (Based on Requirements)

Develop a plan for implementing the processes and the tools in the organization. This plan should describe how to efficiently move from the organization's current state to the release and deployment state. To develop this plan, you need to follow specific project steps.[15] The output of step 21 is consumed by steps 22, 24, 25, and 27.

Typical tasks that are done within this step:

- Set or revise goals
- Identify risks
- Distribute responsibilities and tasks
- Decide when to launch processes and tools
- Plan training and mentoring

Typical templates that are used:

- Process Map and/or Matrix
- Service Map and/or Matrix
- Stakeholder Map and/or Matrix
- Object Map and/or Matrix

Typical BPM CoE roles involved:

- Process eXperts
- Process Architects

Step 22: Process Rollout

During the rollout phase, all areas of change are tested together in the business environment to generate confidence that everything is ready to "go live." During this phase, business users and support teams also receive appropriate training concerning the new processes and the associated systems, organization, and infrastructure.[16] The process rollout should be meticulously executed by using a step-by-step approach and also categorized into levels of importance, preferably based on criteria such as complexity, time, cost, and urgency as well as with clearly defined steps for when the main, supporting, and management process rollouts should occur, and in what sequence. The output of step 22 is consumed by steps 23 and 28.

● 软件许可管理。

发布团队应使用流程、模板和指导以便于能够高效地构建整合的发布包。在采购、分发、安装、移动和控制与获取、构建和测试发布相关的资产和组件过程中需要程序和文档[14]。

步骤21：决定流程实施（基于要求）

制定一个在组织中实施流程和工具的计划。这个计划应该描述如何有效地从组织的当前状态转移到发布和部署状态。要制定这个计划，您需要遵循特定的项目步骤[15]。步骤22、步骤24、步骤25和步骤27使用步骤21的输出。

在此步骤中完成的典型任务：

● 设定或修订目标；

● 识别风险；

● 分配职责和任务；

● 决定何时启动流程和工具；

● 计划培训和指导。

使用的典型模板：

● 流程图及矩阵；

● 服务图及矩阵；

● 利益相关者图及矩阵；

● 目标图及矩阵。

涉及的典型BPM CoE角色：

● 流程专家；

● 流程架构师。

步骤22：流程展示

在展示阶段，所有变化的领域都在业务环境中进行测试，以产生一种信心，即一切都准备好"投入使用"。在此阶段，业务用户和支持团队还将接受有关新流程和相关系统、组织和基础设施的适当培训[16]。应使用一步一步的方法仔细地执行流程展示，并将其按重要性分级，最好是基于复杂性、时间、成本和紧急程度等标准，以及明确定义的步骤主要展示、支持和管理流程应在何时发生以及以何种顺序发生的步骤。步骤22的输出由步骤23和步骤28使用。

Typical tasks that are done within this step:

- Process rollout
- Ensure end-to-end process rollout and consistency
- Bring all processes up to target performance
- Business users and process team training
- Test process capability and process adjustment
- Manage issue management and change-request handling
- Implement all the components of the solution

Typical templates that are used:

- Process Map and/or Matrix
- Service Map and/or Matrix
- Object Map and/or Matrix
- Application Service Map and/or Matrix
- Data Service Map and/or Matrix
- Application Rule Map and/or Matrix
- Data Rule Map and/or Matrix
- Compliance Map and/or Matrix

Typical BPM CoE roles involved:

- Process eXperts
- Process Architects

Quality Gate 4:

- Process rollout
- Ensure process quality
- Ensure process coverage

Step 23: Add Process Rewards

Process reward recognition is not just a nice thing to do for the organization or its employees. Process reward recognition is a communication tool that reinforces and rewards the most important process outcomes that people create for your organization. When you recognize people effectively, you reinforce, with your chosen means of process reward recognition, the actions and behaviors you most want to see people repeat. Therefore, process rewards should be defined and created to incite employee motivation for successful implementation, and as rewards for achieving process and value goals. The output of step 23 is consumed by steps 22 and 24.

Typical tasks that are done within this step:

- Establish criteria for what process performance or process contribution constitutes behavior or actions that are rewarded
- All employees must be eligible for the process reward

在此步骤中完成的典型任务：

- 流程发布；
- 确保端到端流程的发布和一致性；
- 使所有流程达到目标绩效；
- 业务用户和流程团队培训；
- 测试流程能力和流程调整；
- 管理发行管理和变更请求的处理；
- 实施解决方案的所有组件。

使用的典型模板：

- 流程图及矩阵；
- 服务图及矩阵；
- 目标图及矩阵；
- 应用服务图及矩阵；
- 数据服务图及矩阵；
- 应用程序规则图及矩阵；
- 数据规则图及矩阵；
- 合规图及矩阵。

涉及的典型BPM CoE角色：

- 流程专家；
- 流程架构师。

质量门4：

- 流程展示；
- 确保流程质量；
- 确保流程覆盖范围。

步骤23：添加流程奖励

流程奖励不仅仅是为组织或其员工做的一件好事，而是一种沟通工具，它强化和奖励人们为组织创建的最重要的流程结果。当您有效地识别出人员时，您会通过您选择的流程奖励识别的方式，强化您最希望看到人们重复的行动和行为。因此，应定义和创建流程奖励，以激励员工去成功实施的积极性，并作为实现流程和价值目标的奖励。步骤23的输出被步骤22和步骤24使用。

在此步骤中完成的典型任务：

- 为流程绩效或流程贡献构成奖励的行为或行动建立标准；
- 所有员工必须有资格获得流程奖励；

- Implement process rewards into the process performance model
- Build organizational motivation for chasing process rewards to elevate process performance
- The process reward recognition should occur as close to the performance of the actions as possible, so the recognition reinforces behavior the employer wants to encourage.

Typical templates that are used:

- Value Map and/or Matrix
- Stakeholder Map
- Organizational Chart Map
- Performance Map and/or Matrix

Typical BPM CoE roles involved:

- Process eXperts
- Value eXperts

Step 24: Enable Process Performance Measurements

Process performance measurement is the process of collecting, analyzing, and reporting information regarding the process performance of a group of processes or an individual process. Enabling performance measurements for processes on all measureable levels is an essential behavior of any BPM Life-cycle project and directly links to monitoring, reporting, decision making, as well as process evaluation and audits. The output of step 24 is consumed by steps 23 and 25.

Typical tasks that are done within this step:

- Develop measurement metrics for a process performance model
- Define and relate Process Performance Indicators (PPIs) for process levels 3–5
- Enable Process Performance Reporting and Evaluation
- Identify, categorize, and label Strategic, Tactical, and Operational Process Performance Indicators
- Associate and categorize processes the strategic, Tactical, and Operational Process Performance Indicators to the relevant performance goals/objectives
- Create a Performance Model with decision making and reporting that illustrates the connection and relationship between Strategic, Tactical, and Operational Process Performance Indicators and the business goals and objectives.

Typical templates that are used:

- Process Map and/or Matrix
- Measurement and Reporting Map and/or Matrix
- Performance Map and/or Matrix

Typical BPM CoE roles involved:

- 在流程绩效模型中实施流程奖励;
- 建立追求流程奖励的组织动力,以提高流程绩效;
- 流程奖励识别应尽可能接近行动的执行,因此识别会强化雇主希望鼓励的行为。

使用的典型模板:

- 价值图及矩阵;
- 利益相关者图;
- 组织结构图;
- 绩效图及矩阵。

涉及的典型BPM CoE角色:

- 流程专家;
- 价值专家。

步骤24:启用流程绩效度量

流程绩效度量是收集、分析和报告一组流程或单个流程的流程绩效信息的程序。在所有可度量级别上为流程启用绩效度量是任何BPM生命周期项目的基本行为,并直接链接到监控、报告、决策以及流程评估和审计。步骤24的输出由步骤23和步骤25使用。

在此步骤中完成的典型任务:

- 为流程绩效模型制定度量标准;
- 定义和关联流程3～5级的PPI;
- 实现流程绩效报告和评估;
- 识别、分类和标记战略、战术和操作PPI;
- 将战略、战术和操作PPI与相关绩效目标关联并分类;
- 通过决策和报告创建绩效模型,说明战略、战术和操作PPI与业务目的和目标之间的联系和关系。

使用的典型模板:

- 流程图及矩阵;
- 度量和报告图及矩阵;
- 绩效图及矩阵。

涉及的典型BPM CoE角色:

- Process eXperts
- Process Architects
- Value eXperts

Value Gate 4a:

- Process performance measurements
- Performance measurement tools efficiency
- Process efficiency evaluation
- Process reporting and evaluation

Step 25: Define Performance Indicators Based on Value Drivers

Establishing direct links between performance indicators and value drivers is essential for both process-modeling and value-modeling perspectives. Therefore, because value drivers indicate value-generating mechanisms, it is important to define performance indicators and let them be based on predefined value drivers. This enables process owners to control and measure the flow of value within the processes on both the high-level and detailed process landscape. The output of step 25 is consumed by steps 24 and 26.

Typical tasks that are done within this step:

- Define, associate, and relate the Process Performance Indicators based on Value Drivers
- Develop value measurements linked to the process performance measurements
- Enable Value based Reporting and Evaluation
- Create a Value Model with decision making and reporting that illustrates the connection and relationship between performance indicators and value indicators.

Typical templates that are used:

- Process Map and/or Matrix
- Performance Map and/or Matrix
- Value Map and/or Matrix

Typical BPM CoE roles involved:

- Value eXperts
- Process eXperts
- Enterprise Architects

Value Gate 4b:

- Process performance indicators establishment
- Number of process targets reached
- Number of process targets obsolete
- Increase/decrease number of process targets

- 流程专家；
- 流程架构师；
- 价值专家。

价值门4a：
- 流程绩效测量；
- 绩效衡量工具效率；
- 流程效率评估；
- 流程报告和评估。

步骤25：基于价值驱动因素定义绩效指标

在绩效指标和价值驱动因素之间建立直接联系对于流程建模和价值建模观点都是至关重要的。因此，由于价值驱动因素表示价值生成机制，定义绩效指标并使其基于预定义的价值驱动因素是很重要的。这使流程所有人能够控制和测量流程中高级和详细流程布局上的价值流。步骤24和步骤26使用步骤25的输出。

在此步骤中完成的典型任务：
- 根据价值驱动因素定义、联系和关联PPI；
- 开发与流程绩效度量相关的价值度量；
- 实现基于价值的报告和评估；
- 创建一个具有决策和报告功能的价值模型，说明绩效指标和价值指标之间的联系和关系。

使用的典型模板：
- 流程图及矩阵；
- 绩效图及矩阵；
- 价值图及矩阵。

涉及的典型BPM CoE角色：
- 价值专家；
- 流程专家；
- 企业架构师。

价值门4b：
- 建立PPI；
- 达到的流程目标数量；
- 废弃的流程目标数量；
- 增加/减少流程目标数量。

Step 26: Harmonize Terms

The harmonization of process terms across the process landscape has to be continuously evaluated and managed by process owners and teams. Different BPM-oriented organizations and groups today have the tendency to call certain process objects various names, and the same thing goes for the various BPM frameworks, methods, and approaches, such as Six Sigma, Lean, and BPR that use terms in specialized ways. Several business process methodologies have described the use of terms in specific ways. Formal business process languages, like BPML, have semantic definitions that are enforced by the language. Unfortunately, many of these different sources use terms in slightly different ways.[17] We have, therefore, provided basis process terminology and definitions in the BPM ontology chapter. However, it will still be necessary for any organization to tailor these terms, gather additionally needed terms, and establish their own documentation for process terms and definitions to be able to harmonize variants across process groups and process areas to achieve process harmonization (i.e., standardization and integration). The output of step 26 is consumed by steps 25 and 27.

Typical tasks that are done within this step:

- Identify, assess, and establish process-level commonality across the organization
- Identify, assess, and establish process-harmonization opportunities across the organization
- Gather existing process terminology or use the BPM Ontology terminology as a basis to identify relevant terms
- Agree on process terms relevant for the organization
- Ensure process-term harmonization across the organization
- Identify, assess, and establish process-standardization opportunities across the organization

Typical templates that are used:

- Process Map and/or Matrix
- Information Map and/or Matrix
- Service Map and/or Matrix
- Object Map and/or Matrix

Typical BPM CoE roles involved:

- Process eXperts
- Process Architects

Step 27: Establish Process Ownership

As process owners are responsible for the management of processes within the organization, the success of the organization's BPM initiatives depends heavily on implementing good process ownership (see Figure 12). Regardless of the maturity model

步骤26：协调术语

流程所有人和团队必须持续评估和管理整个流程布局中流程术语的协调。今天，不同的以BPM为导向的组织和集团倾向于称某些流程对象为不同的名称，同样的事情也适用于不同的BPM框架、方式和方法，如以特定的方式使用术语的六西格玛、精益和BPR。一些业务流程方法以特定的方式描述了术语的使用。正式的业务流程语言（如BPML）具有由语言强制执行的语义定义。不幸的是，这些不同来源中的许多使用术语的方式略有不同[17]。因此，我们在BPM本体一节中提供了基本的流程术语和定义。但是，对于任何组织来说，仍然有必要修改这些术语，收集额外需要的术语，并为流程术语和定义建立自己的文档，以便能够协调跨流程组和流程区域的变量，以实现流程协调（即标准化和一体化）。步骤26的输出由步骤25和步骤27使用。

在此步骤中完成的典型任务：

- 识别、评估和建立整个组织的流程级共性；
- 识别、评估和建立整个组织的流程协调机会；
- 收集现有流程术语或使用BPM本体术语作为识别相关术语的基础；
- 同意与组织相关的流程术语；
- 确保整个组织的流程术语协调；
- 识别、评估和建立整个组织的流程标准化机会。

使用的典型模板：

- 流程图及矩阵；
- 信息图及矩阵；
- 服务图及矩阵；
- 目标图及矩阵。

涉及的典型BPM CoE角色：

- 流程专家；
- 流程架构师。

步骤27：建立流程所有权

流程所有人管理负责组织内流程，因此组织取得BPM计划的成功在很大程度上取决于实施良好的流程所有权（图12）。无论组织应用的成熟度模型是什么，流程所有权的创建或分配通常都会比现状高一级。然而，为什么这很困难？具有讽

Business Architecture: Value Map, Performance Map & Process Map Alignment Worksheet

Business Competency	Strategic Business Objectives	Ownership	Critical Success Factors	Ownership	Key Performance Indicators	Major Business Process and Performance Measures	Ownership	Activity and Performance Measures
						Process:		Process:
						M1.		M1.
						M2.		M2.
						Process:		Process:
						M1.		M1.
						M2.		M2.
						Process:		Process:
						M1.		M1.
						M2.		M2.
						Process:		Process:
						M1.		M1.
						M2.		M2.

Ref. 18.

FIGURE 12

A table tool that can be used to link process ownership with value maps and performance maps.

业务架构：价值图、绩效图和流程图协调工作表

业务能力	战略业务目标	所有权	关键成功因素	所有权	关键绩效指标	主要业务流程和绩效指标	所有权	活动和绩效指标
						流程：		流程：
						测量1		测量1
						测量2		测量2
						流程：		流程：
						测量1		测量1
						测量2		测量2
						流程：		流程：
						测量1		测量1
						测量2		测量2
						流程：		流程：
						测量1		测量1
						测量2		测量2

图 12　可用于将流程所有权与价值图和绩效图链接起来的图表工具[18]

being applied by an organization, the creation or assignment of process ownership normally occurs one level up from the status quo. However, why is this difficult? Ironically, one of the most neglected areas of process transformation in any kind of change is the definition and assignment of roles and responsibilities. Although there is now a general acknowledgement that people are one of (if not *the* most) critical success factors in any type of business transformation, most organizations are not very accomplished at implementing "people"-oriented changes.[19] In some cases, process owners are current leaders/managers, and in other cases, process owners may be taken from nonleadership positions. Organizational management and structure is an effective tool to use to establish process ownership along with a clear definition of employee requirements and responsibilities. This, at the same time, also incites the need for documenting definitions around process roles, responsibilities, and the who-does-what structure within process-specific teams. The output of step 27 is consumed by step 26.

Typical tasks that are done within this step:

- Specify process ownership responsibility and tasks
- Select process owners
- Implement a process-ownership organization
- Appoint key process roles reporting or working with process owner
- Develop and implement process-improvement initiatives
- Define the process and monitor process performance
- Develop and manage policies and procedures related to the process
- Ensure process adoption, harmonization, standardization, and integration
- Enable process innovation and transformation (link to BPM Change Management and Continuous Improvement)

Typical templates that are used:

- Process Map
- Information Map
- Owner Map and/or Matrix

Typical BPM CoE roles involved:

- Process eXperts
- Process Architects

In the 4th phase of the BPM Life Cycle, we went through a series of steps to execute a successful Release and Deployment Management plan to take the business processes out of the production environment and go live. In the upcoming Run and Maintain phase (see Figure 13), we focus on management of the running process environment in which we will put a lot of effort into monitoring and governing the entire process landscape of the organization.

刺意味的是，在任何类型的变更中，流程转换最被忽视的领域之一就是角色和职责的定义和分配。尽管现在人们普遍承认，在任何类型的业务转型中，人员都是（如果不是最重要的）关键的成功因素之一，但大多数组织在实现"以人为本"的变革方面并不是很成功[19]。

在某些情况下，流程所有人是当前的领导者/经理，而在其他情况下，流程所有人可以从非领导职位上获得。组织管理和结构是建立流程所有权的有效工具，同时明确定义了员工的要求和职责。同时，这也激发了对流程角色、职责以及谁在流程特定团队中执行什么结构的定义进行文档化的需求。步骤27的输出被步骤26使用。

在此步骤中完成的典型任务：
- 指定流程所有权责任和任务；
- 选择流程所有人；
- 实施流程所有权组织；
- 指定关键流程角色报告或与流程所有人合作；
- 制定和实施流程改进计划；
- 定义流程并监控流程绩效；
- 制定和管理与流程相关的政策和程序；
- 确保流程采用、协调、标准化和一体化；
- 实现流程创新和转型（链接到BPM变更管理和持续改进）。

使用的典型模板：
- 流程图；
- 信息图；
- 所有者图及矩阵。

涉及的典型BPM CoE角色：
- 流程专家；
- 流程架构师。

在BPM生命周期的第四个阶段，我们经历了一系列步骤来执行成功的发行和部署管理计划，以将业务流程从生产环境中剥离并上线。在即将到来的运行和维护阶段（图13），我们专注于运行流程环境的管理，在运行流程环境中，我们将投入大量精力来监控和管理组织的整个流程布局。

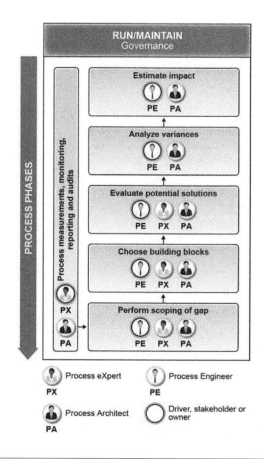

FIGURE 13

The Run/Maintain phase of the BPM Life Cycle.

Ref. 20.

PHASE 5: RUN/MAINTAIN—RUN PROCESSES AND GOVERN PERFORMANCE

The 5th phase, the Process Run and Maintain Phase (see Figure 13), is the phase in which we govern and monitor the active processes that were deployed and implemented during the previous phase. Governance, derived from the Greek verb (kubernáo)—which means to *steer*—is essentially the act of governing what already exists or is in the process of getting developed, deployed/implemented, and/or something that is running.

The *LEADing Practice Way of Process Governance* relates to decisions and guidance that define expectations and direction, grant power, or verify and ensure value identification and creation. It consists of process governance within the entire process life cycle in terms of process analysis, design, construction (build), implementation, and execution (run/maintain), and allows for process monitoring and governance as well as continuous process improvements and optimization disciplines.

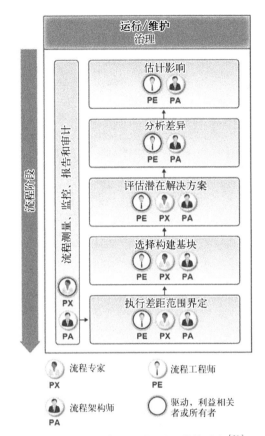

图13 BPM生命周期的运行/维护阶段[20]

3.5.6 第五阶段：运行/维护（运行流程并控制绩效）

第五阶段是流程运行和维护阶段（图13），是我们管理和监控在前一阶段部署和实施的活动流程的阶段。治理（governance），源自希腊动词（kubernáo），意思是引导，本质上是管理已经存在的或正在开发、部署/实施及正在运行的事物的行为。

流程治理的主导实践方式涉及定义期望和方向、授予权力或验证和确保价值识别和创造的决策和指导。它包括整个流程生命周期内的流程治理，包括流程分析、设计、构建（建立）、实施和执行（运行/维护），并允许流程监控和治理以及持续的流程改进和优化规程。

The governance phase also includes the many different relationships among the many practitioners in the mentioned phases to ensure that each task enables specific value identification, creation, and realization in achieving the outlined goals. Process governance involves setting standards and priorities for BPM efforts, identifying process governance leaders, and defining BPM project participant roles—all for the purpose of executing and improving upon an organization's process transformation and innovation strategies. The ultimate goal of both business governance and process governance is to both optimize an organization's business processes and make workflow more efficient and effective by implementing and using the built-in continuous improvement concept during phase 6 of the BPM Life Cycle.

As a part of the Continuous Improvement concept, the process governance steps include the establishment of internal BPM or process centers of excellence or competency centers to share process improvement, best practices, as well as leading practices applied within the organization, and spread awareness of the process standards and priorities. Process governance also works to monitor and document both the successes and shortcomings of an organization's operational execution.

In Business Process Management, an additional purpose of governance is to assure (sometimes on behalf of others in terms of stakeholders) that an organization produces the defined pattern of good results, while avoiding an undesirable pattern of bad circumstances. Therefore, the process governance and Continuous Improvement processes and systems are typically administered by a governance body.

Business process governance is often overseen by teams made up of both business and IT professionals. The daily process governance consists of assuring, on behalf of those governed, the desired business innovation, transformation, and value creation while avoiding an undesirable pattern of high cost, process ineffectiveness, and process inefficiency (low performance). Process governance, therefore, consists of the set of governance gates within the life cycle that ensures quality and value aspects within the various phases and tasks.

Step 28: Process Measurements, Monitoring, Reporting, and Audits

Immediately after going live with the processes, it is important to establish an effective way of monitoring and governing the processes while being able to capture real-time data on measurements for the purpose of reporting process performance (see Figure 14) and performing follow-up audits. The output of step 28 is consumed by step 29.

Typical tasks that are done within this step:

- Specify process measurements
- Select real-time process monitoring and governance
- Capture process performance measurements (see Figure 15)
- Document and performance measurement results for reporting and auditing

治理阶段还包括上述阶段中许多从业者之间的许多不同关系,以确保每个任务在实现所概述的目标时能够实现特定的价值识别、创造和认识。流程治理包括为BPM工作设定标准和优先级,确定流程治理领导者,以及定义BPM项目参与者角色,所有这些都是为了执行和改进组织的流程转型和创新战略。业务治理和流程治理的最终目标是通过在BPM生命周期的第六阶段实施和使用内置的持续改进概念来优化组织的业务流程,并使工作流更加高效和有效。

作为持续改进概念的一部分,流程管理步骤包括建立内部的BPM或卓越流程中心或能力中心,以共享流程改进、最佳实践以及组织内应用的领导实践,并传播对流程标准和优先级的认识。流程治理还可以监控和记录组织的操作执行的成功和不足。

在业务流程管理中,治理的另一个目的是确保(有时代表其他利益相关者)一个组织生成已定义的良好结果模式,同时避免不良情况模式。因此,流程治理、流程和系统的持续改进通常由治理机构管理。

业务流程治理通常由业务和IT专业人员组成的团队进行监督。日常流程治理包括代表被治理者确保所需的业务创新、转型和价值创造,同时避免高成本、流程无效和流程效率低下(低绩效)的不良模式。因此,流程治理由生命周期内的一系列治理组成,这些关口确保了各个阶段和任务中的质量和价值方面。

步骤28:流程测量、监控、报告和审计

在实施流程后,必须立即建立有效的流程监控和管理方法,同时能够获取测量的实时数据,以便报告流程绩效(图14)和执行后续审计。步骤28的输出被步骤29使用。

在此步骤中完成的典型任务:

- 指定流程测量;
- 选择实时流程监控和治理;
- 捕获流程绩效度量(图15);
- 记录和执行测量结果,以便报告和审计。

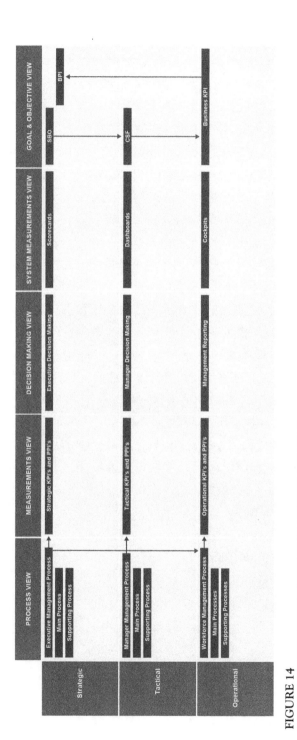

FIGURE 14

How process monitoring links to process measurements and reporting.

Ref. 21.

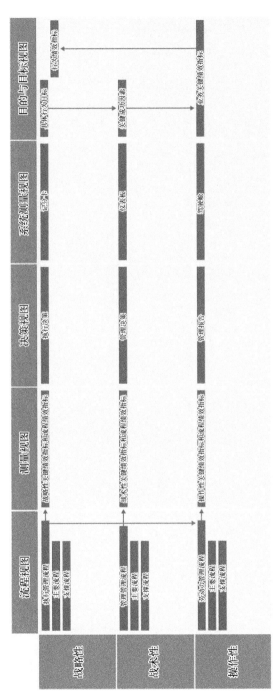

图 14 流程监控如何链接到流程测量和报告[21]

FIGURE 15

An example of a Performance and Value Measurement Template model that can be effectively used to measure process performance.

Ref. 22.

Typical templates that are used:

- Process Map and/or Matrix
- Measurement and Reporting Map and/or Matrix
- Performance Map and/or Matrix
- Role Map and/or Matrix
- Owner Map and/or Matrix
- Application Rule Map and/or Matrix
- Data Rule Map and/or Matrix
- Rule Map and/or Matrix
- Compliance Rule Map and/or Matrix

Typical BPM CoE roles involved:

- Process eXperts
- Process Architects

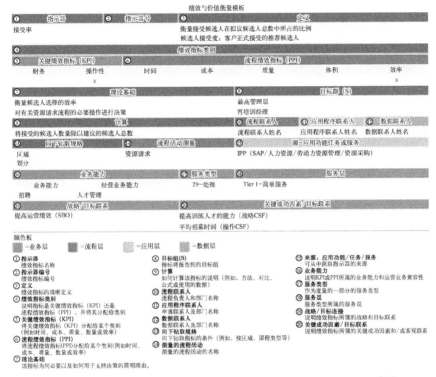

图15　可有效用于测量流程绩效的性能和价值测量模板模型示例[22]

使用的典型模板：

- 流程图及矩阵；
- 测量和报告图及矩阵；
- 绩效图及矩阵；
- 角色图及矩阵；
- 所有者图及矩阵；
- 应用程序规则图及矩阵；
- 数据规则图及矩阵；
- 规则图及矩阵；
- 合规规则图及矩阵。

涉及的典型BPM CoE角色：

- 流程专家；
- 流程架构师。

Step 29: Perform Scoping of Gaps

This includes detailed review, identification, and classification of all running processes in the process portfolio of the organization, and then scoping for performance gaps, irregularities, and other kinds of process performance mishaps and misbehavior. This serves as a staging point for choosing new building blocks to re-engineer existing processes that need to be reconfigured and/or rebuilt. The output of step 29 is consumed by step 30.

Typical tasks that are done within this step:

- Align and associate with defined process goals
- Align with process performance expectations
- Review and document performance gaps

Typical templates that are used:

- Process Map and/or Matrix
- Measurement and Reporting Map and/or Matrix
- System Measurements/Reporting Map and/or Matrix
- Performance Map and/or Matrix
- Role Map and/or Matrix
- Owner Map and/or Matrix

Typical BPM CoE roles involved:

- Process Engineers
- Process eXperts
- Process Architects

Step 30: Choose Building Blocks

As already mentioned in step 3, building blocks are important for the reusability of certain components/aspects. Also in the Run/Maintain phase can building blocks be used, in that the governance and monitoring of value creation is checked. If the value is not realized, alignment of relevant areas between strategic, organizational, process, and technology contexts is initiated. Below in (Figure 16) is an example of typical feedback loops used around business value; among them are business governance, business performance, process monitoring, and IT governance.

In step 30, if business value is not realized as expected, a feedback loop is triggered through either business governance, business performance, process monitoring, and/or IT governance. Building-block concepts are used to enable teams to reuse the artifacts, templates, and models that already have been developed starting in step 3 and through the design, build, and deploy/implement phases. So building blocks from the previously defined process reference content are chosen for the reengineering, process modeling, or process architecture of existing processes to close process performance gaps in Phase 6: Continuous Improvement.

步骤29：执行差距范围

这包括对组织的流程组合中所有正在运行的流程进行详细的审查、识别和分类，然后确定绩效差距、异常情况以及其他类型的流程绩效事故和不当行为的范围。这是选择新构建模块以重新设计需要重新配置及重建的现有流程的一个过渡点。步骤29的输出被步骤30使用。

在此步骤中完成的典型任务：

- 与已定义的流程目标保持一致和关联；
- 符合流程绩效预期；
- 审查和记录绩效差距。

使用的典型模板：

- 流程图及矩阵；
- 测量和报告图及矩阵；
- 系统测量/报告图及矩阵；
- 绩效图及矩阵；
- 角色图及矩阵；
- 所有者图及矩阵。

涉及的典型BPM CoE角色：

- 流程工程师；
- 流程专家；
- 流程架构师。

步骤30：选择构建模块

如步骤3所述，构建模块对于某些组件/方面的可重用性很重要。同样，在运行/维护阶段，也可以使用这些构建模块，以便检查价值创造的治理和监控。如果价值没有实现，战略、组织、流程和技术环境之间的相关领域将开始协调。下面（图16）是围绕业务价值使用的典型反馈循环示例，其中包括业务治理、业务绩效、流程监控和IT治理。

在步骤30中，如果没有按预期实现业务价值，则会通过业务治理、业务绩效、流程监控及IT治理触发反馈循环。构建模块概念用于使团队能够重用已经从步骤3开始并通过设计、构建和部署/实现阶段开发的制品、模板和模型。因此，从先前定义的流程参考内容中选择构建模块用于现有流程的重新设计、流程建模或流程架构，以弥补第六阶段：持续改进中的流程绩效差距。

FIGURE 16

Example of the relationship between business value realized, the areas working and focusing on value creation and realization, such as business governance, business performance, process monitoring, and IT governance, and the related building-block groups, that is, strategic, organizational, process, and technology contexts.

Ref. 23.

Therefore, the output of step 30 is consumed by step 31.
Typical tasks that are done within this step:

- Specification of the value and/or performance gaps
- Description of the pain points
- Identification of relevant building blocks
- Choose building blocks from existing process reference content
- Alignment and unification of building blocks across areas
- Review and document possible solutions

Typical templates that are used:

- Process Map and/or Matrix
- Object Map and/or Matrix
- Service Map and/or Matrix

Typical BPM CoE roles involved:

- Process Engineers
- Process eXperts
- Process Architects
- Business Analysts

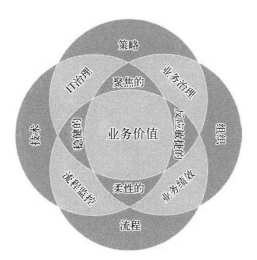

图16 实现的业务价值、工作领域和专注于价值创造和实现的领域（如业务治理、业务绩效、流程监控和IT治理）与相关构建模块组（即战略、组织、流程和技术环境）之间的关系示例[23]

因此，步骤30的输出被步骤31使用。

在此步骤中完成的典型任务：

- 价值及绩效差距说明；
- 描述业务痛点；
- 相关构建模块的识别；
- 从现有流程参考内容中选择构建模块；
- 跨区域的构建模块调整和统一；
- 审查和记录可能的解决方案。

使用的典型模板：

- 流程图及矩阵；
- 目标图及矩阵；
- 服务图及矩阵。

涉及的典型BPM CoE角色：

- 流程工程师；
- 流程专家；
- 流程架构师；
- 业务分析师。

Step 31: Evaluate Potential Solutions

Evaluating a potential solution is within BPM CoE as much about BPM Governance, BPM Portfolio Management, BPM Alignment, as it is about BPM Change Management. The evaluation for implementing new processes and/or reengineering existing processes is more or less solely based on the scoping of the performance gap, the available building blocks, as well as the value and performance expectations dictated by the process owners and stakeholders.

The output of step 31 is consumed by step 32.

Typical tasks that are done within this step:

- Identify performance gap (link to BPM Governance)
- Specify root cause of performance gap (link to BPM Change Management)
- Identify alternatives and potential solutions (link to BPM Portfolio Management)
- Collect and list advantages and disadvantages of potential solutions (BPM Governance and BPM Change Management)
- Compare and align potential solutions to the existing process landscape (link to BPM Alignment)
- Evaluate and decide upon alternatives, if any are proposed (link to BPM Portfolio Management)

Typical templates that are used:

- Process Map and/or Matrix
- Service Map and/or Matrix
- Performance Map and/or Matrix
- Value Map and/or Matrix
- Operating Map and/or Matrix
- Measurement and Reporting
- System Measurements/Reporting Map and/or Matrix

Typical BPM CoE roles involved:

- Process Engineers
- Process eXperts
- Business Analysts
- Process Architects

Step 32: Analyze Variances

In today's optimized organizations, analyzing process variances is a must-do. This is not only about identifying duplication and a potential for integration, it is also important for the standardization of the various processes that should be similar, but are different. While capturing existing processes in BPMN, performing value-stream mapping and statistical analysis provides critical insight into key factors that help improve business processes. Analyzing traditional business processes does not

步骤31：评估潜在解决方案

评估一个潜在的解决方案和评估BPM变更管理一样，都是在BPM CoE中进行的，包括BPM治理、BPM项目组合管理、BPM协调。实施新流程及重新设计现有流程的评估或多或少仅基于绩效差距的范围、可用的构建模块以及流程所有人和利益相关者规定的价值和绩效预期。

步骤31的输出被步骤32使用。

在此步骤中完成的典型任务：

- 确定绩效差距（与BPM治理的链接）；
- 确定绩效差距的根本原因（链接到BPM变更管理）；
- 确定备选方案和潜在解决方案（链接到BPM产品组合管理）；
- 收集并列出潜在解决方案的优缺点（BPM治理和BPM变更管理）；
- 将潜在解决方案与现有流程布局进行比较和协调调整（链接到BPM协调）；
- 评估并决定备选方案（如果有）（链接到BPM项目组合管理）。

使用的典型模板：

- 流程图及矩阵；
- 服务图及矩阵；
- 绩效图及矩阵；
- 价值图及矩阵；
- 操作图及矩阵；
- 测量和报告；
- 系统测量/报告图及矩阵。

涉及的典型BPM CoE角色：

- 流程工程师；
- 流程专家；
- 业务分析师；
- 流程架构师。

步骤32：差异分析

在当今组织优化中，分析流程差异是必须做的。这不仅是为了识别重复和整合的可能性，而且对于各种流程的标准化也很重要，这些流程应该是相似的，但又是不同的。在BPMN中捕获现有流程的同时，执行价值流图和统计分析提供了对帮助改进业务流程的关键因素的决定性洞察。分析传统业务流程并不能提供足够的信息来比较和指定差异。详细的业务流程或工作流程分析是需要的，在这种

provide enough information to be able to compare and specify variances. Detailed business process or workflow analysis is needed, in which one examines processes using various techniques, such as BPR, Six Sigma, Lean, providing alternatives to identify duplication, define variances, reduce time and cost, specify steps, waste, and other factors important to the organization when analyzing variances. It is important to remember that process variances differ by variances of a task and in the way the business process flows.

A common business process might exist in multiple variations in an enterprise, due to different legal requirements in different countries, deviations in the supporting IT infrastructure, or differences in the organizational structure. To explore and control such variability,[24] Weidlich and Weske argue that the notion of a main process, the invariant nucleus of all process variants, might be applied, because the degree of variability of process variants might be explored using the notion of a main (core) process. Such a process captures structural and behavioral aspects that are invariant across all process variants. SAP's business process handbook specifies [25] that a business process variant is a fundamental flow variant of a Business Process that uses the same input and delivers the same measurable outcome. The flow of process steps is defined at business process variant level. To keep level consistency it is necessary that each Business Process has at least one business process variant attached. A business process variant should differ from another at least in one of the following:

1. Flow of documents
2. The specific business objects needed
3. Life-cycle schema of the business objects (status and status transitions)
4. Application to Application/Business to Business (A2A/B2B) message choreography or choreography with direct interactions with other Business Processes
5. A business process variant is not just an alternative User Interface (UI)
6. A business process variant is not just another sequence a user decides to perform tasks on the User Interface (UI).
7. Two Business Process Variants differ in the way the business process flows. The difference is so important that the variants are to be considered separately in a business process analysis. The difference is so fundamental that it typically needs to be treated by special software functionality and not just configuration, if implemented in software.

The above clearly illustrates the importance of analyzing, identifying, tracking, and documenting process variances. To test possible solutions, it is important to analyze different process variants in different process scenarios and setups. The output of step 32 is consumed by both steps 33 and 34.

Typical tasks that are done within this step:

- Perform detailed process analysis (main process, flow, roles, tasks, objects, etc.)
- Undertake process performance simulations and report on output
- Specify process variances

分析中,使用各种技术(如BPR、六西格玛、精益)检查流程,提供识别重复、定义差异、减少时间和成本、指定步骤、损耗的替代方案以及在分析差异时对组织重要的其他因素。重要的是要记住,流程差异因任务的差异和业务流程的流动方式而不同。

由于不同国家的不同法律要求、支持IT基础设施的偏差或组织结构的差异,企业中可能存在多种不同的通用业务流程。为了探索和控制这种差异性[24],Weidlich和Weske认为:可以应用主流程(main process)的概念(即所有流程变体的不变核心),因为可以使用主流程(核心)的概念来探索流程变量的差异程度。这样的流程捕获结构和行为在所有流程变量中都是不变的。SAP的业务流程手册规定[25]:业务流程变体是业务流程的基本流程变体,它使用相同的输入并提供相同的可度量结果。流程步骤的流程在业务流程变量级别定义。为了保持级别的一致性,每个业务流程都必须至少附加一个业务流程变量。业务流程变量应至少在以下方面与其他变量不同。

(1)流程文档;

(2)所需的特定业务对象;

(3)业务对象的生命周期模式(状态和状态转换);

(4)应用程序到应用程序/企业到企业(A2A/B2B)与其他业务流程直接交互的消息编排或编排;

(5)业务流程变量不仅仅是可选的UI;

(6)业务流程变量不仅仅是用户决定在UI上执行任务的另一个序列;

(7)两种业务流程变体在业务流程的流动方式上有所不同。差异是如此重要,以至于在业务流程分析中要单独考虑变量。这种差异是如此基础,以至于如果在软件中实现,通常需要通过特殊的软件功能而不仅仅是配置来处理。

上面清楚地说明了分析、识别、跟踪和记录流程差异的重要性。为了测试可能的解决方案,在不同的流程场景和设置中分析不同的流程变体非常重要。步骤32的输出由步骤33和步骤34使用。

在此步骤中完成的典型任务:

- 执行详细的流程分析(主要流程、流程、角色、任务、对象等);
- 进行流程绩效模拟和输出报告;
- 指定流程差异;

- Detailed and thorough testing of new processes
- Detailed and thorough testing of reengineered processes

Typical templates that are used:

- Process Map and/or Matrix
- Service Map and/or Matrix
- Performance Map and/or Matrix

Typical BPM CoE roles involved:

- Process Engineers
- Process Architects

Value Gate 5:

- Evaluate potential solutions
- Investigate opportunities for higher process quality delivery
- Investigate opportunities for higher Return on Investment (ROI)

Step 33: Estimate Impact

Based on acquired process analysis and testing simulations, it is critical to assess the possible impact that either new and/or reengineered processes have upon many different business aspects, but most importantly, what their execution will mean for the value-generating cycle of the organization (Figure 17).

The purpose of the business impact analysis (BIA) is to identify which business units/departments and processes are essential to the survival of the organization. The BIA will identify how quickly essential business units and/or processes have to return to full operation following a disaster situation. The BIA will also identify the resources required to resume business operations. Business impacts are identified based on worst-case scenarios that assume that the physical infrastructure supporting each respective business unit has been destroyed and all records, equipment, etc. are not accessible for 30 days. Please note that the BIA will not address recovery solutions.[26]

The objectives of the BIA are as follows:

- Estimate the financial impacts for each business unit, assuming a worst-case scenario.
- Estimate the intangible (operational) impacts for each business unit, assuming a worst-case scenario.
- Identify the organization's business unit processes and the estimated recovery time frame for each business unit.

The output of step 33 is consumed by step 34.
Typical tasks that are done within this step:

- Document and prepare new processes in a process-testing portfolio (separating them from active processes)

- 新流程的详细和彻底测试；
- 详细而彻底测试重新设计流程。

使用的典型模板：

- 流程图及矩阵；
- 服务图及矩阵；
- 绩效图及矩阵。

涉及的典型BPM CoE角色：

- 流程工程师；
- 流程架构师。

价值门5：

- 评估潜在解决方案；
- 调查更高流程质量交付的机会；
- 调查提高ROI的机会。

步骤33：评估影响

基于获得的流程分析和测试模拟，评估新流程及重新设计的流程对许多不同业务方面的可能产生至关重要影响，但最重要的是，这些流程的执行对于组织的价值生成周期意味着什么（图17）。

业务影响分析（business impact analysis，BIA）的目的是确定哪些业务单位/部门和流程对组织的生存至关重要。BIA将在灾难情况发生后，确保基本业务部门及流程必须快速地恢复完整运营。BIA还将确定恢复业务运营所需的资源。业务影响根据最坏情况确定，假设支持每个业务单元的物理基础设施已被破坏，所有记录、设备等在30天内无法访问。请注意，BIA不会处理恢复解决方案[26]。

BIA的目标如下：

- 假设最坏情况，估计每个业务部门的财务影响；
- 假设最坏情况，估计每个业务部门的无形（运营）影响；
- 确定组织的业务部门流程和每个业务部门的预计恢复时间框架。

步骤33的输出被步骤34使用。

在此步骤中完成的典型任务：

- 记录并准备流程测试组合中的新流程（将其与活动流程分离）；

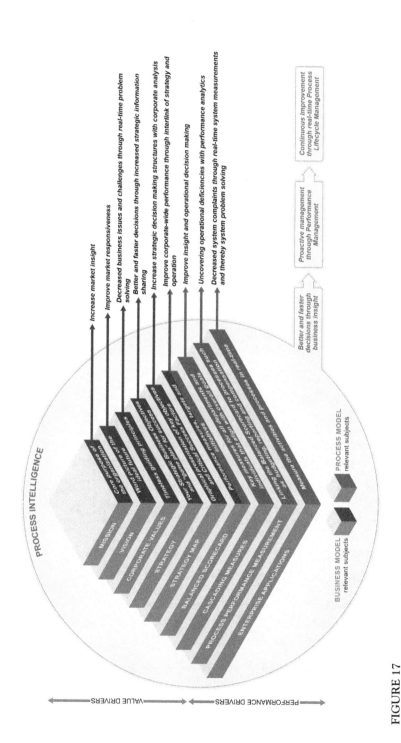

FIGURE 17

How process intelligence should be understood and undertaken by the organization.

Ref. 27.

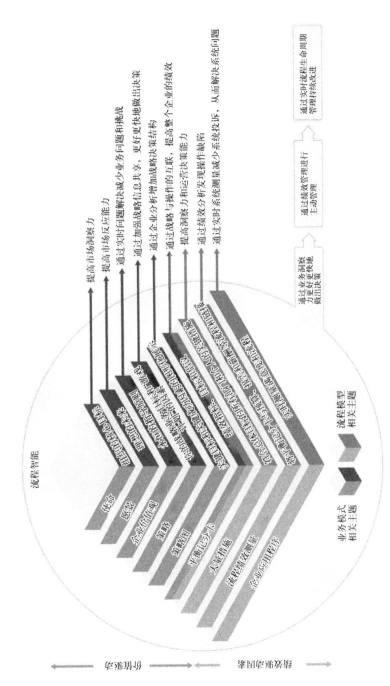

图17 组织应如何理解和执行流程智能[27]

- Document and prepare the reengineering of existing processes in a process-testing portfolio (move them out of the existing process portfolio temporarily while testing is underway)
- Execute thorough process-testing simulations of both newly created processes as well as reengineered processes

Typical templates that are used:

- Process Map and/or Matrix
- Service Map and/or Matrix
- Performance Map and/or Matrix
- Value Map and/or Matrix
- Operating Map and/or Matrix
- Measurement and Reporting
- System Measurements/Reporting Map and/or Matrix
- Risk Map and/or Matrix

Typical BPM CoE roles involved:

- Process Engineers
- Process Architects

Quality Gate 5:

- Investigate possible business impact and consequences
- Estimate impact on process quality
- Estimate impact on process coverage
- Estimate impact on process goals and scope

During the 5th phase of the BPM Life Cycle, the Run and Maintain phase, we have been focusing mainly on process measurements, monitoring, reporting, and audits. As we enter the last phase of the BPM Life Cycle, the Continuous Improvement phase (see Figure 18), we will be focusing on optimizing processes around prioritized improvement areas, such as the Value Model, Revenue Model, Cost Model, Service Model, Operating Model, and the Performance Model in a collaborative business environment using feedback loops to report on process improvements, requests for changes, and to further manage the process landscape of the organization.

PHASE 6: CONTINUOUS IMPROVEMENT—CONTINUOUSLY OPTIMIZE AND DEVELOP PROCESSES

Business Process Improvement (BPI) is a systematic approach to help an organization optimize its underlying processes to achieve results that are more efficient. The methodology was first documented in H. James Harrington's 1991 book, Business Process Improvement.[28] It is the methodology upon which both Process Redesign and Business Process Reengineering are based (see Figure 18). BPI has allegedly

- 记录并准备流程测试组合中现有流程的业务流程再设计（在测试进行期间将其暂时移出现有流程组合）；
- 对新创建的流程和重新设计的流程执行全面的流程测试模拟。

使用的典型模板：

- 流程图及矩阵；
- 服务图及矩阵；
- 绩效图及矩阵；
- 价值图及矩阵；
- 操作图及矩阵；
- 测量和报告；
- 系统测量/报告图及矩阵；
- 风险图及矩阵。

涉及的典型 BPM CoE 角色：

- 流程工程师；
- 流程架构师。

质量门 5：

- 调查可能的业务影响和后果；
- 估计对流程质量的影响；
- 估计对流程覆盖范围的影响；
- 估计对流程目标和范围的影响。

在 BPM 生命周期的第五个阶段，即运行和维护阶段，我们主要关注流程测量、监控、报告和审计。当我们进入 BPM 生命周期的最后一个阶段，即持续改进阶段（图18）时，我们将专注于围绕优先改进领域的优化流程，如价值模型、收入模型、成本模型、服务模型、运营模型，在协作式业务环境中，使用反馈循环报告流程改进、变更请求并进一步管理组织的流程布局。

3.5.7 第六阶段：持续改进（持续优化和发展流程）

BPI 是帮助组织优化其底层流程以实现更高效的结果的系统方法。该方法首先记录在 H. J. Harrington 于 1991 年出版的《业务流程改进》一书中[28]。它是流程再设计和业务流程再设计的基础方法（图18）。据称，BPI 负责将成本和周期时间

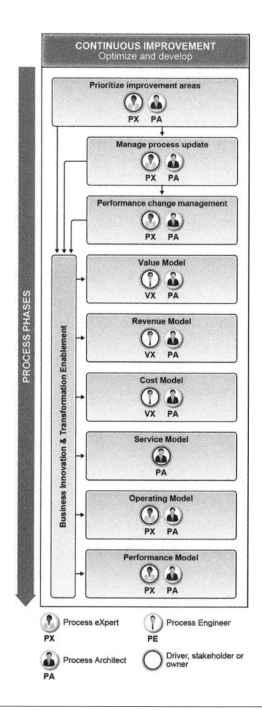

FIGURE 18

The Continuous Development phase of the BPM Life Cycle.

Ref. 29.

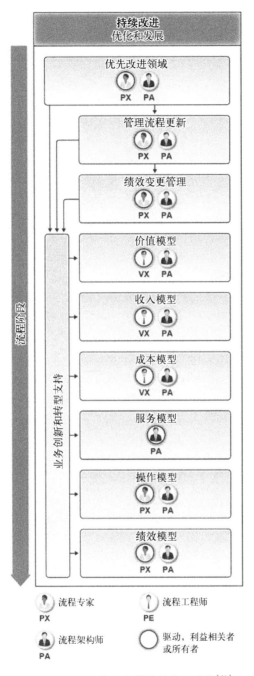

图18 BPM生命周期的持续发展阶段[29]

been responsible for reducing cost and cycle time by as much as 90% while improving quality by over 60%. In the meantime, the idea and concept of Continuous Improvement (CI) or Continuous Process Improvement (CPI) is applied in multiple areas. Many frameworks, methods, and approaches have some sort of CI and or CPI incorporated in one way or another as an ongoing effort to improve products, services, or processes. To mention some:

- Business process reengineering
- Six Sigma
- Theory of Constraints, Lean, Six Sigma
- Kaizen
- Toyota Production System
- Zero Defect
- Mottainai
- Muda
- Total productive maintenance

There has been some criticism of these approaches, however, and there are claims that these methods are more resource intensive and thereby cost-cutting-focused and that the measures come at the expense of fair labor practices and quality products. This criticism furthermore argue that real continuous improvement governance models would have to consider more than just cost drivers. They would have to incorporate both performance/cost as well as value drivers to link to business transformation and innovation aspects.

Continuous Improvement: Performance and Value Drivers

Performance drivers and value drivers can go hand in hand to specify the link to business innovation and transformation enablement. Even though they are not applied by many organizations, applying both have proven over the years to deliver above-average results. They are what we call outperforming organizations. The first time this was researched and proven was in 1984 by Dutton and Thomas.[30] They measured the results in progress ratios. The number they used represents the cost of production after cumulative production doubles. Dutton and Thomas found that those companies across different industries that apply these principles had the ratio typically around 80%. Thus, if a business has a progress ratio of 80% and it costs $100 to produce a unit after producing 100 units, when cumulative production reaches 200 units, it will cost only $80 to produce the same unit. Formally, for some commodity, if Cost(t) is cost at time t, d(t) is the number of doublings of cumulative output of the commodity in time t, and a is the percent reduction in cost for each doubling of cumulative output (note: $1-a$ is the progress ratio), then we have Cost(t) = Cost(0)(1−a)^d.

Here, some of the known industries that have applied the previously mentioned progress ratios (Table 1):

The core principle here is not only to identify performance/cost as well as value drivers, it is to attach them to a continuous feedback loop and thereby reflect

减少90%,同时将质量提高60%以上。同时,CI或CPI的思想和概念在多个领域得到了应用。许多框架、方式和方法都以某种方式合并了某种CI及CPI,不间断的努力去改进产品、服务或流程。这里提到一些:

- BPR;
- 六西格玛;
- 约束理论、精益、六西格玛;
- 改善;
- TPS;
- 零缺陷;
- 珍惜;
- 浪费;
- 全面生产维护。

然而,有人对这些方法提出了一些批评,有人声称,这些方法更加浪费资源密集,因为为了削减成本,这些措施是以牺牲公平劳动实践和优质产品为代价的。这种批评进一步指出,真正的持续改进治理模型将不得不考虑的不仅仅是成本驱动因素。他们必须将性能/成本和价值驱动因素结合起来,以连接到业务转型和创新方面。

持续改进:绩效和价值驱动因素

绩效驱动因素和价值驱动因素可以共同指定到业务创新和转型支持(BITE)的链接。尽管许多组织没有应用这两种方法,但多年来,应用这两种方法的组织都已证明其能够提供高于平均水平的结果。它们就是我们所说的优秀组织。1984年,J. M. Dutton和A. Thomas首次对此进行了研究和验证[30]。他们测量了进步率的结果,它们使用的数字表示累计生产翻倍后的生产成本。J. M. Dutton和A. Thomas发现,应用这些原则的不同行业的公司的比率通常在80%左右。因此,如果一个企业的进步率为80%,那么生产100个单位后生产一个单位的成本为100美元,当累计生产达到200个单位时,生产相同单位的成本仅为80美元。从形式上讲,对于某些商品,如果成本(t)是时间t的花费,$d(t)$是时间t中商品累计产量的两倍,a是累计产量每增加一倍的成本减少的百分比(注:$1-a$是进步率),那么我们就有了成本$(t)=$成本$(0) \times (1-a)^{\wedge}d$。

这里(表1)显示了一些应用了上述进步率的已知行业。

这里的核心原则不仅是确定性能/成本以及价值驱动因素,还将它们附加到连续的反馈循环中,从而反映(优化、改进和创新)流程。因此,CI的目的是识别、

Table 1 *Industry Table Index*

Technology	Period	Year 1 Production	Cumulative Production	Cost Index	Progress Ratio
Ford model T auto	1909–1923	15,741	8,028,000	0.290	87%
Integrated circuits	1962–1968	4 million unites	828 million units	0.047	67%
CFC substitutes	1988–1999	100,000 tons	3,871,000 tons	0.690	93%
Scrubbers	1987–1995	65.8 GW	84.3 GW	0.	89%
Photovoltaic	1971–2000	0.1	1451.4	0.042	72%
Magnetic ballasts	1977–1993	29.4 million	629.3 million	0.897	97%
Electronic ballasts	1986–2001	431	350 million units	0.277	88%
Refrigerators	1980–1998	5.1 million	126.3 million	0.556	88%
Freezers	1980–1998	1.8 million	26.1 million	0.374	78%
Clothes washers	1980–1998	4.4 million	104.7 million	0.536	87%
Electronic clothes	1980–1998	2.5 million	61.0 million	0.557	88%
Gas clothes dryer	1980–1998	0.7 million	18.2 million	0.593	90%
Dishwasher	1980–1998	2.7 million	69.7 million	0.450	84%
Room air conditioner	1980–1998	2.4 million	63.3 million	0.478	85%
Selective window coatings	1992–2000	4.8 million m^2	157.4 million m^2	0.394	83%

(optimization, improvement, and innovation) processes. The purpose of continuous improvement (CI), therefore, is the identification, reduction, elimination, and innovation of suboptimal processes (efficiency). This already goes beyond most approaches that are result/effect-driven and thereby focuses on effectiveness. The emphasis of CI is on incremental, continuous steps rather than giant leaps (Evolution). Continuous Improvements are thereby based on many, small changes rather than the radical optimization project that is more likely handled through a project. The change should come from the operation (the workers themselves) and enable business model changes both in the areas of revenue, value, and service model, as well as the cost, performance, and operating model. Such a feedback loop to the business model domains is more likely to succeed in enabling change. Practically speaking, process improvement is, therefore, an aspect of organizational development (OD) in which a series of actions are taken by a process owner to identify, analyze, and improve existing business processes within an organization to meet new goals and objectives, such as increasing profits and performance, reducing costs, and accelerating schedules. These actions often follow a specific methodology or strategy to increase the likelihood of successful results. Process improvement may include the restructuring of company training programs to increase their effectiveness. Process improvement is also a method to introduce process changes to improve the quality of a product or service to better match customer and consumer needs.

Continuous Business Process Improvement, however, is about taking the cycle of the process optimization phases and steps to another level of detail and efficiency.

表1　行业表索引

工业技术	时期/年	第一年产量	累计产量	成本指数	进步比率/%
福特T型汽车	1909 ~ 1923	15 741	8 028 000	0.290	87
集成电路	1962 ~ 1968	4 000 000单位	828 000 000单位	0.047	67
CFC替代品	1988 ~ 1999	100 000吨	3 871 000吨	0.690	93
洗涤器	1987 ~ 1995	65.8 GW	84.3 GW	0	89
光伏发电	1971 ~ 2000	0.1	1 451.4	0.042	72
磁性镇流器	1977 ~ 1993	29 400 000	629 300 000	0.897	97
电子镇流器	1986 ~ 2001	431	350 000 000单位	0.277	88
冰　箱	1980 ~ 1998	5 100 000	126 300 000	0.556	88
冷冻机	1980 ~ 1998	1 800 000	26 100 000	0.374	78
洗衣机	1980 ~ 1998	4 400 000	104 700 000	0.536	87
电子服装	1980 ~ 1998	2 500 000	61 000 000	0.557	88
燃气干衣机	1980 ~ 1998	700 000	18 200 000	0.593	90
洗碗机	1980 ~ 1998	2 700 000	69 700 000	0.450	84
室内空气调节器	1980 ~ 1998	2 400 000	63 300 000	0.478	85
选择性窗涂层	1992 ~ 2000	4 800 000平方米	157 400 000平方米	0.394	83

减少、消除和创新次优流程(效率)。这已经超越了大多数由结果/效果驱动的方法,因此侧重于有效性。CI的重点是渐进的、连续的步骤,而不是巨大的飞跃(进化)。因此,CI是基于许多微小的变化,而不是更可能通过项目处理的根本优化项目。变更应该来自运营(工人本身),并使业务模型能够在收入、价值和服务模型领域以及成本、绩效和运营模型方面进行变更。这种对业务模型领域的反馈循环更有可能成功地实现变更。因此,实际上,流程改进是组织开发(organizational development, OD)的一个方面,流程所有人采取一系列行动来识别、分析和改进组织内现有的业务流程,以实现新的目的和目标,如提高利润和绩效、降低成本和加速评级表。这些行动通常遵循特定的方法或策略,以增加结果成功的可能性。流程改进可能包括重组公司培训计划以提高其有效性。流程改进也是引入流程更改以提高产品或服务质量以更好地满足客户和消费者需求的方法。

　　然而,持续的BPI是将流程优化阶段和步骤的周期提升到另一个详细和有效率的级别。接收、评估和执行流程变更请求,以支持变更的流程及业务目的和目

Process change requests are received, evaluated, and carried out in support of changing process and/or business objectives and goals, and the simulation and performance measurements are then reported to process owners, stakeholders, and decision makers who are involved with the BPM Life Cycle. This ensures a chain reaction of quality assurance, evaluation, and decision making on behalf of the continuous feedback loop and collaborative work done by all participants involved.

The 6th and final phase of the BPM Life Cycle (see Figure 18), the continuous improvement phase, is the phase in which the processes are managed in terms of their effectiveness, efficiency, incidents/issues, and process change request fulfillments, etc. This is also when the organization improves the existing process operations and evaluates, adjusts, alters, amends, changes, corrects, eliminates, enhances, increases, modifies, optimizes, and/or excludes specific process parts within the process portfolios and landscapes. This phase is all about the link between business process operation and business, and thereby BPM Innovation and Transformation Enablement (see Figure 19). This interlinks with the improvement and optimization of the six business model domains.

Step 34: Prioritize Improvement Areas

Improvement areas have to be defined, documented, and selected through collaborative efforts between business, IT, and technology units of the organization (see Figure 19). As processes are heavily dependent on so many different aspects, it is important to select areas that correlate and align directly with the overall process strategy and the established process goals (Figure 20).

The first step of any process improvement initiative is to take stock of as many organizational processes as possible, because everything is connected to everything else in the value chain from concept to customer. Considering all organizational processes will force the team to think about the interdependencies between individuals, departments, vendors, and customers, all of whom may influence the process.

The next step is to determine which process, if improved, would have the greatest positive impact on the organization (i.e., would most likely contribute to the fulfillment of the organization's goals). There are five steps involved in the selection of choosing which processes to prioritize:

1. *List success criteria:* Success criteria are those measures, ranging from most tangible (e.g., financial measures) to least tangible (e.g., strategic measures), that indicate the larger organization is on the right strategic path. In this case, the organization has determined that hit ratio, combined ratio, and compliance are the three criteria that demonstrate it is performing according to its strategic plan.

2. *Weighted success criteria:* Success criteria are then weighted relative to each other using an index from 0.5 to 1.5, in which 0.5 indicates the criterion has the least weight, and 1.5 indicates the criterion has the most weight. In our example, the organization has deemed new business to be a critical success

标,然后将模拟和绩效度量报告给流程所有人、利益相关者和参与BPM生命周期的决策者。这确保了质量保证、评估和决策的连锁反应,代表着持续的反馈循环和所有参与者完成的协作工作。

BPM生命周期的第六阶段(图18)也是最后阶段,即持续改进阶段,是指流程在其有效性、效率、事故/问题和流程变更请求履行等方面进行管理的阶段。这也是组织改进现有流程操作、评估、调整、更改、修改、改变、修正、消除、增强、增加、改善、优化及排除流程组合和布局中的特定流程部分。这个阶段是关于业务流程操作和业务之间的联系,从而使BPM创新和转型(图19)。这与六个业务模型领域的改进和优化相互关联。

步骤34:确定改进区域的优先级

必须通过组织的业务部门、IT部门和技术部门之间的协作工作来定义、记录和选择改进领域(图19)。由于流程在很大程度上依赖于这么多不同的方面,选择与整个流程战略和既定流程目标直接相关和协调的领域非常重要(图20)。

任何流程改进计划的第一步都是尽可能多地评估组织流程,因为从概念到客户,价值链中的所有东西都与其他东西相连。考虑到所有组织流程将迫使团队考虑个人、部门、供应商和客户之间的相互依赖性,所有这些人都可能影响流程。

下一步是确定哪一个流程,如果改进,将对组织产生最大的积极影响(即最有可能有助于实现组织的目标)。选择优先处理哪些流程涉及以下五个步骤。

(1)列出成功标准:成功标准是指从最有形的(如财务指标)到最无形的(如战略指标)的指标,这些指标表明较大的组织处于正确的战略路径上。在这种情况下,组织已经确定命中率、组合比率和合规性是三个标准,证明它是根据其战略计划执行的。

(2)加权成功标准:然后使用指数0.5 ~ 1.5对成功标准进行相对加权,其中0.5表示标准的权重最小,1.5表示标准的权重最大。在我们的示例中,组织认为新业务是一个关键的成功因素。因此,命中率(1.5)是相对于合规性(1.0)和组合比

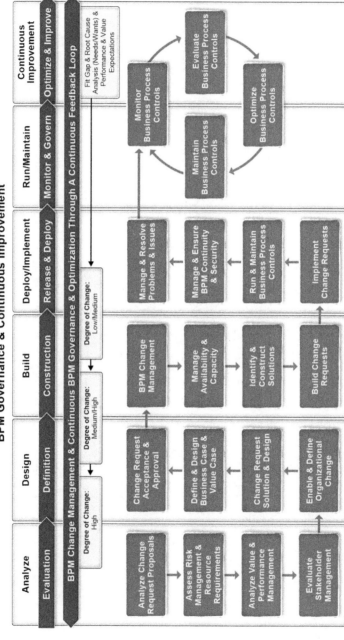

FIGURE 19

A high-level view example of a BPM Governance model that can be used to structure and organize the activities performed and to continuously monitor, govern, and administer the flow of business processes within the organization.

Ref. 31.

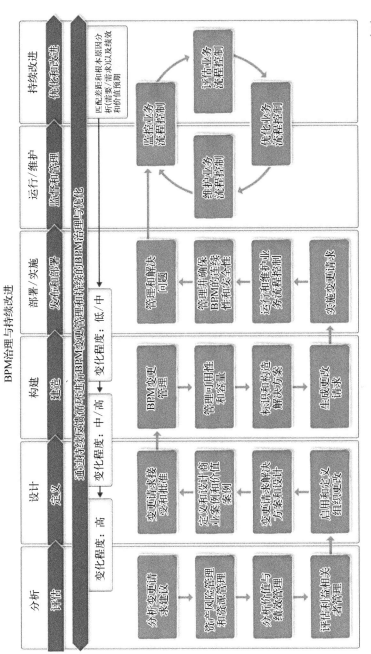

图19　BPM治理模型的高级视图示例，可用于组织和组织所执行的活动，并持续监控、管理和执行组织内的业务流程流[31]

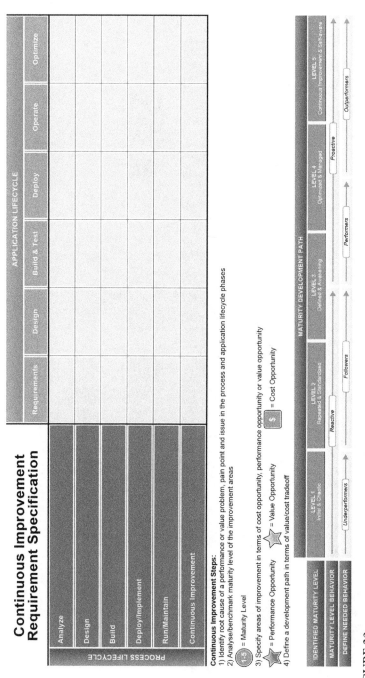

FIGURE 20

An example model of Continuous Improvement and Requirement Specification between the BPM Life Cycle and the Application Life Cycle. This model relates only to automated processes in the process landscape, and will aid the user in identifying the maturity levels as well as performance, value, and cost opportunities associated with processes throughout all six life-cycle phases.

Ref. 32.

持续改进需求规范

	应用程序生命周期					
	要求	设计	构建和测试	部署	操作	优化
分析						
设计						
构建						
部署/实施						
运行/维护						
持续改进						

BPM 生命周期

持续改进步骤：
1）确定绩效或价值问题的根本原因、流程和应用生命周期阶段的痛点和问题
2）分析/基准改进领域的成熟度水平
3）在成本机会、绩效机会方面指定改进领域
4）根据价值/成本权衡定义并开发路径

(+5) ＝成熟度

☆ ＝价值机会

[5] ＝成本机会

成熟度发展路径

确定的成熟度水平
成熟度行为
定义所需行为

1级 初始与非正式的	2级 重复的与标准化的	3级 被定义的和被认知的	4级 优化与管理	5级 持续改进与利益相关者
表现不佳者	反应性	追随者	表演者	积极主动
				超越大市场

图20 BPM 生命周期和应用生命周期之间持续改进需求规范和需求规范的示例模型。此模型仅与流程布局中的自动化流程相关，并将帮助用户识别成熟度级别以及在整个六个生命周期阶段中与流程相关的绩效、价值和成本机会[32]

factor. As such, the hit ratio (1.5) is the most important strategic measure relative to compliance (1.0) and combined ratio (0.5). They are all important, however; this is simply a relative weighting.

3. *List processes:* List the names of the processes from the process master.

4. *Assign anchors:* A number from 1 to 5 (each, an "anchor") is inserted in each cell indicating the strength of correlation between each process and each success criterion.

5. *Determine score and rank:* The resulting scores, which are the products of the relative weights times the anchors summed across each process, provide a ranking, based on the success criteria, which indicates what specific processes should be given improvement priority. In our example, it has been determined that the state filings process (with a score of 13) most contributes to the fulfillment of the organization's strategic objectives.[33]

The output of step 34 is consumed by step 35.
Typical tasks that are done within this step:

- Define and document steps for improving process structures around each business model domain
- Gather and list advantages and disadvantages for reengineering processes around each business model domain
- Establish process ownership and responsibilities around improvement steps
- Clarify process roles around improvement steps
- Reevaluate and reestablish process goals and requirements around each business model domain

Typical templates that are used:

- Process Map and/or Matrix
- Service Map and/or Matrix
- Information Map and/or Matrix
- Performance Map and/or Matrix
- Value Map and/or Matrix
- Operating Map and/or Matrix
- Measurement and Reporting
- Risk Map and/or Matrix

Typical BPM CoE roles involved:

- Process eXperts
- Business Analysts
- Transformation eXperts
- Process Architects

Quality Gate 6:

- Identify and determine areas to improve upon
- Initiate process quality improvements

率(0.5)最重要的战略措施。然而,它们都很重要,这只是一个相对的权重。

（3）列出流程：列出流程主数据中流程的名称。

（4）分配锚：在每个单元格中插入一个从1到5的数字（每个数字都是一个锚），指示每个流程和每个成功标准之间的关联强度。

（5）确定分数和等级：结果分数是相对权重乘以每个流程中锚的总和的乘积,根据成功标准提供一个等级,表明哪些特定流程应被给予改进优先级。在我们的例子中,已经确定国家归档流程（得分13）最有助于实现组织的战略目标33。

步骤34的输出被步骤35使用。

在此步骤中完成的典型任务：

- 定义并记录改善每个业务模型领域流程结构的步骤；
- 收集并列出在每个业务模型领域再设计流程的优缺点；
- 围绕改进步骤建立流程所有权和责任；
- 围绕改进步骤明确流程角色；
- 重新评估和重新建立每个业务模型域周围的流程目标和需求。

使用的典型模板：

- 流程图及矩阵；
- 服务图及矩阵；
- 信息图及矩阵；
- 绩效图及矩阵；
- 价值图及矩阵；
- 操作图及矩阵；
- 测量和报告；
- 风险图及矩阵。

涉及的典型BPM CoE角色：

- 流程专家；
- 业务分析师；
- 转换专家；
- 流程架构师。

质量门6：

- 确定并决定需要改进的领域；
- 启动流程质量改进；

- Initiate process coverage improvements
- Initiate process goals improvements
- Update process scope documentation and deliver evaluation report to stakeholders

Step 35: Manage Process Update

Process updates have to be continuously updated and managed by a dedicated process unit/team. This is important because process portfolios are usually quite expansive and cumbersome to manage, so it is essential that a process portfolio is efficiently managed when undergoing regular updates. The output of step 35 is consumed by step 36.

Typical tasks that are done within this step:

- Supervise and execute Change Management

Typical templates that are used:

- Process Map and/or Matrix
- Service Map and/or Matrix
- Information Map and/or Matrix
- Performance Map and/or Matrix
- Operating Map and/or Matrix

Typical BPM CoE roles involved:

- Process eXperts
- Process Architects

Step 36: Performance Change Management

Change management relevant for BPM and how it is handled across the entire BPM Life Cycle is handled in the chapter BPM Change Management. As requirements, the organizations, competitors, or the environment are constantly changing, adapting to outside changes is a challenge faced by nearly all organizations. The changes do not only impact the strategic aspects, the business models, the employees, and the way an organization utilizes technology, but the degree of outside change also influences the organization's ability to maintain control of their work. So while the organization manages the changes befalling the organization, it must also actively manage the change of their business processes. Developing an effective change road map that is integrated into the business process management life cycle and the BPM CoE change and issue management is imperative for the change effectiveness if your initiative is to avoid "the valley of despair" (see BPM Change Management chapter). Purpose and objectives of the BPM Change Management concept are to respond to both process change requests as well as the BPM client's changing requirements while maximizing value and reducing incidents, disruptions, and rework. When executing change management principles

- 启动流程覆盖改进;
- 启动流程目标改进;
- 更新流程范围文档并向利益相关者提交评估报告。

步骤35:管理流程更新

流程更新必须由专门的流程单位/团队持续更新和管理。这一点很重要,因为流程组合通常非常庞大,管理起来很麻烦,所以在进行定期更新时,必须有效地管理流程组合。步骤35的输出被步骤36使用。

在此步骤中完成的典型任务:

- 监督和执行变更管理。

使用的典型模板:

- 流程图及矩阵;
- 服务图及矩阵;
- 信息图及矩阵;
- 绩效图及矩阵;
- 操作图及矩阵。

涉及的典型BPM CoE角色:

- 流程专家;
- 流程架构师。

步骤36:绩效变更管理

与BPM相关的变更管理,以及如何在整个BPM生命周期中进行处理,请参见BPM变更管理一节。随着需求、组织、竞争对手或环境的不断变化,适应外部变化是几乎所有组织都面临的挑战。这些变化不仅影响战略方面、商业模式、员工和组织使用技术的方式,而且外部变化的程度也影响组织保持对其工作控制的能力。因此,当组织管理组织之前的变更时,它还必须积极地管理其业务流程的变更。如果您的计划是避免"绝望之谷",那么开发一个整合了BPM生命周期和BPM CoE变更以及问题管理的有效变更路线图对于变更的有效性是必不可少的(请参阅BPM变更管理章节)。BPM变更管理概念的目的和目标是响应流程变更请求以及BPM客户的变更需求,同时使价值最大化并减少事件、中断和返工。在执行变更

for the purpose of managing changes to process performance, it is important to apply both BPM CoE Change Management to the process portfolio, as well as BPM Change Management throughout the life cycle. This must go hand-in-hand with the process reports on performance, their gaps, and suggested solutions and alternatives. It also has to include the expected business impact parameters as well as outline both value and performance expectations. The output of step 36 is consumed by step 37.

Typical tasks that are done within this step:

- Identify performance gaps
- Specify stakeholder value and or performance expectation (BPM Requirements Management)
- Document improvements
- Planning with link to process portfolio, program, and project management
- Business organizational changes that need to be channeled through the Process Portfolio Management channel
- Investigate BPM Continuous Improvement feedback loop in terms of degree of change (low, medium, or high)
- Clarify value and performance expectations
- Structural changes that need to be channeled through the BPM CoE management; the people-side of change that needs to be channeled through the business change management group

Typical templates that are used:

- Process Map and/or Matrix
- Service Map and/or Matrix
- Information Map and/or Matrix
- Performance Map and/or Matrix
- Value Map and/or Matrix
- Operating Map and/or Matrix
- Measurement and Reporting
- Risk Map and/or Matrix

Typical BPM CoE roles involved:

- Process eXperts
- Transformation eXperts
- Enterprise Architects
- Process Architects

Step 37: Business Innovation and Transformation Enablement

Business Innovation and Transformation Enablement (BITE) principles go hand-in-hand with Performance Change Management described earlier. BITE principles, however, do not only emphasize performance, but also go into detail around

管理原则以管理流程绩效的变更时，重要的是在整个生命周期内将BPM CoE变更管理应用于流程组合，以及BPM变更管理。这必须与有关绩效、差距、建议的解决方案和替代方案的流程报告同步进行。它还必须包括预期的业务影响参数，并概述价值和绩效预期。步骤36的输出被步骤37使用。

在此步骤中完成的典型任务：

- 确定绩效差距；
- 明确利益相关者的价值及绩效期望（BPM需求管理）；
- 文档改进；
- 计划与流程组合、计划和项目管理链接；
- 需要通过流程组合管理渠道进行的业务组织变更；
- 根据变化程度（低、中或高）调查BPM持续改进反馈循环；
- 明确价值和绩效预期；
- 需要通过BPM CoE管理进行引导的结构变更，需要通过业务变更管理组进行引导的变更的人员方面。

使用的典型模板：

- 流程图及矩阵；
- 服务图及矩阵；
- 信息图及矩阵；
- 绩效图及矩阵；
- 价值图及矩阵；
- 操作图及矩阵；
- 测量和报告；
- 风险图及矩阵。

涉及的典型BPM CoE角色：

- 流程专家；
- 转换专家；
- 企业架构师；
- 流程架构师。

步骤37：业务创新和转型支持

业务创新和转型支持原则与前面描述的绩效变更管理密切相关。然而，业务创新和转型支持原则不仅强调性能绩效，还详细介绍建议的流程更改及其对组织的价值、收入、成本、服务和运营模型以及绩效模型的影响。步骤37的输出由步骤

suggested process changes and their impact on the organization's value, revenue, cost, service, and operating model as well as the performance model. The output of step 37 is consumed by steps 37a–f, depending on which business model domain the focus is placed. Please note that moving onward to any—or all—of the six different business model domains is not only allowed, but is highly recommended and should be done at all times during phases of process improvement, optimization, and development.

Doing this is also strongly advised because the vast majority of all of the existing and in-use processes in the process landscape plays an important role in the organization. It is often estimated that a significant number of active processes are likely to have a dramatic impact and effect upon the workflow within, execution of, and results delivered by any of the six different business model domains. Therefore, careful precaution and consideration has to be maintained when choosing to eliminate and/or reengineer any of the active processes in the process landscape.

The structure around Business Innovation and Transformation Enablement for the BPM Life Cycle focuses on two directions; the Process Way of thinking (innovation and transformation principles) and the Process Way of working (the sequence of actions taken). The 8 distinct interrogatives that can be used effectively for process analysis, design, construction, and monitoring, as well as for Continuous Improvement are as follows:

1. *Where:* The location/area of the process (i.e., where it is located or resides in the context of the process landscape)
2. *When:* Time of the process (i.e., the timing/time at execution or the length [time consumption] of the process)
3. *Whence:* The source of the process (what exactly is the source of the process, i.e., manual labor sequence, automation through software, etc.)
4. *How:* The manner of the process (how it behaves)
5. *What:* The context of the process (i.e., in which context does it have relevance)
6. *Why:* The reason for the process' existence (i.e., why does it exist, why do we have it/use it)
7. *Who:* Is the process of personal relevance (i.e., does it have an actor)
8. *Whether:* What are the choices, alternatives, and options of the process (i.e., should we choose another path)

The classification form of the 8 distinct process interrogatives can be structured and efficiently organized as shown in the example of Figure 21.

Typical templates that are used:

- Process Map and/or Matrix
- Service Map and/or Matrix
- Performance Map and/or Matrix
- Value Map and/or Matrix
- Operating Map and/or Matrix

37a ~ 步骤37f使用,具体取决于放置的是哪个业务模型域。请注意,不仅允许在六个不同业务模型域中的任何一个或全部中活动,而且强烈建议在流程改进、优化和开发阶段始终进行活动。

我们也强烈建议这样做,因为流程布局中绝大多数现有和使用中的流程在组织中起着重要作用。通常估计,大量的活动流程可能会对六个不同业务模型域中的任何一个域的工作流、执行和结果产生巨大的冲击和影响。因此,在选择消除及重新设计流程布局中的任何活动流程时,必须保持谨慎的预防和考虑。

围绕业务创新和BPM生命周期的转型支持的结构集中在两个方向:流程思维方式(创新和转型原则)和流程运行方式(采取的行动顺序)。可有效用于流程分析、设计、施工和监控以及持续改进的8个不同问题如下。

(1)位置:流程的位置/区域(即位于流程全景的位置);

(2)时间:流程的时间[即执行的时长/时间或流程的长度(时间使用)];

(3)来源:流程的来源(流程的确切来源是什么,即手动劳动顺序、通过软件实现自动化等);

(4)方式:流程的方式(行为方式);

(5)内容:流程的环境(即它在哪个环境中具有相关性);

(6)原因:流程存在的原因(即为什么它存在,为什么我们拥有/使用它);

(7)人物:是与个人关联的流程(即它有角色吗);

(8)可能的选择:流程的选择、备选方案和选项是什么(即我们是否应该选择其他路径)。

如图21的示例所示,8个不同流程疑问句的分类形式可以结构化并有效地组织起来。

使用的典型模板:

- 流程图及矩阵;
- 服务图及矩阵;
- 绩效图及矩阵;
- 价值图及矩阵;
- 操作图及矩阵;

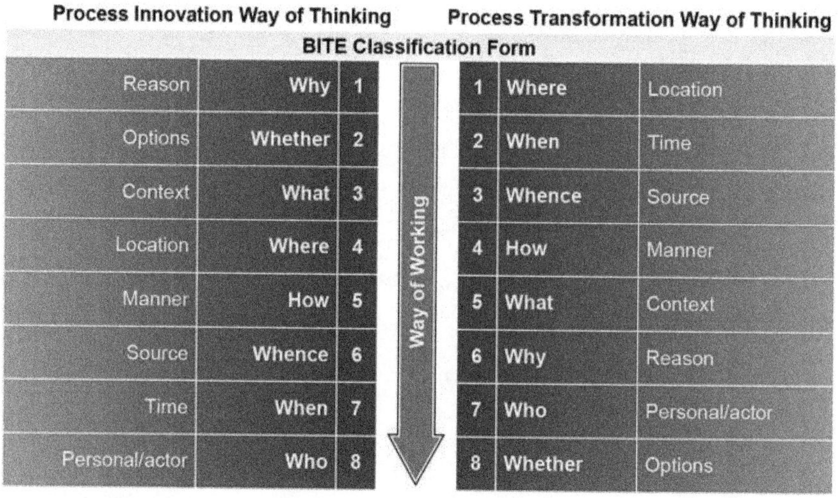

Process Innovation Way of Thinking   Process Transformation Way of Thinking

BITE Classification Form

Reason	Why	1	1	Where	Location
Options	Whether	2	2	When	Time
Context	What	3	3	Whence	Source
Location	Where	4	4	How	Manner
Manner	How	5	5	What	Context
Source	Whence	6	6	Why	Reason
Time	When	7	7	Who	Personal/actor
Personal/actor	Who	8	8	Whether	Options

Way of Working

FIGURE 21

An example of a model that shows the 8 interrogatives related to process reference content. This particular example also shows the direction of and approach toward Process Innovation and Process Transformation (both of them being a way of thinking and a way of working around processes).

Ref. 34.

- Cost Map and/or Matrix
- Revenue Map and/or Matrix
- Measurement and Reporting
- Risk Map and/or Matrix

Typical BPM CoE roles involved:

- Transformation eXperts

Value Gate 6:

- Business Innovation and Transformation Enablement (BITE) opportunities

Step 37a: Value Model

Process performance changes are closely connected to the organization's value model because all processes are designed to deliver value in one way or another (see Figure 22). In most instances, it is necessary to continuously improve processes by relating them to changes in the value model, such as, for instance, updated or newly defined value drivers and value indicators. The outcome of step 37 is consumed by step 37a.

流程创新思维方式　　　　　　　　　流程转换思维方式

基地设施试验设备分类表

原因	为什么	1		1	哪里	位置
选项	是否	2		2	什么时间	时间
上下文	什么	3		3	何处	来源
位置	哪里	4	运行方式	4	如何	方式
方式	如何	5		5	什么	上下文
来源	何处	6		6	为什么	原因
时间	什么时间	7		7	谁	个人/参与者
个人/参与者	谁	8		8	是否	选项

图21　显示与流程参考内容相关的8个疑问句的模型示例。这个特殊的例子也显示了流程创新和流程转换的方向和方法（两者都是一种思维方式和围绕流程工作的方式）[34]

- 成本图及矩阵；
- 收入图及矩阵；
- 测量和报告；
- 风险图及矩阵。

涉及的典型BPM CoE角色：

- 转换专家。

价值门6：

- 业务创新和转型支持机会。

步骤37a：价值模型

流程绩效变化与组织的价值模型密切相关，因为所有流程都是以某种方式交付价值（图22）。在大多数情况下，有必要通过将流程与价值模型中的变化（例如，更新或新定义的价值驱动因素和价值指标）联系起来，不断改进流程。步骤37的结果由步骤37a使用。

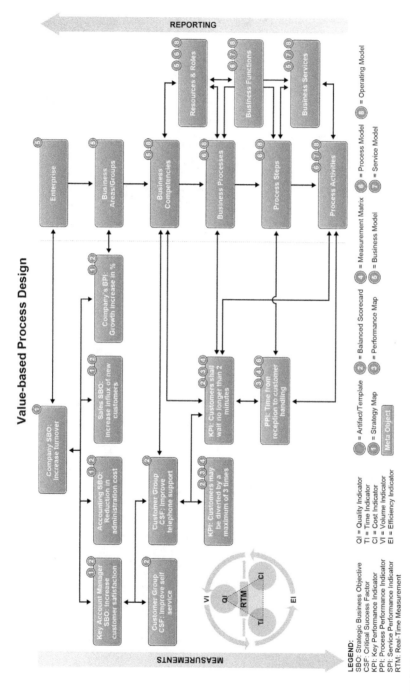

FIGURE 22

An example of a value-based process design in which the business processes of the organization have been designed, constructed, and fully integrated with the organization's value model.

Ref. 35.

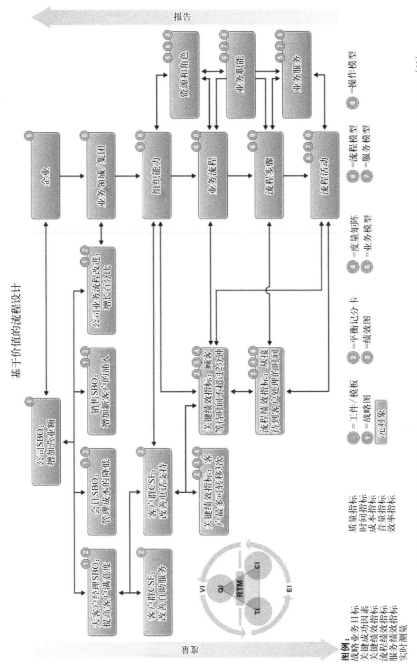

图22　一个基于价值的流程设计示例，其中组织的业务流程已经设计、构建并与组织的价值模型完全整合[35]

Typical tasks that are done within this step:

- Align and unify all strategic, tactical, and operational processes to the strategic, tactical, and operational key Performance Indicators of the organization
- Include process performance changes to the value flow model
- Relate business processes to process performance indicators and measurements
- Relate process steps and process activities to service performance indicators and measurements
- Relate business processes to business competencies, business resources and roles, and business functions
- Relate process steps and activities to business functions and business services

Typical templates that are used:

- Process Map and/or Matrix
- Service Map and/or Matrix
- Performance Map and/or Matrix
- Value Map, Matrix, and/or Model
- Risk Map and/or Matrix

Typical BPM CoE roles involved:

- Value eXperts
- Business Analysts
- Enterprise Architects
- Process Architects

Step 37b: Revenue Model

When main processes undergo change requests, it is important to realign them with the revenue model, because they are the driving force of processes that are specifically involved with value-generation flows of the organization. The outcome of step 37 is consumed by step 37b.

Typical tasks that are done within this step:

- Optimize and improve main (value-generating) processes in the process landscape
- Focus on communication and analysis around processes that help identify customers, market to those customers, and generate sales to those customers
- Relate business processes, process steps, and process activities to revenue flows
- Relate business process owners, process roles, and process measurement to revenue flows

Typical templates that are used:

- Process Map and/or Matrix
- Service Map and/or Matrix
- Performance Map and/or Matrix
- Value Map and/or Matrix

在此步骤中完成的典型任务：

- 根据组织的战略、战术和运行KPI,协调和统一所有战略、战术和运行流程；
- 将流程绩效变更纳入价值流模型；
- 将业务流程与流程绩效指标和度量相关联；
- 将流程步骤和流程活动与服务绩效指标和度量相关联；
- 将业务流程与业务能力、业务资源和角色以及业务职能联系起来；
- 将流程步骤和活动与业务职能和业务服务联系起来。

使用的典型模板：

- 流程图及矩阵；
- 服务图及矩阵；
- 绩效图及矩阵；
- 价值图、矩阵及模型；
- 风险图及矩阵。

涉及的典型BPM CoE角色：

- 价值专家；
- 业务分析师；
- 企业架构师；
- 流程架构师。

步骤37b：收入模型

当主要流程接受变更请求时,重要的是将其与收入模型重新调整,因为它们是专门涉及组织价值生成流的流程的驱动力。步骤37的结果由步骤37b使用。

在此步骤中完成的典型任务：

- 优化和改进流程布局中的主要（创造价值）流程；
- 专注于围绕有助于识别客户、向这些客户推销并为这些客户带来销售的流程进行沟通和分析；
- 将业务流程、流程步骤和流程活动与收入流相关联；
- 将业务流程所有人、流程角色和流程度量与收入流相关联。

使用的典型模板：

- 流程图及矩阵；
- 服务图及矩阵；
- 绩效图及矩阵；
- 价值图及矩阵；

- Revenue Map, Matrix, and/or Model
- Risk Map and/or Matrix

Typical BPM CoE roles involved:

- Value eXperts
- Business Analysts
- Enterprise Architects
- Process Architects

Step 37c: Cost Model

Processes are an essential part of production facilities regardless of industry, product, or general process practice principles. Creating new processes or reengineering existing processes within the established process portfolio enables organizations to reduce production cost, save time, reduce manpower, and automate previously manual labor through automated processes (see Figure 23). The outcome of step 37 is consumed by step 37c.

Typical tasks that are done within this step:

- Link operational processes to the cost model and cost profiles
- Define cost drivers and opportunities for cost reduction
- Examine cost reduction opportunities in operational processes
- Relate business processes, process steps, and process activities to cost flows
- Relate process roles, process measurement, and process owners to

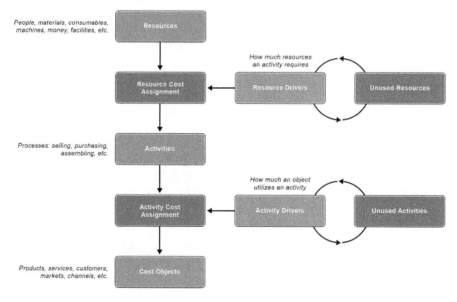

FIGURE 23

An example of how processes relate and integrate with an activity-based costing model.

- 收入图、矩阵及模型；
- 风险图及矩阵。

涉及的典型BPM CoE角色：

- 价值专家；
- 业务分析师；
- 企业架构师；
- 流程架构师。

步骤37c：成本模型

流程是生产设施的重要组成部分，与工业、产品或一般工艺实践原则无关。在已建立的流程组合中创建新流程或重新设计现有流程，使组织能够通过自动化流程降低生产成本、节省时间、减少人力和自动化以前的人工（图23）。步骤37的结果由步骤37c使用。

在此步骤中完成的典型任务：

- 将运营流程与成本模型和成本概况联系起来；
- 定义成本驱动因素和降低成本的机会；
- 检查运营流程中的成本降低机会；
- 将业务流程、流程步骤和流程活动与成本流相关联；
- 将流程角色、流程度量和流程所有人相关联；

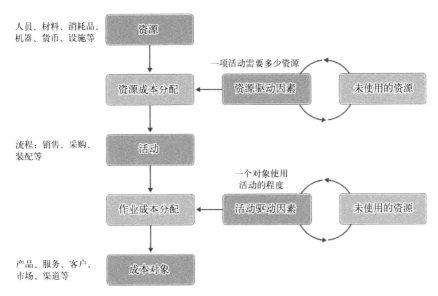

图23　流程如何与基于活动的成本计算模型关联和整合的示例

- High-cost types
- Medium-cost types; and
- Low-cost types

Typical templates that are used:

- Process Map and/or Matrix
- Service Map and/or Matrix
- Performance Map and/or Matrix
- Value Map and/or Matrix
- Cost Map, Matrix, and/or Model
- Risk Map and/or Matrix

Typical BPM CoE roles involved:

- Value eXperts
- Business Analysts
- Enterprise Architects
- Process Architects

Step 37d: Service Model

Business processes deliver business services (Service Provider) to one or more customers (whether internal or external) who then consume the services delivered to them (Service Consumer), thereby making the Service Model and the entire Service-Oriented Architecture landscape of the organization heavily dependent on a stable, reliable, and high-performance process landscape (see Figure 25).

Business services are directly affected by process-change requests and optimizations, because changes in processes will ultimately change the behavior of how services are being delivered by an organization. Therefore, it is essential to continuously update the service models to reflect any changes to existing processes and/or the introduction of new processes to allow the service models to perform at maximum efficiency (see Figure 24). The outcome of step 37 is consumed by step 37d.

Typical tasks that are done within this step:

- Associate and link changes to process steps and activities to business, application, data, platform, and infrastructure services
- Associate and link changes to events and gateways to business, application, data, platform, and infrastructure services
- Associate and link changes to process flow, process roles, and process owners to business, application, data, platform, and infrastructure services

Typical templates that are used:

- Process Map and/or Matrix
- Service Map, Matrix, and/or Model
- Performance Map and/or Matrix

- 高成本类型；
- 中等成本类型；
- 低成本类型。

使用的典型模板：

- 流程图及矩阵；
- 服务图及矩阵；
- 绩效图及矩阵；
- 价值图及矩阵；
- 成本图、矩阵及模型；
- 风险图及矩阵。

涉及的典型BPM CoE角色：

- 价值专家；
- 业务分析师；
- 企业建筑师；
- 流程架构师。

步骤37d：服务模型

业务流程将业务服务（服务提供商）提供给一个或多个客户（无论是内部还是外部），然后使用提供给他们的服务（服务消费者），从而使组织的服务模型和整个面向服务的体系结构依赖于稳定、可靠和高性能的流程布局（图25）。

业务服务直接受流程变更请求和优化的影响，因为流程的变化最终将改变组织提供服务的行为。因此，必须不断更新服务模型以反映对现有流程的任何更改及引入新流程，以使服务模型以最高效率执行（图24）。步骤37的结果由步骤37d使用。

在此步骤中完成的典型任务：

- 将流程步骤和活动的变更关联并链接到业务、应用程序、数据、平台和基础架构服务；
- 将事件和网关的更改关联并链接到业务、应用程序、数据、平台和基础架构服务；
- 将流程、流程角色和流程所有人的变更关联并链接到业务、应用程序、数据、平台和基础架构服务。

使用的典型模板：

- 流程图及矩阵；
- 服务图、矩阵及模型；
- 绩效图及矩阵；

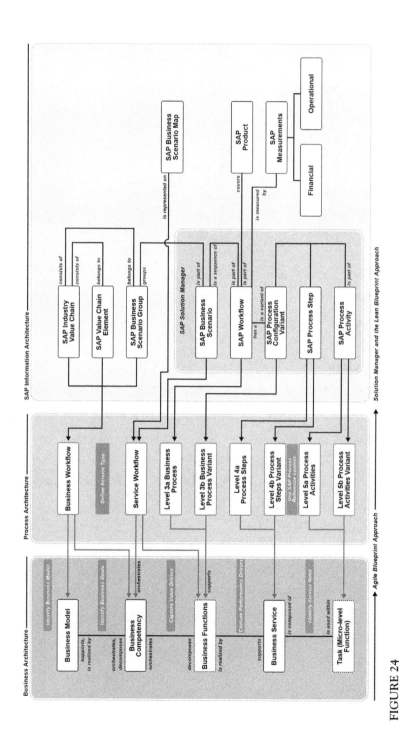

FIGURE 24

Example of how process steps (and variants thereof) deliver business services.

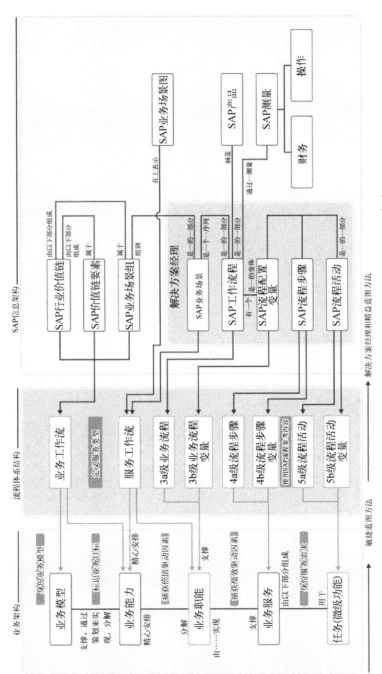

图24 流程步骤（及其变体）如何提供业务服务的示例 [36]

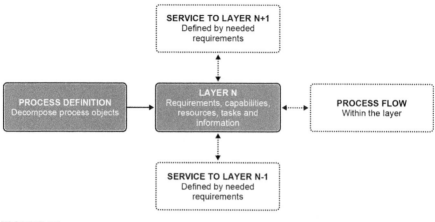

FIGURE 25

An example of a Layered Process Architecture model that shows how the execution of processes delivers services across layers within an organization based on layer requirements.

Ref. 37.

- Value Map and/or Matrix
- Risk Map and/or Matrix

    Typical BPM CoE roles involved:

- Service eXperts
- Business Analysts
- Enterprise Architects
- Process Architects.

Step 37e: Operating Model

The operating model (see Figure 26) focuses on process standardization and integration across the organization, and, because of this, changes made to any of the existing processes of the process portfolio need to be updated with the current operating model to reflect changes made and how services are being delivered as either strategic, tactical, or operational services. The outcome of step 37 is consumed by step 37e.

Typical tasks that are done within this step:

- Investigate opportunities for establishing processes around:
    - *Coordination*—low process standardization but high process integration
    - *Unification*—both high standardization and integration
    - *Diversification*—businesses requiring low standardization and low integration
    - *Replication*—high standardization but low integration
- Identify and categorize business processes around

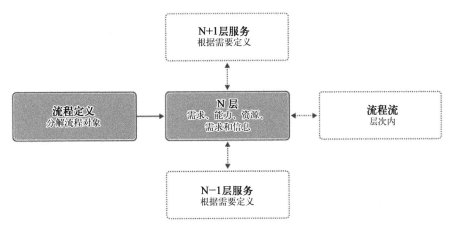

图25 一个分层流程架构模型的示例,它显示了流程的执行如何根据分层需求跨组织内的层提供服务[37]。

- 价值图及矩阵;
- 风险图及矩阵。

涉及的典型BPM CoE角色:

- 服务专家;
- 业务分析师;
- 企业架构师;
- 流程架构师。

步骤37e:操作模型

操作模型(图26)侧重于整个组织的流程标准化和一体化,因此,对流程组合的任何现有流程所做的更改需要用当前的操作模型更新以反映所做的更改是如何以战略、战术或运行的形式交付服务的。步骤37的结果由步骤37e使用。

在此步骤中完成的典型任务如下。

- 调查建立流程的机会。
  - 协调——低流程标准化,但高流程一体化;
  - 统一——高标准化和一体化;
  - 多样化——需要低标准化和低一体化的业务;
  - 复制——高标准化但低一体化。
- 识别和分类业务流程。

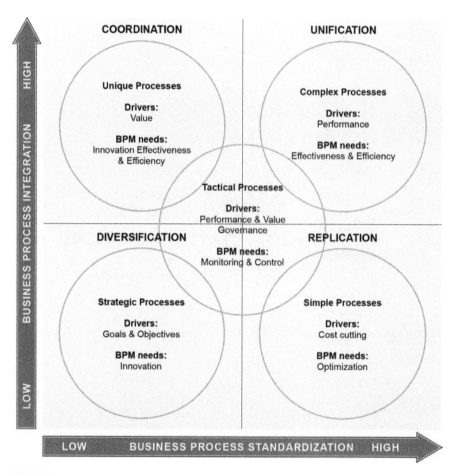

FIGURE 26

An Operating Model focusing on business process integration and business process standardization.

Ref. 38.

- business areas and groups
- business functions
- business roles and owners
- resources/actors
- business competencies
- business rules and compliance
- service construct
- service areas and groups
- service owners
- process areas and groups

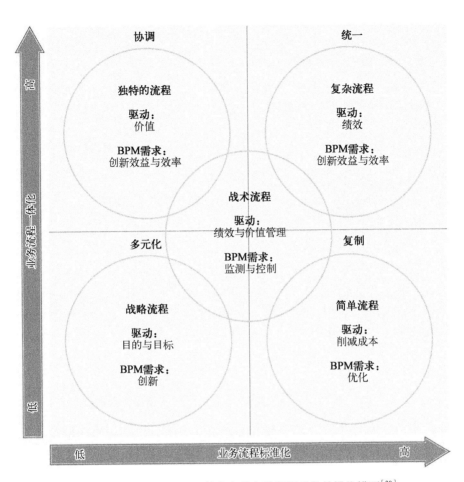

图26　专注于业务流程一体化和业务流程标准化的操作模型[38]

- 业务领域和团体；
- 业务职能；
- 业务角色和所有者；
- 资源/参与者；
- 业务能力；
- 业务规则和合规性；
- 服务构建；
- 服务区域和团体；
- 服务所有者；
- 流程域和流程组；

- process owners
- application/system owners
- data owners
- platform owners
- infrastructure owners
- Associate and connect the process types (i.e., main, management, and support) to
  - business areas and groups;
  - capabilities and functions;
  - business roles and owners;
  - resources/actors;
  - business competencies;
  - business rules and compliance;
  - service construct;
  - service areas and groups;
  - service owners;
  - process areas and groups;
  - business processes;
  - process owners;
  - application/system owners;
  - data owners;
  - platform owners; and
  - infrastructure owners
- Associate each process owner with
  - business and process areas and groups
- Create an Operating Model to illustrate the relationship between process owners and business and process areas and groups

Typical templates that are used:

- Process Map and/or Matrix
- Service Map and/or Matrix
- Performance Map and/or Matrix
- Value Map and/or Matrix
- Operating Map, Matrix, and/or Model
- Risk Map and/or Matrix

Typical BPM CoE roles involved:

- Process eXperts
- Business Analysts
- Information Architects
- Technology Architects
- Enterprise Architects
- Process Architects

- 流程所有人；
- 应用程序/系统所有者；
- 数据所有者；
- 平台所有者；
- 基础设施所有者。
- 关联并连接流程类型（即首要、管理和支持）。
  - 业务领域和团体；
  - 能力和功能；
  - 业务角色和所有者；
  - 资源/行动者；
  - 业务能力；
  - 业务规则和合规性；
  - 服务构建；
  - 服务区域和团体；
  - 服务所有者；
  - 流程区域和组别；
  - 业务流程；
  - 流程所有人；
  - 应用/系统所有者；
  - 数据所有者；
  - 平台所有者；
  - 基础设施所有者。
- 将每个流程所有人相关联。
  - 业务和流程领域和团体。
- 创建操作模型以说明流程所有人与业务、流程区域和组别之间的关系。

使用的典型模板：

- 流程图及矩阵；
- 服务图及矩阵；
- 绩效图及矩阵；
- 价值图及矩阵；
- 操作图、矩阵及模型；
- 风险图及矩阵。

涉及的典型BPM CoE角色：

- 流程专家；
- 业务分析师；
- 信息架构师；

Step 37f: Performance Model

Processes are constantly being monitored for performance gaps and issues (see Figure 27), and therefore, changes made to processes must be directly related to the performance model. The outcome of step 37 is consumed by step 37f.

Typical tasks that are done within this step:

- Relate, monitor, and measure strategic, tactical, and operational processes to Strategic Business Objectives (SBOs)
- Relate, monitor, and measure strategic, tactical, and operational processes to Critical Success Factors (CSFs)
- Relate, monitor, and measure strategic, tactical, and operational processes to Key Performance Indicators (KPIs)
- Develop, monitor, and measure Process Performance Indicators (PPIs) around operational processes
- Develop, monitor, and measure Service Performance Indicators (SPIs) around operational processes
- Connect and link process areas, process groups, process measurement, and process owners to
  - Strategic, tactical, and operational Key Performance Indicators (KPIs)
- Connect and link process steps, process activities, and process owners to
  - Strategic, tactical, and operational Process Performance Indicators (PPIs); and
  - Strategic, tactical, and operational Service Performance Indicators (SPIs)

Typical templates that are used:

- Process Map and/or Matrix
- Service Map and/or Matrix
- Performance Map, Matrix, and/or Model
- Value Map and/or Matrix
- Risk Map and/or Matrix

Typical BPM CoE roles involved:

- Process eXperts
- Value eXperts
- Business Analysts
- Enterprise Architects
- Process Architects

CONCLUSION

The organizational requirements of implementing a usable and effective BPM Life Cycle in any organization is a very demanding task in itself. Even more difficult is the need to structure the life cycle in a way that fully, and in a very detailed and explicit manner, revolves around accomplishing not only process-related goals, but

- 技术架构师；
- 企业架构师；
- 流程架构师。

步骤37f：绩效模型

不断监控流程的绩效差距和问题，因此，流程所做的更改必须与绩效模型直接相关。步骤37f使用步骤37的结果。

在此步骤中完成的典型任务：

- 将战略、战术和操作过程与战略业务目标（SBO）联系起来，并对其进行监控和度量；
- 关联、监控和衡量CSF的战略、战术和运营流程；
- 关联、监控和衡量KPI的战略、战术和运营流程；
- 围绕运营流程制定、监控和衡量PPI；
- 围绕运营流程制定、监控和衡量SPI；
- 连接并链接流程区域、流程组、流程测量和流程所有人；
- 战略、战术和运营KPI；
- 连接并链接流程步骤、流程活动和流程所有人；
- 战略、战术和运营PPI；
- 战略、战术和运营SPI。

使用的典型模板：

- 流程图及矩阵；
- 服务图及矩阵；
- 绩效图、矩阵及模型；
- 价值图及矩阵；
- 风险图及矩阵。

涉及的典型BPM CoE角色：

- 流程专家；
- 价值专家；
- 业务分析师；
- 企业架构师；
- 流程架构师。

3.5.8　结论

在任何组织中实施可用且有效的BPM生命周期的组织要求本身就是一项非常艰巨的任务。更加困难的是，需要以一种完全、非常详细和明确的方式构建生命

PROCESS/INPUT
Pain Point Situation & Effect Analysis Form

Process or Product Name:

Responsible/Owner:

Prepared by:

Date (original):

Date (revised):

Process Step/Input:	Potential Pain Point Situation:	Potential Pain Point Effects:	Seve-rity (0-10):	Potential Weakness Causes:	Occu-rrence (0-10):	Current Weakness Cluster Controls:	Detec-tion: (0-10):	Risk Priority Number:	Actions Recom-mended:	Res-ponse:	Actions Taken:

General Ranking Guidelines
0-1: N/A or Limited   2-3: Low   4-6: Average   7-8: High   9-10: Critical

FIGURE 27

The Process/Input Pain Point Situation & Effect Analysis Form is a powerful tool to help monitor, identify, and suggest changes to process pain points.

Ref. 39.

流程／输入
痛点情况及效果分析表

过程或产品名称：
负责人/所有者：
编制人：
日期（原件）：
日期（修改）：

过程要素/输入：	潜在的痛点情况：	潜在的痛点影响：	严重程度(0~10)：	潜在的弱点原因：	出现次数(0~10)：	当前弱点集群控制：	检测(0~10)：	风险优先级编号：	建议采取的行动：	回复：	所采取的行动：	Severity 0-10	Occurrence 0-10	Detection 0-10	Risk Priority Number (RPN)

通用排名指南
0-1:不适用或有限　2-3:低　4-6:平均　7-8:高　9-10:关键

图27　流程/输入痛点情况和效果分析表是一个功能强大的工具，可帮助监控、识别和建议处理痛点的变化[39]

more importantly, business objectives, goals, and strategy. Because, in a nutshell, processes are essentially a sequenced flow of steps and activities that have been specifically designed to achieve a defined business objective and eventually allow for the fulfilment of the strategy of the organization. Thus, processes act as a chain reaction of actions that are indirectly responsible for fulfilling the strategy of an organization, and ultimately, that is the goal.

Most BPM and process life cycles focus almost exclusively on process-oriented solutions and goals, rather than also incorporating business-related challenges. That is where many organizations go about it the wrong way. As mentioned earlier, processes are but a tool to fulfill the goals of the business, and with that in mind, it is important to maintain a strong focus on business objectives and goals when designing the structure and the steps involved with the BPM Life Cycle. Process goals have to serve the needs of the business, and designing a tight collaboration between process objectives and goals, and business objectives and goals, is of the utmost importance.

The work done by the process roles (experts, engineers and architects) in various programs and projects along with their specific tasks always relate to a specific business goal, the roles, tasks, and deliverables that link to the business goals, and can be managed within the process life cycle. This is done through managing the link to strategy on a portfolio, program, and project level, and managing the deliverables/artefacts assigned to the specific roles and their tasks within the process life cycle. As the process management life cycle has already been aligned with BPM Portfolio Management, the key touch points can be highlighted for specific focus on governance and value delivery. This clearly illustrates that the BPM Life Cycle has been specifically designed for creating process solutions that also focus on reaching business objectives and goals as well as solving business issues and challenges (see step 1: "Identify critical business factors", step 2: "Describe process goals" as well as the Business Innovation and Transformation Enablement steps around steps 37a–f).

Going through a detailed and highly analytical design and construction phase to create a useful and effective BPM Life Cycle should be top priority in most organizations. The sheer benefits and many different advantages of doing so far outweigh the number of resources required to complete the production of the life cycle. Let us take a glance at some of the many benefits and advantages you can harvest from creating and using an effective BPM Life Cycle.

Let's start with a few examples of advantages from a **strategic point of view:**

- *Industry and trends:* The identification of industry changes and key external trends can be a great asset for adapting existing processes to meet these changes. Such changes, whether internal or external, should be seen as opportunities and drivers for optimizing existing processes or simply acting as a catalyst for the creation of new processes—not only to adapt to, for instance, a changing market, but certainly also to benefit from new business opportunities and ventures.

周期,不仅要实现与流程相关的目标,更重要的是实现业务目的、目标和战略。简而言之,流程本质上是一系列步骤和活动的顺序流程,这些步骤和活动专门用于实现既定的业务目标,并最终实现组织战略。因此,流程充当行动的连锁反应,间接负责实现组织的战略,并最终实现目标。

大多数BPM和流程生命周期几乎只关注面向流程的解决方案和目标,而不是合并与业务相关的挑战。这是许多组织错误的做法。如前所述,流程只是实现业务目标的工具,记住,在设计结构和与BPM生命周期相关的步骤时,保持对业务目的和目标的强烈关注是很重要的。流程目标必须服务于业务需求,设计流程目的和目标以及业务目的和目标之间的紧密协作至关重要。

流程角色(专家、工程师和架构师)在各种计划和项目中完成的工作及其特定任务始终与特定业务目标,链接到业务目标的角色、任务和可交付成果相关,并且可以在流程生命周期内进行管理。这是通过管理产品组合,程序和项目级别的战略链接,以及管理在流程生命周期内分配给特定角色及其任务的可交付成果/人工制品来完成的。由于流程管理生命周期已与BPM组合管理产品保持一致,可以突出关键接触点,以便特别关注治理和价值交付。这清楚地表明,BPM生命周期专门用于创建流程解决方案,该流程解决方案还专注于实现业务目的和目标,以及解决业务问题和挑战[请参阅步骤1:"识别关键业务因素(Identify critical business factors)",步骤2:"描述流程目标(Describe process goals)"以及步骤37a ~ 步骤37f的业务创新和转型支持]。

在大多数组织中,通过一个详细而高度分析的设计和构建阶段来创建一个有用且有效的BPM生命周期应该是最优先考虑的。到目前为止,这样做的纯粹好处和许多不同的好处都超过了完成生命周期生产所需的资源数量。让我们来看看您通过创建和使用有效的BPM生命周期可以获得的许多好处和优势。

让我们从战略角度出发,举几个优势的例子。

- 行业和趋势:识别行业变化和关键外部趋势对于调整现有流程以满足这些变化是一项巨大的资产。这些变化,无论是内部的还是外部的,都应被视为优化现有流程的机会和驱动因素,或者只是作为创造新流程的催化剂,不仅要适应不断变化的市场,而且一定要从新的业务机会和风险中获益。

- *Understanding and commitment:* A BPM Life Cycle helps to secure a common organizational understanding and commitment of owners, stakeholders, and management teams across organizational boundaries.
- *Proactive organization:* It will help jump start and enable a continuously proactive organization in which all share a common goal—the goal of creating or reengineering processes to reach business objectives and business goals, and thereby realize business strategies.

Advantages from an **organizational point of view:**

- *Ownership, roles, and responsibilities:* Every individual involved with the life cycle is assigned both tasks and responsibilities for carrying out particular objectives. This makes it easier to distribute the appropriate tasks and assignments to the appropriate experts across the organization. Process ownership, stakeholders, and decision makers are also involved in all of the context of the BPM Life Cycle.
- *Communication, collaboration, and feedback loop:* Because ownership, roles, and responsibilities have been placed accordingly, organizational communication across business units and effective collaboration within project teams become much more efficient because everybody knows what's going through regular status reporting, evaluations, audits, and measurements (in terms of process testing and simulation). The feedback loop ensures clear communication about improvements and optimization of existing processes, and also closes bottlenecks and encourages participation and involvement by all parties.

And advantages from a **process-oriented point of view:**

- *Overview and structure:* Creates a single point of reference for all future steps and activities that evolve around process creation and reengineering. The visual landscape and point of reference of the high-level BPM Life Cycle gives everyone involved a clear view of all the steps and activities that are to be done. These range from setting process goals, searching for existing process reference content, adjusting, matching, and/or creating new processes, and inciting high performance and good results through the distribution and assignment of process rewards.
- *Detail and granularity:* It is not only possible, but is highly recommended, to create further levels of detail from the perspective of the high-level view of the BPM Life Cycle. Note that there are literally thousands of tasks involved in all of the steps of the high-level BPM Life Cycle, and all the tasks involved need to be described in detail and with an analytical, determined approach.
- *Process innovation:* It is important to note that the BPM Life Cycle any organization creates should always adhere to innovation principles around processes on all levels. Innovation creates value that was previously unavailable, and process innovation is the key to opening up the gates to new - or better yet—the creation of entirely new markets and/or business opportunities. Process innovation should always be an integrated part of the six different business model domains (i.e., value, revenue, cost, service, operating, and performance models).

- 理解和承诺：BPM生命周期有助于确保跨组织边界的所有者、利益相关者和管理团队对组织的共同理解和承诺。

- 主动组织：它将有助于启动并实现一个持续主动性的组织，其中所有人都有一个共同的目标，即创建或重新设计流程以实现业务目的和业务目标，从而实现业务战略。

从组织角度看的优势。

- 所有权、角色和责任：生命周期中涉及的每个人都被分配了执行特定目标的任务和责任。这使得将适当的任务和差事分配给整个组织中适当的专家更加容易。流程所有人、利益相关者和决策者也参与了BPM生命周期的所有环境。

- 沟通、协作和反馈循环：因为所有权、角色和责任都已相应地分配，跨业务部门的组织沟通和项目团队内的有效协作变得更加高效，因为每个人都知道通过定期的状态报告、评估、审计和测量（在流程测试和模拟方面）会发生什么。反馈循环可确保有关改进和优化现有流程的清晰沟通，并消除瓶颈和鼓励各方积极参与。

从面向流程的角度来看，其优势如下。

- 概述和结构：为所有未来步骤和活动创建一个单一的参考点，这些步骤和活动围绕流程创建和重新设计而发展。高级BPM生命周期的视觉景观和参考点使每个参与的人都能清楚地看到要完成的所有步骤和活动。这些范围包括设定流程目标，搜索现有流程参考内容，调整、匹配及创建新流程，以及通过分发和分配流程奖励来激发高绩效和良好结果。

- 细节和粒度：不仅有可能，而且强烈建议从BPM生命周期的高级视图的角度创建更高层次的细节。请注意，在高级BPM生命周期的所有步骤中，实际上都涉及数千个任务，并且所有涉及的任务都需要详细描述，并采用分析、确定的方法。

- 流程创新：重要的是要注意，任何组织创建的BPM生命周期都应始终遵循所有级别流程的创新原则。创新创造了以前无法获得的价值，而流程创新是打开通向新的或更好的大门的关键，也是创造全新市场及商业机会的关键。流程创新应始终是六个不同业务模型领域（即价值、收入、成本、服务、运营和绩效模型）的集成整合部分。

- *Process transformation:* Transformation principles have been well-known factors around process change since business process reengineering [40] was first introduced in the early 1990s. Going through a cycle of process reengineering allows for detailed and continuous process optimization and improvements while adhering to process goals and requirements. And, just as with process innovation, process transformation should also always be an integrated part of the six different business model domains (i.e., value, revenue, cost, service, operating, and performance models).

End Notes

1. The BPM Life Cycle Model, LEADing Practice Business Process Reference Content [#LEAD-ES20005BP].
2. See note 1 above.
3. Document Engineering: Analyzing and Designing Documents for Business Informatics and Web Services", R.J. Glushko and T. McGrath, The MIT Press, (2005).
4. Process Life Cycle Verb Taxonomy Model, LEADing Practice Business Process Reference Content [#LEAD-ES20005BP].
5. BPM Building Blocks Model, LEADing Practice Business Process Reference Content [#LEAD-ES20005BP].
6. Layered Process Architecture, LEADing Practice Process Architecture Reference Content [#LEAD-ES40004BP].
7. See note 1 above.
8. "What is Business Process Design and Why Should I Care?", Jay Cousins and Tony Stewart, RivCom Ltd, (2002)
9. Process Workflow Connection Diagram Model, LEADing Practice Business Process Reference Content [#LEAD-ES20005BP].
10. See note 1 above.
11. Process Decomposition & Composition Model, LEADing Practice Decomposition & Composition Reference Content [#LEAD-ES30001AL].
12. "Standardization or Harmonization? You need Both!", Albrecht Richen and Ansgar Steinhorst, (2008).
13. See note 1 above.
14. "ITIL; A. Guide To Release And Deployment Management", UCISA.
15. "Implementing a Process in an Organization", Rational Software Corporation, (2001).
16. "Process roll-out and nurturing", ProcessNet srl, (2014).
17. "Business Process Change", Paul Harmon, Morgan-Kaufmann, (2003).
18. Value Map, Performance Map & Process Map Alignment Worksheet Model, LEADing Practice Business Architecture Reference Content [#LEAD-ES40002PGBCPSI].
19. K.A. Long, "Process Roles — Who are the Process Owners?" Business Rules Journal, 13, no. 9 (September 2012).
20. See note 1 above.
21. Process Monitoring & Measurements Model, LEADing Practice Business Process Reference Content [#LEAD-ES20005BP].
22. Performance & Value Measurement Model, LEADing Practice Measurement Reference Content [#LEAD-ES20014PG].

- 流程转换：自20世纪90年代初首次引入业务流程再设计[40]以来，转换原则一直是流程更改的众所周知因素。通过流程再设计的循环，可以在遵循流程目标和要求的同时进行详细和持续的流程优化和改进。而且，与流程创新一样，流程转换也应该始终是六个不同业务模型领域（即价值、收入、成本、服务、运营和绩效模型）的整合部分。

参考文献

［ 1 ］ The BPM Life Cycle Model, LEADing Practice Business Process Reference Content［#LEAD-ES20005BP］.

［ 2 ］ See note 1 above.

［ 3 ］ "Document Engineering: Analyzing and Designing Documents for Business Informatics and Web Services", R.J. Glushko and T. McGrath, The MIT Press, (2005).

［ 4 ］ Process Life Cycle Verb Taxonomy Model, LEADing Practice Business Process Reference Content［#LEAD-ES20005BP］.

［ 5 ］ BPM Building Blocks Model, LEADing Practice Business Process Reference Content［#LEAD-ES20005BP］.

［ 6 ］ Layered Process Architecture, LEADing Practice Process Architecture Reference Content［#LEAD-ES40004BP］.

［ 7 ］ See note 1 above.

［ 8 ］ "What is Business Process Design and Why Should I Care?", Jay Cousins and Tony Stewart, RivCom Ltd, (2002).

［ 9 ］ Process Workflow Connection Diagram Model, LEADing Practice Business Process Reference Content［#LEAD-ES20005BP］.

［ 10 ］ See ote 1 above.

［ 11 ］ Process Decomposition & Composition Model, LEADing Practice Decomposition & Composition Reference Content［#LEAD-ES30001AL］.

［ 12 ］ "Standardization or Harmonization? You need Both!", Albrecht Richen and Ansgar Steinhorst, (2008).

［ 13 ］ See note 1 above.

［ 14 ］ "ITIL；A. Guide To Release And Deployment Management", UCISA.

［ 15 ］ "Implementing a Process in an Organization", Rational Software Corporation, (2001).

［ 16 ］ "Process roll-out and nurturing", ProcessNet srl, (2014).

［ 17 ］ "Business Process Change", Paul Harmon, Morgan-Kaufmann, (2003).

［ 18 ］ Value Map, Performance Map & Process Map Alignment Worksheet Model, LEADing Practice Business Architecture Reference Content［#LEAD-ES40002PGBCPSI］.

［ 19 ］ K.A. Long, "Process Roles — Who are the Process Owners?" Business Rules Journal, 13, no. 9 (September 2012).

［ 20 ］ See note 1 above.

［ 21 ］ Process Monitoring & Measurements Model, LEADing Practice Business Process Reference Content［#LEAD-ES20005BP］.

［ 22 ］ Performance & Value Measurement Model, LEADing Practice Measurement Reference Content［#LEAD-ES20014PG］.

23. Business Value Model, LEADing Practice Value Model Reference Content [#LEAD-ES20007BCPG].
24. Structural and Behavioural Commonalities of Process Variants, Matthias Weidlich and Mathias Weske, Hasso-Plattner-Institute, University of Potsdam, German.
25. Business process modeling notation, Jan Mendling, Matthias Weidlich, Mathias Weske, Springer, 2010.
26. "Business Impact Analysis", ISACA.
27. Process Intelligence Model, LEADing Practice Business Process Reference Content [#LEAD-ES20005BP].
28. "Business Process Improvement: The Breakthrough Strategy for Total Quality, Productivity, and Competitiveness", H. James Harrington, McGraw-Hill, (1991).
29. See note 1 above.
30. Treating Progress Functions as a Managerial Opportunity, John M. Dutton and A. Thomas, The Academy of Management Review 9, no. 2 (April 1984): 235–247.
31. BPM Governance Model, LEADing Practice Business Process Reference Content [#LEAD-ES20005BP].
32. Process & Application Lifecycle Continuous Improvement Model, LEADing Practice Lifecycle Management Reference Content [#LEAD-ES30002AL].
33. "Business Process Improvement", Rob Berg, (2008).
34. An example of a Transformation & Innovation Interrogative Model around Business Processes, LEADing Practice Enterprise Transformation & Innovation Reference Content [#LEAD-ES60ETI].
35. Value-based Process Design Model, LEADing Practice Business Process Reference Content [#LEAD-ES20005BP].
36. Service & Process Activity Model, LEADing Practice Process Architecture Reference Content [#LEAD-ES40004BP].
37. Layered Process Architecture Model, LEADing Practice Process Architecture Reference Content [#LEAD-ES40004BP].
38. Process Operating Model, LEADing Practice Process Architecture Reference Content [#LEAD-ES40004BP].
39. Business Process Reference Content: An example of a BPM Continuous Improvement Tool.
40. "Reengineering Work: Don't Automate, Obliterate", Michael Hammer, Harvard Business Review, (1990).

[23] Business Value Model, LEADing Practice Value Model Reference Content［#LEAD-ES20007BCPG］.

[24] Structural and Behavioural Commonalities of Process Variants, Matthias Weidlich and Mathias Weske, Hasso-Plattner-Institute, University of Potsdam, German.

[25] Business process modeling notation, Jan Mendling, Matthias Weidlich, Mathias Weske, Springer, 2010.

[26] "Business Impact Analysis", ISACA.

[27] Process Intelligence Model, LEADing Practice Business Process Reference Content［#LEAD-ES20005BP］.

[28] "Business Process Improvement: The Breakthrough Strategy for Total Quality, Productivity, and Competitiveness", H. James Harrington, McGraw-Hill, (1991).

[29] See note 1 above.

[30] Treating Progress Functions as a Managerial Opportunity, John M. Dutton and A. Thomas, The Academy of Management Review 9, no. 2 (April 1984): 235−247.

[31] BPM Governance Model, LEADing Practice Business Process Reference Content［#LEAD-ES20005BP］.

[32] Process & Application Lifecycle Continuous Improvement Model, LEADing Practice Lifecycle Management Reference Content［#LEAD-ES30002AL］.

[33] "Business Process Improvement", Rob Berg, (2008).

[34] An example of a Transformation & Innovation Interrogative Model around Business Processes, LEADing Practice Enterprise Transformation & Innovation Reference Content［#LEAD-ES60ETI］.

[35] Value-based Process Design Model, LEADing Practice Business Process Reference Content［#LEAD-ES20005BP］.

[36] Service & Process Activity Model, LEADing Practice Process Architecture Reference Content［#LEAD-ES40004BP］.

[37] Layered Process Architecture Model, LEADing Practice Process Architecture Reference Content［#LEAD-ES40004BP］.

[38] Process Operating Model, LEADing Practice Process Architecture Reference Content［#LEAD-ES40004BP］.

[39] Business Process Reference Content: An example of a BPM Continuous Improvement Tool.

[40] "Reengineering Work: Don't Automate, Obliterate", Michael Hammer, Harvard Business Review, (1990).

The Chief Process Officer: An Emerging Top Leadership Role

Mathias Kirchmer, Peter Franz, Mark von Rosing

INTRODUCTION

More organizations are establishing business process management (BPM) as a discipline to move their strategies into operational execution with certainty. This is particularly important in the increasingly dynamic and connected business environment. As with any other management discipline, BPM was established through the process of process management. The use of BPM has become a key driver for optimization, cost cutting, effectiveness, and enterprise transformation, especially in an environment of external forces and drivers initiating constant change. As a result, many companies are beginning to develop a dedicated role to lead these initiatives. This emerging top executive role, which manages all process initiatives, is called the chief process officer (CPO).[1] The CPO oversees process management so that it increases performance and ensures value creation[2] by executing the business process strategy across organizational boundaries, such as departments or divisions.

THE EMERGING ROLE OF THE CPO

Process modelling, process optimization, and process innovation have been critical parts of organizations since the Industrial Revolution.[3] Today, with the increasing pace of business change, organizations are experiencing pressure to make continuous process improvements a part of daily operations. Given the increasing importance of processes as the key enablers in the transfer of the business strategy into execution, operational excellence,[4] optimization, and standardization, a top management role has emerged. As with any other important area, good leadership is needed; hence, an appropriate management role is required. In this environment, the ability to rally resources and drive collaboration effectively is vital, especially in diverse organizations, making process leadership not only a management but a C-level imperative.

This new top executive, which we see emerging in various organizations, is the CPO.[5] The title may not always be "chief process officer," but a champion of process is an important executive position at many enterprises. Some process management leaders hold titles such as "chief transformation officer" or "vice president of business process management." However, as the vice president of Gartner Group stated[6]: "Regardless of what the title is, it's for whoever is championing and leading business process improvement."[7]

3.6 CPO：新兴的高级领导角色

Mathias Kirchmer, Peter Franz, Mark von Rosing

3.6.1 介绍

越来越多的组织将BPM作为一个规程，以确保战略始终与实际操作结合。这在日益变化且相互关联的业务环境中尤为重要。与其他管理规程一样，BPM是通过流程管理的流程建立的。BPM的使用已经成为优化、成本削减、有效性和企业转型的关键驱动因素，尤其是在外部力量和驱动因素引发持续变化的环境中。因此，许多公司开始设立一个专门的角色来领导这些行动。这一新兴的最高执行官角色被称为CPO[1]，负责管理所有流程方案。CPO通过跨组织边界（如分部或分派）执行业务流程战略来监督流程管理，从而提高绩效并确保价值创造[2]。

3.6.2 新兴角色CPO的作用

自工业革命以来，流程建模、流程优化和流程创新一直是组织的关键部分[3]。如今，随着业务变化的步调越来越快，企业也随之面临压力，因此要将持续的流程改进作为日常运营的一部分。鉴于流程作为关键推动因素在将业务战略转变为执行、卓越运营[4]、优化和标准化过程中越来越重要，因此出现了高层管理角色。与其他重要领域一样，需要良好的领导。在这种环境下，有效地聚集资源和推动协作的能力至关重要，尤其是在不同的组织中，流程领导不仅属于管理层，而且是C级的必要条件。

我们看到这个新的高层管理人员出现在不同的组织中，他就是CPO[5]。这个头衔不一定总是首席流程官，在许多企业中，这是充当流程捍卫者的一个重要执行职位。一些流程管理领导者拥有"首席转型官"或"业务流程管理副总裁"等头衔。然而，正如Gartner Group的副总裁所言[6]："无论头衔是什么，都是为支持和领导业务流程改进的人而设的[7]。"

Managing processes appropriately gives an organization the capability to successfully deal with a volatile business environment. The CPO enables the journey of an organization to the next-generation enterprise. He or she develops an integrated view of the organization across organizational boundaries, helping business people to see the power of information technology (IT) and IT people to understand the business challenges.

The transparency created under the leadership of the CPO facilitates other values, including quality and efficiency, agility and compliance, external integration of the company and internal alignment of the employees, and innovation and conservation where appropriate. The result is a lasting competitive advantage, empowered through BPM-enabled transformation under the leadership of the CPO.

The CPO creates a process-centric organization and culture across the more or less functional organization of an existing company; we often refer to this value proposition offering as the "BPM value flow."[8] He or she integrates function-driven and process-driven decision-making and management, as well as enables an end-to-end process view focused on value planning, value identification, and value creation[9] for clients. This overall management approach of the CPO, resulting in company-wide process governance,[10] is shown in Figure 1.

Where do you find such a CPO? In many cases, it is an enlightened chief information officer (CIO) who recognizes that with certain trends,[11] such as "the cloud" or "software as a service," the key assets in the organization are the processes. BPM is the means to get business value out of such technology trends. Some organizations even show that transition openly and move from the CIO role to the "chief process and information officer" (CPIO). Also, a chief operating officer (COO) could become a

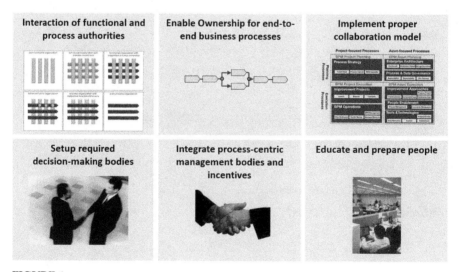

FIGURE 1

Enabling process-centric management of functional organizations.

　　适当地管理流程使组织能够成功地处理不稳定的业务。CPO实现了一个组织到下一代企业的旅程。他或她开发了跨组织边界的组织整合视图,帮助业务人员了解IT和IT人员的能力,了解业务挑战。

　　在CPO领导下创建的透明度有助于实现其他价值,包括质量和效率、灵活性和合规性、公司外部整合和员工内部协调,以及适当的创新和节约。其结果是持久的竞争优势,在CPO的领导下,并通过BPM的能力实现转变。

　　CPO在现有公司的职能组织中或多或少的创建了一个以流程为中心的组织和文化,我们通常将此价值主张产品称为BPM价值流(BPM value flow)[8]。他或她将功能驱动和流程驱动的决策和管理整合在一起,并支持端到端流程视图,重点关注客户的价值规划、价值识别和价值创造[9]。CPO的这种总体管理方法导致了公司范围的流程治理[10],如图1所示。

　　您在哪里找到这样一个CPO? 在许多情况下,有见识的首席信息官(CIO)认识到,在某些趋势下[11],如"云"或"软件即服务",组织中的关键资产是流程。BPM是从这些技术趋势中获取业务价值的手段。一些组织甚至公开表明这种转变,并从CIO角色转变为首席流程和信息官(CPIO)。

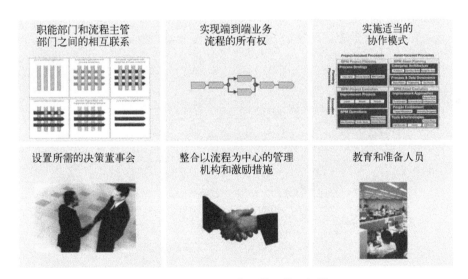

图1　实现以流程为中心的职能组织管理

CPO if he or she recognizes that BPM is more than just efficiency improvement. The role frequently develops on the back of a large business transformation program, with the "transformation director" stepping into this role to create sustainable value. Other organizations need to build the role of a CPO from scratch, co-existing with the CIO and COO. The best solution depends on the specific situation of an organization.

KEY TASKS OF THE CPO

The tasks of the CPO can be directly deduced from the process of process management that he or she owns. Figure 2 shows an overview of the process of process management.[12]

There are five groups of tasks:

- General integration tasks
- Project-related planning tasks
- Project-related execution tasks
- Asset-related planning tasks
- Asset-related execution tasks.

The CPO drives a cross-functional culture in which process owners, process experts, and the various knowledge workers know how they fit into the overall end-to-end process and what that means for their work. The CPO is the overall contact for all process-related topics and provides input in strategic business planning.

The core of the project-related planning task is the management of a BPM strategy. This includes the identification of the high-impact, high-opportunity processes of an organization on which BPM initiatives focus, the identification of required capabilities and capability gaps, as well as the definition of the overall

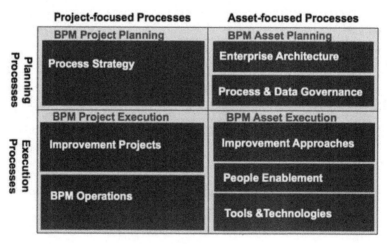

FIGURE 2

Overview of the process of process management.

此外，如果首席运营官（COO）认识到BPM不仅仅是提高效率，那么他或她可能会成为CPO。这一角色经常在大型企业转型计划的支持下发展，而"转型主管"也会加入这一角色，以创造可持续价值。其他组织需要从头开始构建CPO的角色，与CIO和COO共同存在。最佳方案取决于组织的具体情况。

3.6.3 CPO的关键任务

CPO的任务可以直接从其拥有的流程管理流程中推导出来。图2显示了流程管理流程的概述[12]。

有五组任务：

- 一般的整合任务；
- 项目相关的计划任务；
- 项目相关的执行任务；
- 资产相关的计划任务；
- 资产相关的执行任务。

CPO推动一种跨职能的文化，在这种文化中，流程所有人、流程专家和各种知识工作者知道他们如何适应整个端到端流程，以及这对他们的工作意味着什么。CPO是所有流程相关主题的整体联系人，并提供战略业务规划的输入。

与项目相关的计划任务的核心是管理BPM策略。这包括识别BPM计划重点关注的组织的高影响、高机会流程，识别所需能力和能力差距，以及定义整个流程管理议程。流程管理议程包括高级业务案例，允许对项目进行优先级排序。

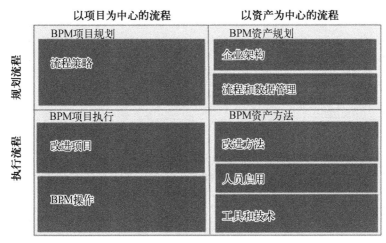

图2 流程管理的流程的概述

process management agenda. The process management agenda includes high-level business cases that allow the prioritization of projects.

Project-related execution tasks include the launch and oversight of improvement initiatives, as well as the ongoing BPM operations. Typical operation tasks are the organization of a value-realization approach that enforces the ongoing focus on value, even after the project has concluded, or the procurement of administrative parts of the BPM Centre of Excellence (CoE) as externally delivered managed services.

The organization of an enterprise architecture (EA) approach and the appropriate process governance are key asset-related planning tasks. The governance-related tasks are crucial for a successful BPM discipline. They include the definition of governance-related roles, process owners, and procedures as well as the organization of governance bodies,[13] such as for the process owners to make cross-process decisions.

Asset-related execution tasks include the development of capabilities in various improvement methods, such as a process transformation approach, so that they can be applied in improvement projects. A key task is to organize people enablement initiatives, such as change management (ongoing or as part of a project) or the launch of BPM communities. Last but not least, the direction and oversight for tool- and technology-related assets, such as repository tools, process execution systems, social platforms, and process intelligence tools, comes from the CPO.

POSITIONING OF THE CPO IN THE ORGANIZATION

The positioning of the CPO (or another more relevant name for the role) in an organization depends heavily on the nature of the specific organization. It has to be defined in the context of the existing organizational structure and market dynamics. However, there are several aspects to consider when positioning this role:

- The CPO provides input in the overall business strategy based on his or her cross-functional view and drives the execution of this strategy. The CPO requires appropriate access to the board and the position needs to provide the necessary standing in the organization.
- The CPO works closely with the other process owners, who are in general very senior executives. He or she needs to have the standing and positioning to be successful in this context.
- The CPO leads the BPM core team, which is often a center of excellence. The CPO's positioning needs to give his or her reports the appropriate standing and development perspectives.

These points motivate most organizations to have the CPO role report directly to the management board or at least to another C-level position. This has been confirmed in an empirical research study.[14] It is consistent with our observations working with large and mid-sized organizations around the world.

The typical positioning of the CPO is shown in Figure 3. It also describes the CPO's integration in various governance bodies.

与项目相关的执行任务包括启动和监督改进计划,以及正在进行的BPM操作。典型的运营任务是组织一种价值实现方法,即使在项目结束后,也要对价值进行持续关注,或者获取作为外部交付的管理服务的BPM CoE的行政部分。

企业架构(enterprise architecture,EA)方法的组织和适当的流程治理是与资产相关的关键计划任务。治理相关的任务对于成功的BPM规则至关重要。它们包括与治理相关的角色、流程所有人和流程的定义,以及治理机构的组织[13],如流程所有人做出跨流程决策。

与资产相关的执行任务包括各种改进方法能力的开发,以便将其应用于改进项目中,如流程转换方法。关键任务是组织人员发挥主动性,如变更管理(正在进行或作为项目的一部分)或启动BPM社区。最后同样重要的是,工具和技术相关资产(如存储库工具、流程执行系统、社会平台和流程智能工具)的方向和监督来自CPO。

3.6.4　CPO在组织中的定位

CPO(或角色的另一个更相关名称)在组织中的定位在很大程度上取决于特定组织的性质。它必须在现有组织结构和市场动态的背景下进行定义。但是,在定位此角色时需要考虑以下几个方面。

- CPO根据其跨职能的观点提供总体业务战略的输入,并推动该战略的执行。CPO需要适当地进入董事会,该职位需要在组织中提供必要的地位。
- CPO与其他流程所有人密切合作,他们通常都是非常高级的管理人员。在这种情况下,他或她需要具备成功的地位和定位。
- CPO领导BPM核心团队,该团队通常是卓越中心。CPO的定位需要在报告中明确适当的地位和发展视角。

这些要点促使大多数组织直接向管理委员会或至少向另一个C级职位报告CPO角色。这已在一项实证研究中得到证实[14]。这与我们与世界各地大中型组织合作的观察结果一致。

CPO的典型定位如图3所示。它还描述了CPO在各种治理机构中的整合。

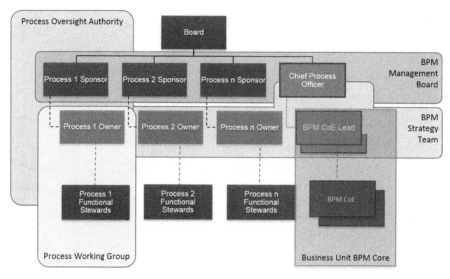

FIGURE 3

Typical positioning of the chief process officer.

The CPO is an emerging role targeting the creation of value by focusing on transferring the business strategy into execution, quickly and with minimal risk. We expect this role to become more and more important in organizations, similar to the rise of the CIO about in the latter part of the 1980s.

CONCLUSION

As organizations respond to the need for greater market responsiveness, they recognize the need for a more robust process management discipline that treats the processes in the organization as an important asset. As with any other major asset, processes need nurturing and caring with the right level of priority. The emergence of a CPO is therefore an inevitable, important enabler in achieving this goal.

End Notes

1. Franz, Kirchmer, *The Chief Process Officer – A Role to Drive Value* (London, Philadelphia: Accenture Whitepaper, 2012-2).
2. Kirchmer, "How to Create Successful IT Projects with Value-Driven BPM," in *CIO Magazine Online* (February 27th 2013).
3. Scheer, *Business Process Engineering – Reference Models of Industrial Enterprises* (Berlin e.a: Springer, 2nd edition, 1995).
4. Kirchmer, *High Performance through Process Excellence – From Strategy to Execution with Business Process Management* (Berlin, e.a: Springer, 2nd edition, 2011).
5. Jost, "Vom CIO zum CPO," *Harvard Business Manager*, (2004): 88–89.
6. Bruce Robertson, Gartner Group Inc. Nov-2013.

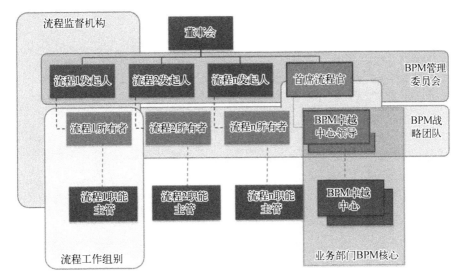

图3　CPO的典型定位

CPO是一个新兴的角色,其目标是通过专注于将业务战略转化为执行,快速且风险最小地创造价值。我们期望这个角色在组织中变得越来越重要,类似于20世纪80年代后期CIO的崛起。

3.6.5　结论

当组织响应更大的市场适应能力需求时,他们认识到需要一个更强大的流程管理规则,将组织中的流程视为一项重要资产。与任何其他主要资产一样,流程需要以正确的优先级进行培育和关心。因此,CPO的出现是实现这一目标的一个不可避免的重要推动者。

参考文献

［1］ Franz, Kirchmer, The Chief Process Officer — A Role to Drive Value (London, Philadelphia: Accenture Whitepaper, 2012–2).

［2］ Kirchmer, "How to Create Successful IT Projects with Value-Driven BPM," in CIO Magazine Online (February 27th 2013).

［3］ Scheer, Business Process Engineering — Reference Models of Industrial Enterprises (Berlin e.a: Springer, 2nd edition, 1995).

［4］ Kirchmer, High Performance through Process Excellence — From Strategy to Execution with Business Process Management (Berlin, e.a: Springer, 2nd edition, 2011).

［5］ Jost, "Vom CIO zum CPO," Harvard Business Manager, (2004): 88–89.

［6］ Bruce Robertson, Gartner Group Inc. Nov-2013.

7. Morris, *Architect, Design, Deploy, Improve (ADDI) – A BPMS Development Methodology* (Chicago: Wendan Whitepaper, 2014).

8. LEADing Practice Value Modelling Reference Content.

9. Franz, Kirchmer, *Value-Driven Business Process Management – The Value-Switch for Lasting Competitive Advantage* (New York, e.a: McGraw-Hill, 2012-1).

10. Kirchmer, Hofmann, "Value-Driven Process Governance – Wettbewerbsvorteile durch die richtige Processorganisation,". in *IM+io Fachzeitschrift fuer Innovation, Organisation und Management* (Germany, 03/2013).

11. Scheer, "Tipps fuer den CIO: Vom Tekki zum Treiber neuer Business Modelle," in *IM+IO – Das Magazin fuer Innovation, Organisation und Management* (Sonderausgabe, Dezember 2013).

12. Kirchmer M, Franz P, *The BPM-Discipline – Enabling the Next Generation Enterprise. BPM-D Executive Training Documentation* (Philadelphia, London).

13. Kirchmer, Lehmann, Rosemann, zur Muehlen, Laengle, *Research Study – BPM Governance in Practice* (Philadelphia: Accenture Whitepapers, 2013).

14. Ibid.

［ 7 ］ Morris, Architect, Design, Deploy, Improve (ADDI) — A BPMS Development Methodology (Chicago: Wendan Whitepaper, 2014).

［ 8 ］ LEADing Practice Value Modelling Reference Content.

［ 9 ］ Franz, Kirchmer, Value-Driven Business Process Management — The Value-Switch for Lasting Competitive Advantage (New York, e.a: McGraw-Hill, 2012−1).

［10］ Kirchmer, Hofmann, "Value-Driven Process Governance — Wettbewerbsvorteile durch die richtige Processorganisation,". in IM+io Fachzeitschrift fuer Innovation, Organisation und Management (Germany, 03/2013).

［11］ Scheer, "Tipps fuer den CIO: Vom Tekki zum Treiber neuer Business Modelle," in IM+IO — Das Magazin fuer Innovation, Organisation und Management (Sonderausgabe, Dezember 2013).

［12］ Kirchmer M, Franz P, The BPM-Discipline — Enabling the Next Generation Enterprise. BPM-D Executive Training Documentation (Philadelphia, London).

［13］ Kirchmer, Lehmann, Rosemann, zur Muehlen, Laengle, Research Study — BPM Governance in Practice (Philadelphia: Accenture Whitepapers, 2013).

［14］ Ibid.

iBPM—Intelligent Business Process Management

Nathaniel Palmer

The impact of new technologies, the mandate for greater transparency, and the ongoing aftershocks of globalization have collectively removed nearly any trace of predictability within the enterprise environment. In light of this, delivering sustainable competitive advantage can no longer be found through simply scale and efficiency, but rather requires to ability interpret and adapt in real-time to streams of information flows, make sense of these, rapidly translating these into effective responses designed for precision rather than repeatability. The ability today is best delivery through Business Process Management (BPM). BPM is a discipline involving any combination of modeling, automation, execution, control, measurement, and optimization of business activity flows in support of enterprise goals, spanning systems, employees, customers, and partners within and beyond the enterprise boundaries.

Today, we are in what many see as the third phase of BPM, and what is commonly distinguished as *"Intelligent BPM"* or *iBPM*. The subject of this chapter, iBPM builds upon the previous two generations, yet extends into directions previously out of reach. Among that previously out of reach is what frames the core of the notion of intelligence. Not that BPM previously was dumb, per se, but rather it was blind. Whereas previous generations of BPM offered limited ability to make sense of business activity flows, iBPM System ("IBPMS") are distinguished foremost by a "sense and respond" orientation. In this way we can group and categorize the three phases of BPM in terms of the synergistic combination of three groups of capabilities.

The first phase ("Phase One") offered the ability to separate application from the processes (business logic) that they support, similar to introduction of the relational database systems (RDBMS) and the ability to separate data from applications. The advent of the RDBMS was essential to transforming how business applications were designed and built. In the same way, the introduction the BPMS offered the ability for the first time to manage processes as separate assets, in a similar manner to how we manage data. Just a few years ago, "Phase Two" of BPM introduced new capabilities that support adaptable, goal-driven process models by maintaining the intelligence for how to access information and application resources without having to bind this into a rigid process model. Today, "Phase Three" builds on the first two generations, adding visibility into the process and what is happening in the real-time, as well as through integrated analytics what will likely occur in the near future.

The Complete Business Process Handbook. http://dx.doi.org/10.1016/B978-0-12-799959-3.00016-1

3.7　iBPM——智能业务流程管理

Nathaniel Palmer

　　新技术的影响、任务的透明化以及全球化的持续发展共同消除了企业环境中几乎所有可预测性的可能。有鉴于此,仅仅通过规模和效率将无法再获得可持续的竞争优势,而是需要能够实时解释、理解、适应这些信息流,并迅速将其转化为旨在精确而非可重复性的有效响应。BPM是一个涉及建模、自动化、执行、控制、度量和业务活动流优化组合的规程,用于支持企业目标、跨越系统、员工、客户以及企业边界内外的合作伙伴。因此,现在的能力是通过BPM进行最佳交付。

　　今天,我们正处于许多人所认为的BPM的第三阶段,以及通常被称为"智能BPM"或iBPM(Intelligent BPM)的阶段。本节的iBPM建立在前两代的基础上,但延伸到了先前无法触及的目标。其中之前无法实现的核心是智能。事实上,并非BPM本身以前是愚蠢的,而是本质上它是盲目的。虽然上一代的BPM提供的理解业务活动流的能力有限,但iBPM系统(iBPMS)的主要特点是感知和响应定位。通过这种方式,我们可以根据三组功能的协同组合对BPM的三个阶段进行分组和分类。

　　第一阶段("阶段一")提供了将应用程序与它们支持的流程(业务逻辑)分离的能力,引入类似于关系数据库系统(relational database systems,RDBMS)和将数据与应用程序分离。RDBMS的出现对于转换业务应用程序的设计和构建方式至关重要,以同样的方式,BPMS的引入首次提供了将流程作为独立资产进行管理的能力,其方式与我们管理数据的方式类似。就在几年前,BPM的"阶段二"引入了新的功能,通过智能维护如何访问信息和应用程序资源,从而支持适应性强、目标驱动的流程模型,而无须将其绑定到严格的流程模型中。今天,"阶段三"建立在前两代的基础上,增加了对流程和实时发生的事情的可视性,以及对不久的将来可能发生的事情进行整合分析。

Within the first several years of the market for BPM products being established, applications that addressed these first two phases of capability dominated. Although having added value and offering improved maturity of BPM practice, these approaches also limited the potential market for BPM software. However, for most organizations the vast majority of their business processes is dynamic, not standardized, and thus requires the business systems (e.g., deployed software) that support them to adapt quickly to changes within the business environment. As a business technology, the greatest value of process management software is delivered, not through automation and integration alone, but by introducing a layer between users and existing IT infrastructure to allow business systems to adapt and keep pace with the constant change found in most business environments. Fully realizing the ability offered through orchestration, however, requires the situational awareness necessary to adapt business systems to a changing business environment—the ability to sense and respond. By taking the lid off the black box of automation, Phase Three of BPM offers a framework for continuously validating and refining an understanding of business performance drivers, and adapting business systems accordingly.

To achieve the visibility and feedback to expose what is going on within a process will require a new level of transparency of processes and operations that is sure to present cultural and human factors challenges. However, this is nothing new for BPM. BPM is only slightly about technology. It is, instead, mostly about the business and the people. What is indeed new, however, and at the center of the Phase Three opportunity, is the ability now to adapt systems continuously to match the ever-changing business environment. The model most frequently referenced throughout this chapter, this continuous loop of visibility and adaptability, offers one of the first real leverage points for transforming business through adaptability.

THE EVOLUTION OF INTELLIGENT BPM

To understand the opportunities offered by Intelligent Business Process Management, it is helpful to consider the phases of maturation that solutions have gone through since the late 1990s. Prior to and during technology expansion of the mid-to late-1990s, the management of business processes was typically limited to the repetitive sequencing of activities, with rigid, "hard-wired" application-specific processes set out in documentation and later specified programmatically in custom-coded or conventional Enterprise Resource Planning (ERP) systems. Any more sophisticated degree of workflow management generally imposed a significant integration burden, frequently accounting for 60–80% of the project cost with little opportunity for reuse. Still, integration was typically limited to retrieval of data or documents, similarly hard wired with one-to-one connection points.

These early process management initiatives often focused on integrating and automating repetitive processes, generally within standardized environments. Whether focused on Straight-Through Processing transactions or a discrete process, such as Account Activation, these are applications in which the flow and sequence of activities is predetermined and immutable. The role of exception handling here is to allow human intervention to quickly resolve or correct a break in the flow of an otherwise standard process.

在建立BPM产品市场的最初几年内,解决前两个阶段能力的应用程序占主导地位。尽管这些方法增加了价值并提高了BPM实践的成熟度,但也限制了BPM软件的潜在市场。然而,对于大多数组织来说,其绝大多数业务流程是动态的,而不是标准化的,因此需要一个支持它们快速适应业务环境变化的业务系统(如已部署的软件)。作为一种业务技术,流程管理软件的最大价值不是通过自动化和一体化实现的,而是通过在用户和现有IT基础设施之间引入一个层次,使业务系统能够适应并跟上大多数业务环境中不断变化的步伐。然而,要完全实现通过协调提供的能力,需要具备必要的情境意识,以便使业务系统适应不断变化的业务环境,即感知和响应能力。通过揭开自动化的黑匣子,BPM的第三阶段提供了一个框架,用于不断验证和完善对业务性能驱动因素的理解,并相应地调整业务系统。

为了实现可视性和反馈,以揭示流程中正在发生的事情,需要对流程和操作提高透明度,这肯定会带来文化和人为因素方面的挑战。然而,这对于BPM来说并不是什么新鲜事。BPM只是稍微涉及了技术,相反,它主要是关于业务和人员。然而,真正的新兴之处,也是第三阶段的核心,是现在能够不断调整系统,以适应不断变化的业务环境。本节中最常引用的模型(即可视性和适应性的连续循环),为通过适应性转换业务提供了第一个真正的杠杆点。

3.7.1 iBPM的发展

为了理解iBPM提供的机会,考虑自20世纪90年代末以来的解决方案中所经历的成熟阶段是很有帮助的。在20世纪90年代中期到后期的技术扩展之前和期间,BPM通常局限于具有严格的特定于应用程序的流程活动的硬连接,这些流程在文档中列出,随后在自定义编码或传统ERP系统中以编程方式指定。任何更复杂程度的工作流管理通常都会带来很大的整合负担,往往占项目成本的60% ~ 80%,而且几乎没有重用的机会。不过,整合通常仅限于数据或文档的检索,类似于硬连接的一对一的点连接。

这些早期的流程管理计划通常在标准化环境中侧重于一体化和自动化重复流程。无论是专注于直通式事务处理,还是专注于离散流程(如账户激活),这些应用程序中的活动流程和顺序都是预先确定且不可变的。异常处理在这里的作用是允许人工干预快速解决或纠正标准流程中的中断。

By the end of the 1990s, however, BPM had emerged as an identifiable software segment, a superset of workflow management, distinguished in part by allowing process management independent of any single application. This was enabled by managing application execution instructions separate from process flows, so processes could be defined without limitation to a single application, as well as through support for variable versus hard-wired process flow paths.

The first wave of BPM deployments were typically aimed at bridging the island of automation described above, such as closing gaps in existing ERP deployments. Early BPM solutions were differentiated by integration-centric functionality, such as application adapters, data transformation capabilities, and product-specific process definitions (e.g., an order-to-cash process). Eventually, the introduction of standards, such as Web Services and advances in the development tools within BPM suites, lowered the cost and complexity of data integration. This began to shift the fundamental value proposition of BPM from discrete capabilities to enabling the management of business logic by business process managers, without threatening the integrity of the application logic (the infrastructure that is rightfully managed and protected by IT personnel).

The availability of standards-based protocols significantly lowered the burden on BPM adopters for building and maintaining integration infrastructure, freeing time and resources to focus on process and business performance, rather than being consumed with plumbing issues. Over time, this facilitated a refocus of process management software from that of automation and integration to orchestration and coordination, bringing BPM into the realm of business optimization. The environment in which the modern enterprise operates is dynamic, requiring the business systems that support it to be so as well. This means that systems must be able to easily adapt to changing circumstances of the enterprise. Phase Two of the BPM opportunity was presented through making orchestration a reality—the ability to connect abstracted application capabilities across orchestrated business processes, thereby transforming existing automation infrastructure into reusable business assets. What separates orchestration from automation is presented by a fundamental shift in perspective, from thinking of processes as a flow of discrete steps, to understanding processes in terms of goals and milestones.

FROM AUTOMATION TO ORCHESTRATION: THE REALIGNMENT OF BPM AROUND SERVICE-ORIENTED ARCHITECTURES

Orchestration allows systems to mirror the behavior of the rest of the enterprise environment (one defined in terms of objectives rather than scripts). Over the last decade, orchestration has introduced a visible shift in the axis of business computing. As organizations realize the opportunities presented by orchestration, it offers (arguably mandates) a wholesale rethinking of the role of applications and information systems.

Orchestration has already had a visible impact on the direction of the BPM market, enabled by standards protocols (notably eXtensible Markup Language (XML)

　　然而,到20世纪90年代末,BPM已经成为一个可识别的软件、一个工作流管理的集合,部分区别在于允许流程管理独立于任何单个应用程序。这是通过管理独立于流程的应用程序执行指令实现的,因此可以定义流程而不局限于单个应用程序,也可以通过支持变量与硬连接流程路径来定义流程。

　　第一批BPM部署通常旨在弥合上述自动化孤岛,如缩小现有ERP部署中的差距。早期的BPM解决方案通过以整合为中心的功能[如应用程序适配器、数据转换功能和特定于产品的流程定义(如订单到现金流程)]加以区分。最终,标准的引入(如Web服务和BPM套件中开发工具的进步)降低了数据整合的成本和复杂性。这开始将BPM的基本价值主张从分散的功能转变为支持业务流程经理管理业务逻辑,而不威胁应用程序逻辑(由IT人员正确管理和保护的基础架构)的完整性。

　　基于标准协议的可用性大大降低了BPM采用者在构建和维护集成基础架构方面的负担,从而腾出了时间和资源来关注流程和业务绩效,而不是浪费在管理路径问题上。随着时间的推移,这促进了流程管理软件从自动化和一体化到协作和协调的重新关注,使BPM进入了业务优化领域。现代企业运行的环境是动态的,要求支持它的业务系统也是动态的。这意味着系统必须能够轻松地适应企业不断变化的环境。BPM机会的第二阶段是通过使流程协作成为现实,即跨流程协作的业务流程连接抽象的应用程序功能,从而将现有的自动化基础架构转换为可重用的业务资产来实现的。协调与自动化的区别在于认知方面根本上的转变,从将流程视为离散步骤的流程,到从目标和里程碑的角度理解流程。

3.7.2　从自动化到流程协作：围绕面向服务架构重新调整BPM

　　流程协作允许其借鉴企业环境的其余部分的行为(一个根据目标而不是脚本定义的)。在过去的十年中,流程协作在业务计算方面引入了一个明显的转变。当组织意识到协调所带来的机遇时,它提供了(可以说是授权)对应用程序和信息系统角色的全面反思。

　　业务流程已经对BPM市场的发展方向产生了明显的影响,这是由标准协议[特别是可扩展标记语言(extensible markup language, XML)及简单对象访问协议(simple object access protocol, SOAP),通用描述、发现和集成整合(universal

and the core Web Services stack of Simple Object Access Protocol (SOAP), Universal Description, Discovery, and Integration (UDDI), and Web Services Description Language (WSDL)); the emergence of Service-Oriented Architectures (SOA) has provided a new level of flexibility and simplicity in resolving integration issues. In fact, it has to such an extent that it almost seems redundant to discuss in the context of the forward-looking perspective of modern BPM.

Indeed, most "adaptability pundits" would find the discussion of SOA as "propeller head" anathema, something only the geekiest techies should worry about. Yet, that is why it is so relevant to the adaptability discussion. Because, previously (e.g., prior to SOA), performing the most basic changes to underlying integration configurations, such as a change in the structure of a document by its sender or the set of information required by the requester, would have required taking running processes and/or systems off line, then having a programmer manually code, test, and deploy each of the changes.

Now we can nearly take for granted that the underlying systems of record are decoupled from how we access them—that access is enabled through a services layer rather than a programmatic interface that requires integration at the code level (i.e., "tightly-coupled"). What SOA provides for BPM and other software environments is a common means for communicating between applications, such that connections do not need to be programmed in advance, as long as the BPM environment "knows" where to find information and how to access it. This is critical to dynamic processes in which the specific information, activities, and roles involved with a process may not be predetermined but identified as the process progresses.

Of course, this does require, however, that the information and infrastructure sought to be accessed is exposed as services. For example, core system capabilities can be exposed as containerized Web services with a WSDL description, able to be invoked by any Web services-compliant application, or increasingly with a representational state transfer (REST)ful[1] interface allowing integration points and data variables to be defined at design time, but resolved at runtime, eliminating the inherent problems of hard-wired integration.

APPLY SOA STRATEGIES TO INTEGRATING UNSTRUCTURED INFORMATION

Although the evolution of SOA has dramatically improved the accessibility of structured information through standardized interfaces, access to unstructured information can be far more challenging. Consider for a moment where customer information resides in your organization. The answer is most likely "everywhere"—records, transactions, profiles, project data, recent news, and other sources of structured and "semi-structured" information (such as correspondence and other documents without uniform representation). For many organizations, it would take years to rationalize all the places where customer data might be found. However, by instead knowing where to find the various instances of each of their data sets and how each is described, they can be left intact yet used for multiple purposes.

description, discovery, and integration, UDDI)和 Web 服务描述语言(Web services description language, WSDL)的核心 Web 服务堆栈] 支持的, SOA 在解决整合问题方面提供了新的灵活性和简单性。事实上, 在现代 BPM 前瞻性视角的背景下进行讨论似乎是多余的。

事实上, 大多数应用型专家(adaptability pundits)会认为 SOA 作为 "螺旋推进器" 式的讨论是令人生厌的, 这是只有极客技术人员才应该担心的事情。然而, 这就是它与适应性讨论如此相关的原因。因为之前(如 SOA)执行最基本的底层集成配置的变化(如由其发送者或请求者所需的信息改变文档的结构)需要采取流程或系统以脱机方式运行, 然后由一个程序员手工编写代码、测试和部署每一个变化。

现在, 我们几乎可以想当然地认为, 记录的底层系统与我们访问它们的方式是分离的, 即通过服务层而不是需要在代码级别整合的编程接口(即 "紧密耦合")来启用访问。SOA 为 BPM 和其他软件环境提供的是在应用程序之间进行通信的一种常见方式, 这样, 只要 BPM 环境 "知道" 在哪里查找信息以及如何访问信息, 就不需要预先对连接进行编程。这对动态流程至关重要, 在动态流程中, 与流程相关的特定信息、活动和角色可能不是预先确定的, 而是随着流程的进展而确定的。

当然, 这确实也需要将寻求访问的信息和基础设施作为服务公开。例如, 核心系统功能可以公开为具有 WSDL 描述的容器化 Web 服务, 可以由任何 Web 服务兼容的应用程序调用, 或者越来越多地使用表示状态传输(REST)全部接口[1], 允许在设计时定义集成点和数据变量, 但是在运行时解决、消除了硬连接集成的固有问题。

3.7.3 应用 SOA 策略集成非结构化信息

尽管 SOA 的发展通过标准化接口极大地提高了对结构化信息的可访问性, 但对非结构化信息的访问可能更具挑战性。考虑一下客户信息在您的组织中的位置是什么, 答案很可能是 "无处不在": 记录、交易、概况、项目数据、最近的新闻以及其他结构化和 "半结构化" 信息来源(如通信和其他没有统一表示的文件)。对于许多组织来说, 要使所有可能找到客户数据的地方合理化需要数年时间。但是, 通过知道在何处查找每个数据集的不同实例以及如何描述每个数据集, 可以保持这些实例的完整性, 同时还可以用于多种目的。

Leveraging Content As a Service

Following the same strategy as that which was presented above as being used by SOA for accessing structured information, a relatively new standard called "Content Management Interoperability Services" (CMIS) enables a services approach to "content middleware." CMIS does this by exposing and offering a means to control diverse unstructured document-based information stored within CMIS-compliant content repositories, whether the material is in either internally or externally managed sources. The result is that, as content is captured or otherwise introduced to a process, it can be automatically categorized and indexed based on process state and predefined rules and policies.

This strategy presents a virtual repository of both content and metadata that describes how and where content is managed at various stages of its life cycle such that Metadata are exposed to the system and process nodes where it can be used, but remains invisible to users. Users, instead, are presented with the appropriate content and format based on their identity and the current state of the process.

REALIZING ADAPTABILITY: SHIFTING FROM EVENT-DRIVEN TO GOAL-DRIVEN

The notion of orchestration has changed the role of BPM from that of a transit system, designed to shuttle data from one point to another over predefined routes, to that of a virtual power user that "knows" how to locate, access, and initiate application services and information sources. In contrast with more easily automated system-to-system processes and activities, "knowledge worker" processes characteristic of manual work involve a series of people-based activities that may individually occur in many possible sequences.

This transition in computing orientation can be described as the shift from *event-driven* in which processes are defined in terms of a series of triggers, to *goal-driven* in which processes are defined in terms of specific milestones, outcomes (goals), and constant cycles of adaptations required to achieve them. In event-driven computing, systems respond to a specific event—a request for information is received and the appropriate information is sent, or a process step is complete and so the results are recorded and the next step is initiated. In most cases, the nature of event-driven computing requires explicit scripting or programming of outcomes. In goal-driven processes, however, things are far more complex. A process that has only 20–30 unique activities, a relatively small number for most knowledge-worker processes, may present over 1000 possible permutations in the sequencing of activities. This, of course, presents too many scenarios to hard-code in advance within linear process flows, or to create a single process definition. This fact helps explain the difficulty traditionally faced in the automation of these types of goal-driven processes. Rather, this capability is enabled through the application of goals, policies, and rules, while adjusting the flow of the process to accommodate outcomes not easily identifiable.

操控内容作为服务

遵循与SOA用于访问结构化信息相同的策略,一个称为"内容管理互操作性服务(content management interoperability services, CMIS)标准协议"的相对较新的标准实现了"内容中间件"的服务方法。CMIS通过公开和提供一种方法来控制存储在符合CMIS的内容存储库中的各种非结构化基于文档的信息,不管这些信息是在内部还是外部管理的来源中。其结果是,当内容被捕获或以其他方式引入流程时,可以根据流程状态和预定义的规则和策略自动对其进行分类和索引。

此策略提供了一个包含内容和元数据的虚拟存储库,描述了在内容生命周期的各个阶段如何以及在何处对内容进行管理,从而将元数据公开给可以使用它的系统和流程节点,但对用户仍然不可见。相反,根据用户的身份和流程的当前状态,向用户提供适当的内容和格式。

3.7.4　实现适应性:从事件驱动转向目标驱动

流程协作的概念将BPM的角色从一个传输系统的角色(设计为通过预定义的路由将数据从一个点传送到另一个点)转变为一个"知道"如何定位、访问和启动应用程序服务和信息源的虚拟高级用户的角色。与更容易自动化的系统到系统的过程和活动相比,手工工作的知识工作者过程涉及一系列基于人的活动,这些活动可能以许多可能的顺序单独发生。

计算方向上的这种转换可以描述为从事件驱动的转换(其中流程是根据一系列触发器定义的)转换到目标驱动[其中流程是根据特定的里程碑、结果(目标)和实现它们所需的不断调整周期来定义的]。在事件驱动计算中,系统响应特定的事件,包括接收到信息请求并发送适当的信息,或者完成一个流程步骤,然后记录结果并启动下一步。在大多数情况下,事件驱动计算的本质要求对结果进行显式脚本或编程。然而,在目标驱动的过程中,事情要复杂得多。一个流程只有20~30个独特的活动,对于大多数知识工作者流程来说,这是一个较小的数量,在活动的排序中可能会出现超过1 000种可能的排列。当然,这提供了太多的场景,无法在线性流程流中预先硬编码,或者创建单个流程定义。这一事实有助于解释这些类型的目标驱动过程的自动化传统上面临的困难。相反,这种能力是通过目标、策略和规则的应用来实现的,同时调整流程流以适应不容易识别的结果。

Goal-Driven Scenarios

In many cases, each subsequent step in a process is determined only by the outcome and other circumstances of the preceding step. In addition, there may be unanticipated parallel activities that occur without warning, and may immediately impact the process and future (even previous) activities. For these reasons and the others described above, managing goal-driven processes requires the ability to define and manage complex policies and declarative business rules—the parameters and business requirements that determine the true "state" of a process. Goal-driven processes cannot be defined in terms of simple "flow logic" and "task logic," but must be able to represent intricate relationships between activities and information, based on policies, event outcomes, and dependencies (i.e., "context").

Such a case is the admission of a patient for medical treatment. What is involved is in fact a process, yet the specific sequence and set of activities mostly does not follow a specific script, but rather is based on a diagnostic procedure that likely involves applying a combination of policies, procedures, other rules, and the judgment of health care workers. Information discovered in one step (e.g., the assessment of a given condition) can drastically alter the next set of steps, and in the same way a change in "patient state" (e.g., patient goes into heart failure) may completely alter the process flow in other ways. What is needed to successfully execute such a process is either a "super user," who knows both the medical protocols to make a successful diagnosis and the system protocols to know where and how to enter and access the appropriate information, or, alternatively, BPM can exist as the virtual user layer. BPM could provide a single access point for the various roles involved, meanwhile assuming the burden of figuring out where and how access required information. Yet what really differentiates this as a goal-driven system is the ability to determine the sequence of a process based on current context. For example, a BPM system can examine appropriate business rules and other defined policies against the status of a process or activity to determine what step should occur next and what information is required.

Often the flow and sequencing of a goal-driven process is determined largely by individual interpretation of business rules and policies. For example, a nurse who initiates a patient-admitting process will evaluate both medical protocol and the policies of the facility where the health care services are administered. Similarly, an underwriter compiling a policy often makes decisions by referring to policy manuals or his own interpretation of rules and codes. As a result, what may be an otherwise "standard" process will be distinguished by exceptions and pathways that cannot be determined in advance, but at each step each activity must nonetheless adhere to specific rules and policies. For further reading and information on Goal-Driven modelling and scenarios, please see the chapter on Value-Oriented Process Modelling.

PHASE THREE: INTELLIGENT BPM

The first two phases of BPM have laid a solid foundation for enabling adaptable business systems, by allowing business logic (processes, policies, rules, etc.) to be defined and managed within a separate environment, as well as using an open approach to

目标驱动方案

在许多情况下,流程中的每个后续步骤仅由前面步骤的结果和其他情况决定。此外,可能会有未预料到的相关活动,它们在没有警告的情况下发生,并可能立即影响流程和未来(甚至以前的)的活动。出于这些原因以及上面描述的其他原因,管理目标驱动的流程需要能够定义和管理复杂的策略和声明性业务规则,以及确定流程真正"状态"的参数和业务需求。目标驱动的流程不能用简单的"流程逻辑"和"任务逻辑"来定义,但必须能够根据策略、事件结果和依赖关系(即语境)来表示活动和信息之间的复杂关系。

这种情况就像是一位病人入院接受治疗。所涉及的实际上是一个流程,但具体的活动顺序大多不遵循特定的脚本,而是基于可能涉及应用政策、程序、其他规则和医疗工作者判断的诊断程序。在一个步骤中发现的信息(如对给定条件的评估)可以极大地改变下一组步骤,同样,"患者状态"(如患者出现心力衰竭)的改变也可以以其他方式完全改变流程流。成功执行这样一个流程所需要的是一个"超级用户",他既知道成功诊断所需的医疗协议,又知道在何处以及如何输入和访问适当信息的系统协议,或者,BPM可以作为虚拟用户层存在。BPM可以为所涉及的各种角色提供一个单一的访问点,同时假设要计算访问所需信息的位置和方式。然而,真正区别于目标驱动系统的是能够根据当前语境确定流程的顺序。例如,BPM系统可以根据流程或活动的状态检查适当的业务规则和其他定义的策略,以确定接下来应该执行什么步骤以及需要什么信息。

通常,目标驱动流程的流和顺序在很大程度上取决于对业务规则和策略的个别解释。例如,启动患者入院流程的护士将评估医疗协议和医疗服务管理机构的政策。类似地,编制保单的承销商通常通过参考保单手册或自己对规则和代码的解释来做出决策。因此,可能一个"标准"流程将通过例外情况和路径加以区分,但却无法提前确定,但在每个步骤中,每个活动都必须遵守特定的规则和政策。有关目标驱动的建模和场景的进一步阅读和信息,请参阅4.4节以价值为导向的流程建模"。

3.7.5　第三阶段: iBPM

BPM的前两个阶段通过允许在单独的环境中定义和管理业务逻辑(流程、策略、规则等)以及使用开放的方式与其他系统(Web 服务)通信为实现可适应的业务系统奠定了坚实的基础。这提供了一个适应性级别,使得BPM采用者能够

communicating with other systems (Web Services). This has provided a level of adaptability that allows BPM adopters to respond to changes in the operating environment with far greater agility than ever before. This shift towards goal-oriented computing has laid the path for Phase Three BPM, which combines integration and orchestration with the ability to continuously validate and refine the business users' understanding of business performance drivers, and allowing them to adapt business systems and process flows accordingly. The effect of Phase Three BPM is to "take the lid off" what has for years been a black box shrouding automation.

With the third phase of BPM, visibility combines with integration and orchestration to enable business process owners and managers to discover the situation changes that require adaptation. Phase Three of BPM offers a framework for continuously validating and refining an understanding of business performance drivers, and adapting business systems accordingly. This should represent a new and significantly greater level of interest and adoption of BPM software, by attracting organizations seeking to optimize business performance, rather than integrating and automating systems and tasks.

Part of the recent evolution toward iBPMS strategies and technology is the inclusion of more sophisticated reporting capabilities within the BPM environment itself. This is both enabled and in many ways necessitated by the greater flexibility of the architectures introduced with the BPM suites that provide BPM Phase Two capabilities. With these environments, the ability to support nonsequential, goal-driven models is greatly increased, requiring more feedback (reporting) to enable successful execution of this type of less deterministic process models.

With few exceptions, reporting on process events and business performance was previously done only after a process had been executed, or otherwise within a separate environment disjointed from the process. This obviously prevented any opportunity to impact the direction of a process, but was based on a limitation of the management process as well as system and software architectures. Specifically with regard to BPM, process models were most commonly defined as proprietary structures, and in many cases compiled into software. Thus, changes either required bringing down and recompiling an application, or were otherwise limited to discrete points in the process (such as exceptions and yes/no decision points).

Adaptability Begins with Reading Signals

Successful adaptation, to move in the right direction, requires the ability to accurately assess the full and relevant context of status. Indeed, the more flexible and adaptable the systems are, the greater the requirement for visibility. In the same manner, the greater the ability to monitor the signals that define business performance, the more value can be found in the ability to adapt processes and systems accordingly. To illustrate the distinction of orchestration over automation, the metaphor of rail transportation—which moves across a predictable path and direction (quite literally "set in stone") has been used (arguably overused) to illustrate *automation*, contrasted with *orchestration*, described in terms of a car or other personal

以前所未有的灵活性响应操作环境的变化,向面向目标计算的转变为第三阶段的BPM铺平了道路,该阶段将整合和协调业务用户,持续验证和改进对业务性能驱动因素的理解的能力,并允许他们相应地调整业务系统和流程。第三阶段BPM的效果是"揭开一个覆盖自动化的黑匣子的盖子"。

在BPM的第三个阶段,可视性与整合、协调相结合,使业务流程所有人和经理能够发现需要调整的变化情况。BPM的第三阶段提供了一个框架,用于不断验证和改进对业务性能驱动因素的理解,并相应地调整业务系统。通过吸引组织寻求优化业务绩效,而不是将系统和任务进行整合和自动化,这应该代表了一种新的、更高层次对BPM软件的兴趣和应用。

iBPMS战略和技术的最新发展的一部分是在BPM环境中包含更复杂的报告功能。这在许多方面都是可以实现的,并且BPM第二阶段功能的BPM套件所引入的架构中,为提供更大的灵活性所必需的。有了这些环境,支持非顺序、目标驱动的模型的能力大大提高,需要更多的反馈(报告),才能成功执行这种不太确定的流程模型。

除了少数的例外,以前只有在执行流程之后或者在与流程分离的单独流程中才能报告流程事件和业务绩效。这显然阻止了任何影响流程方向的可能性,但这是基于管理流程以及系统和软件架构的限制。特别是在BPM方面,流程模型通常被定义为专有结构,并且在许多情况下被编译成软件。因此,所需的更改可能会导致应用程序停机和重新编译,也可能仅限于流程中的离散点(如异常和是/否决策点)。

适应性从阅读信号开始

成功的适应的前提是要朝着正确的方向发展,这就需要能够准确地评估完整和相关的地位和背景。实际上,系统的灵活性和适应性越强,对可见性的要求就越高。同样地,监控定义业务性能信号的能力越强,相应地调整流程和系统的能力就越有价值。为了说明协调与自动化之间的区别,可以使用铁路运输作为比喻,可以用铁路运输在可预测的路径和方向上移动(字面意思是"固定不变的")来比喻(可以说是过度使用)自动化,可以用汽车或其他个人运输工具来比喻协调。后者

transportation. The latter offers a vehicle to deliver passengers to a desired destination by understanding the rules of the road and milestones along the way, but does not require scripting every single inch along the way. In fact, it would be nearly impossible to do so given the unpredictability of such factors as traffic and road conditions. Driving through traffic is entirely about *sense-and-respond* in which adaptation is happening real-time—you brake, accelerate, steer left, and so forth. All of these actions are in response to a constant stream of signals and event data.

Thus, what separates personal transportation from rail travel is not only the ability to deviate from the rigidly fixed path, but also the need for visibility. This is an overly simplified, and again arguably overused, metaphor, but it nonetheless offers a tangible concept for why "Intelligent BPM" is indeed a substantively different animal from its predecessor, the currently ubiquitous dumb BPM. The bottom line is that if you cannot see what is ahead of you, you cannot respond accordingly, and prior to achieving the capabilities of "Phase Three BPM," your processes are blind. By way of example, driving a car is an immensely data-driven exercise, even if it is largely tactile and observational data. At least this is the case today, in the absence of widespread adoption of *Google's driverless car*. Yet of course there too, it would be extremely data-intensive and literally data-driven, even if the majority of this data is visible only on a machine-to-machine basis. Similarly, to allow users to drive their process to their self-selected destinations requires access to a rich data environment that connects to the process. As we seek further capability in this area, although it may become more machine and data driven, it may well be that the user increasingly will actually require less ability to see the data as the BPM software "sees."

INTELLIGENT BPM LEVERAGES BIG DATA

Since its inception, IT has been defined by the architecture of the relational database (RDMS). The advances seen in computing, even in the evolution of Internet architecture, were essentially a derivative of the relational database. This has evolved over the years since 1975, and everything from monolithic packaged software to comparatively simple and agile applications have been built on this model. Today, however, we are amidst an inflexion point, moving to the postrelational era, perhaps more aptly named the "Big Data Era."

The intelligence that comes from capturing event data (signal detection), as well as driving greater understanding and innovation through simulation, is a Big Data scenario. This cannot be done with a narrow lens on structured data, nor can it be limited to internal, single-company boundaries (nonetheless organizational or departmental constraints). Successful adaptation requires the ability to manage complex multi-enterprise systems, expanding the window of analysis for strategy beyond the single company or business unit. Through partner value chains, customer interaction, outsourcing, offshoring, peer production, and other extended ecosystems of interdependent entities, Intelligent BPM processes often extend well beyond discrete transactions between suppliers or customers to create the extended enterprise.

提供了一种车辆,通过了解道路规则和沿途的里程碑,将乘客送到理想的目的地,但不需要在沿途每一英寸都编写脚本。事实上,考虑到交通和道路条件等因素的不可预测性,几乎不可能做到这一点。在交通中驾驶完全是关于感觉和反应的,在这种感觉和反应中,当您刹车、加速、左转等时,适应就会实时发生。所有这些动作都是对恒定的信号流和事件数据的响应。

因此,将个人运输与铁路运输分开的不仅是偏离固定路线的能力,而且还需要能见度。这是一个过度简化的,也可以说是过度使用的比喻,但它仍然提供了一个切实的概念,解释为什么iBPM确实与它的前身(当前普遍存在的无言BPM)有本质上的不同。底线是如果看不到前方是什么,就不能做出相应的反应,在实现"第三阶段BPM"的能力之前,流程是盲目的。举个例子,驾驶汽车是一项庞大的数据驱动的运动,即使它主要只是触觉和观测数据。至少现在是这样,因为谷歌的无人驾驶汽车还没有被广泛采用。当然,这也将是数据极其密集和真正的数据驱动,即使这些数据的大部分只在机器对机器的基础上可见。同样,要允许用户将流程驱动到自己选择的目标,需要访问连接到流程的丰富数据环境。随着我们在这一领域寻求进一步的能力,尽管它可能会变得更加机器化和数据驱动,但很可能是用户越来越实际地需要更少的能力来查看BPM软件"看到"的数据。

3.7.6 iBPM利用大数据

自创建以来,它就由关系数据库(RDMS)的体系结构定义。计算方面的进步,甚至是互联网架构的发展,本质上都是关系数据库的衍生产品。这是自1975年以来经过多年发展而来的,从单片封装软件到相对简单和敏捷的应用程序,所有的一切都建立在这个模型之上。然而,今天,我们正处在一个拐点之中,进入后关系时代,或许更恰当地称之为"大数据时代"。

通过模拟捕获事件数据(信号检测)以及推动更好的理解和创新所带来的智能是大数据情景。这不能通过结构化数据的狭窄镜头来完成,也不能局限于内部的单一公司边界(尽管是组织或部门限制)。成功的适应需要能够管理复杂的多企业系统、扩展战略分析窗口、超越单个公司或业务部门。通过合作伙伴价值链、客户互动、外包、离岸外包、同行生产以及其他相互依存实体的扩展生态系统,智能的BPM流程通常远远超出了供应商或客户之间的离散交易,从而实现了扩展企业。

This level of collaboration requires standard conventions and mutually understood meaning of data exchanged between stakeholders, but without rigid structure or formalization. In this way, the postrelational shift to Big Data has largely paralleled, and in many ways is driven by, the same conditions and requirements behind adaptability. The movement in both cases is to expand beyond the limits of relational data structures and capture the richer context that defines business events. One of the most frequently discussed Big Data initiatives is *Hadoop*, which is essentially a flat-file document database or file system very reminiscent of the early database hierarchical architecture. It is parallel to the "*NoSQL*" movement, which while it might sound pointedly anti-SQL, actually stands for "Not Only SQL" in the spirit of postrelational flexibility for enabling greater reach and performance than otherwise possible with traditional relational databases.

A core driver behind Big Data is the vast growth of digitized information, in particular that which begins and exists throughout its life cycle in digital format. It's not just that the data volumes are big, but that the broader spectrum of data must be managed differently. The data management goal is no longer about trying to create a monolithic structure, but rather to arrange data in ways such that the semantics—the understanding of the meaning of the data and the interrelationships within it—are accessible.

In an era when an aberrant Tweet can in a matter of minutes costs shareholders millions,[2] it is the meta-context of business events across a spectrum of structured, unstructured, and semistructured information that defines the larger perspective of business activity. The impact of mobile and social capabilities in enterprise systems, as well as external social networks, is having a very real material impact on business. It has become a critical (even if comparatively smaller but clearly growing) piece of the business event stream. It also has advanced the de-materialization of work as well as personal/professional demarcation. Is *LinkedIn* a "work tool" or personal site? Clearly, it is both. Yet for most organizations an increasing amount of work is conducted through otherwise "unsanctioned" channels, such as *LinkedIn*, *Twitter*, and other social sources, offering either the potential wellspring of value-adding business events and event data, or otherwise process "dark matter" outside of the purview of the traditional business IT environment. Social media has allowed not simply individuals, but businesses and brands, to connect directly to consumers. We now have volumes of case studies of missteps and miscalculations with the personae of corporate brands across social networks.

What is less visible, but arguably more important, is the leverage of these tools for successful adaptable business strategy and processes, as we saw with the Proctor and Gamble (P&G) use of social media, in which the company explicitly exploits these channels to engage customers and obtain market advantage. This brings us back to the value of business event management and the speed of adaptation. All business events have an implicit half-life and utility curve. Whatever business you are in, the value represented by the response to an event diminishes over time. In every case, this is based on a utility curve, not a straight line. Responding twice as fast is more than twice as valuable. The faster the response, the greater the business value realized.

　　这种协作水平要求利益相关者之间相互理解交换数据的标准和含义,但没有严格的结构或形式化。通过这种方式向大数据的后关系转移,在很大程度上是平行的,并且在许多方面是由适应性背后的相同条件和需求驱动的。这两种情况下的移动都是为了超越关系数据结构的限制,并捕获定义业务事件的更丰富的背景。Hadoop 是最常被讨论的大数据计划之一,它本质上是一个平面文件(文档数据库或文件系统),这很像早期的数据库层次结构。它与 "NoSQL" 类似,虽然听起来可能有针对性地反 SQL,但实际上代表 "Not Only SQL",本着后关系灵活性的精神,与传统关系数据库相比,它能够实现更大的范围和更高的性能。

　　大数据背后的一个核心驱动力是数字化信息的巨大增长,特别是在其整个生命周期中以数字格式开始并存在的信息。不仅仅是数据量很大,更广泛的数据范围必须以不同的方式进行管理。数据管理的目标不再是试图创建一个整体结构,而是以语义、对数据含义的理解以及其中的相互关系的方式来安排数据。

　　在这样一个时代,一条异常的推特(twitter)可以在几分钟内让股东们损失数百万美元[2],它是跨越一系列结构化、非结构化和半结构化信息的业务事件的元信息,它定义了业务活动的更大视角。移动和社交功能在企业系统以及外部社交网络中的影响正在对业务产生非常实际的重大影响,它已经成为业务事件流中的一个关键部分(即使较小但明显增长)。它还促进了工作的非物质化以及个人/专业划分。LinkedIn 是 "工作工具" 还是个人网站? 很明显,两者都是。然而,对于大多数组织来说,越来越多的工作是通过其他 "未经批准" 的渠道进行的,如 LinkedIn、Twitter 和其他社会资源,提供潜在的价值源泉业务事件和事件数据,或者在传统业务 IT 环境的范围之外处理 "暗物质"。社交媒体不仅允许个人,而且允许企业和品牌直接与消费者联系。我们现在有大量的案例研究,研究了在社交网络中对企业品牌角色的失误和误算。

　　不太明显但可以说更重要的是,这些工具对成功的适应性业务战略和流程的利用,正如我们在宝洁使用社交媒体时看到的那样,在社交媒体中,公司明确利用这些渠道来吸引客户并获得市场优势,这使我们回到了业务事件管理的价值和适应的速度。所有业务事件都有一个隐式的半衰期和效用曲线。无论您从事什么业务,响应事件所代表的价值都会随着时间的推移而降低。在任何情况下,这都是基于效用曲线,而不是直线。以两倍的速度作出反应的价值多于两倍,响应越快,实现的业务价值越大。

Delays in response ("Latency") can be divided into two distinct groups: "Infrastructure Latency," or delay presented by the system in delivering notification of the event, and "Decision Latency," or the period of time between when the business-event data are captured and when it is responded to. For example, the delay between the time from when a customer submits a complaint or (per the scenario above) tweets about a bad experience—from that moment until when it is within an actionable, reporting framework (e.g., when it becomes an actual *signal*), that is Infrastructure Latency.

The time between when the signal is received until the moment someone responds, first deciding then acting, is Decision Latency. Recall the first organizational capability as necessary to foster rapid adaptation offered by Reeves and Deimler, *the ability to read and act on signals of change*, the speed of this is largely a matter of Decision Latency (Figure 1).

Regardless of the specific circumstances involved, the value of the response is greater closest to the moment of the complaint, diminishes over time, and, after a certain period in time, any response is going to be of little value. There is not a single set of hard metrics for all organizations, or all events, but in every case, predictable value is gained from the ability to capture an event. It could be related to a sales opportunity, field maintenance, or terrorist threat; in every case the faster the response, the greater the value.

Faster Adaptation Is Not Necessarily Faster Decisions

It can be assumed that the ability to take action on a specific event will always involve some delay. Yet there is a similar inevitability that the value lost because of that delay will follow a utility curve, not a straight line. Thus, the greatest source of value will always come from faster notification and actionability, rather than simply faster decision-making.

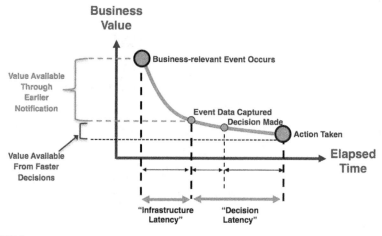

FIGURE 1

Time-based value of business event response.

延迟响应（延迟）可分为两个不同的组：基础设施延迟（或系统在传递事件通知时显示的延迟）和决策延迟（或从捕获业务事件数据到响应业务事件数据之间的时间段）。例如，从客户提交投诉或（根据上面的场景）推特到在可操作的报告框架内（如当它变成实际信号时），这段时间之间的延迟就是基础设施延迟。

从接收到信号到有人响应的时间，首先是决定，然后是行动，这段时间就是决策延迟。回想一下Reeves和Deimler提供的培养快速适应能力所必须的第一种组织能力，阅读并根据变化信号采取行动的能力，这种速度在很大程度上取决于决策延迟（图1）。

无论具体情况如何，回复的价值一定要更接近于投诉的那一时间点，其效果随着时间的推移而减少，在一段时间后，任何回复都将变得毫无价值。对于所有组织或所有事件，并没有单一的硬性度量标准，但是在每种情况下，可预测的价值都是从捕获事件的能力中获得的。它可能与销售机会、现场维护或恐怖主义威胁有关。在每种情况下，响应越快，价值越大。

更快地适应并不一定意味着更快的决策

可以假定，对特定事件采取行动的能力总是会涉及一些延迟。然而，也有一个相似的必然性，即由于延迟而损失的值将遵循效用曲线，而不是直线。因此，最大的价值来源总是来自更快的通知和可操作性，而不仅仅是更快的决策。

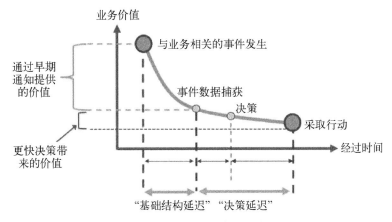

图1　基于时间的业务事件响应价值

The value of faster decisions (automating the function of knowledge workers in the decision-making process) offers little value, particularly when compared to the cost of poor decisions made in haste. Because of the greater the delay in notification and actionability, there is greater pressure on making decisions sooner rather than losing further value. Yet the opportunity lies in reducing Infrastructure Latency. By getting actionable information into the hands of knowledge workers sooner, iBPM systems offer a predictable source of business value and clear differentiation from passive systems (i.e., notification only, without the ability to facilitate a response.)

THE VALUE OF SOCIAL MEDIA TO INTELLIGENT BPM

Understanding the time-based impact of business event response illustrates the critical capability and value offered by Intelligent BPM from the ability to move, to signals and actions, to the edge points of interaction. The sooner signals are received and acted on, the tighter the response loop, and the greater the value derived from that business event. Yet this comes not simply as the ability to read events generated from within social media, but also the means or mode through which events are "socialized."

Consider the revolution of *Facebook* in recent years, which was not about enabling me to post stuff about myself for the world to see. That was an established set of capabilities and, frankly, the world hardly needed another outlet for this. If nothing else, *MySpace* sufficiently addressed this, which in 2009 far exceeded *Facebook's* reach and community size. Rather, it was the introduction of the "Like" button, which rapidly transformed the site's orientation (and ultimately that of social networks overall) from self-publishing to collaboration. Within a few months of introducing the Like button, *Facebook* overtook *MySpace* in community size, and soon after sealed its fate altogether.

The "Like" button was (and is) about tapping into an event pipeline and then enriching these events with personal contributions (if nothing else, indicating your like for them). For all intents and purposes, *Facebook* users are tagging that event, adding value to it across its life cycle, just as presented in the notion of the business event management framework. This new model of collaboration has transformed the now decades-old metaphor of threaded discussions. *Facebook* is now the collaboration metaphor for everything from *LinkedIn* to *Salesforce Chatter* to a growing number of iBPMS product vendors. Yet it is not merely a tribute to the success of *Facebook*, which although quite admirable in areas such as the leverage of agile software development methods and speed of community growth, has otherwise been a lackluster example of adaptability (note initial public offering (IPO) struggles and user backlash, which would likely have been largely diminished by following the strategies described in this chapter.)

Rather, the enduring success of the *Facebook Wall* metaphor for collaboration reflects the importance of business events and business event management to

与匆忙做出的糟糕决策的成本相比,更快决策的价值(使决策过程中知识工作者的功能自动化)几乎没有价值。因为通知和可操作性的延迟越大,制定决策的压力就越大,而不是失去更多的价值。然而,机会在于减少基础设施延迟,通过尽快将可操作的信息交到知识工作者的手中,iBPM 系统提供了可预测的商业价值来源,并明显区别于被动系统(即仅通知,无法促进响应)。

3.7.7　社交媒体对 iBPM 的价值

基于时间的业务事件响应说明了 iBPM 提供的关键功能和价值,从移动能力、信息和操作能力到交互的边缘点。接收和执行信息越快、响应循环越紧,从该业务事件派生的价值越大。然而,这不仅是因为阅读社交媒体中产生的事件的能力,而且是事件"社会化"的手段或模式。

考虑一下近年来 Facebook 的革命,这不是"让我发布关于我自己的东西,让全世界都看到",这是一套既定的能力,坦率地说,世界几乎不需要另外的出路。如果不出意外,MySpace 已经充分解决了这个问题,这在 2009 年远远超出了 Facebook 的影响范围和社区规模。相反,是"Like"按钮的引入,让 Facebook 迅速改变了网站的方向(最终是整个社交网络的方向),从自我出版到协作。在推出"Like"按钮的几个月内,Facebook 在社区规模上超越了 MySpace,并且不久完全封锁了它的命运。

"Like"按钮曾是(并且现在依旧是)打开一个事件的通道,然后通过个人贡献(如果没有其他内容,则表示您对它们的喜欢)丰富这些事件。出于所有的意图和目的,Facebook 用户都在为该事件添加标签,在其整个生命周期中为其增加价值,正如在业务事件管理框架的概念中所展示的那样。这种新的合作模式已经改变了过去几十年来对往来式讨论的隐喻。从 LinkedIn 到 Salesforce Chatter,再到越来越多的 iBPMS 产品供应商,Facebook 现在成了各种产品的协作隐喻。然而,这不仅仅是对 Facebook 成功的一种赞扬,尽管 Facebook 在敏捷软件开发方法的杠杆作用和社区增长的速度等领域相当令人钦佩,但在适应性方面却是一个黯淡的例子。[请注意,首次公开募股(IPO)的斗争和用户的强烈反对,这些情况很可能会因遵循本节所述的策略而大大减少。]

相反,Facebook Wall 对于协作的持久成功,反映了业务事件和业务事件管理对 iBPM 的重要性。社交媒体为业务流程提供的最大价值是能够在事件上进行协

Intelligent BPM and enabling adaptability. The greatest value that social media offers to business processes is the ability to collaborate on events—including but not limited to spontaneous collaboration, but by also tagging and grouping events (e.g., "subscribing") across their life cycle. For further reading and information on social media and BPM, please read the chapter on Social Media and Business Process Management.

The Ability to Mobilize

Reeves and Deimler[3] cite "the ability to mobilize" as a critical organizational practice for enabling adaptability. It was largely presented in the same context as described in the proceeding section, specifically for stakeholders to connect and collaborate informally, but surrounding business events. Yet a more literal interpretation is the ability to extend processes to mobile touch-points as a critical precept of Intelligent BPM.

The fact of the matter is that work is already mobile. The iPad, iPhones, and other smartphones and tablets combined are reaching the point of saturation. Mobile is fast becoming the dominant means of personal computing, connected to the cloud and unleashing a flood of big data. This presents a wellspring of new opportunities for engaging customers and empowering knowledge workers, yet they can also quickly overwhelm existing IT resources. Will you be able to evolve your enterprise applications fast enough to keep pace? Probably not. Yet this is where iBPMS is providing the ability to manage the event life cycle across applications, offering both those connection points to existing applications and the ability to manage business events independently.

CONCLUSION

Change remains the one dependable constant in the business environment, yet the speed of change is greater now than perhaps any time before. Where the challenge may have once been simply to keep pace, today the megatrends of Social, Mobile, Cloud, and Big Data collectively have redefined the IT landscape seemingly overnight. Effectively leveraged, these present a wellspring of new opportunities for engaging customers and empowering knowledge workers. Yet few enterprise systems in place today are positioned to support this. Rather, it is what we see with Intelligent BPM and iBPM systems as the ability to support dynamic and adaptive work patterns is what will enable the type of collaborative work needed to thrive on adaptability in a data-driven work.

We conclude with five basic rules of thumb to follow to foster adaptability, to make mobile and cloud work for you today, and future-proof your IT investments for tomorrow. First, begin by considering the goal-driven processes within your enterprise, and consider how current architecture both obstructs and supports these processes (e.g., identify the obstructions to benchmark your need for adaptability.) Next, assess your ability to manage processes as collections of business events, and

作,包括但不限于自发协作,还可以在事件的生命周期中标记和分组事件(如"订阅")。有关社交媒体和BPM的进一步阅读和信息,请阅读有关社交媒体和BPM的章节。

组织能力

Reeves和Deimler[3]将组织能力作为实现适应性的关键组织实践。它主要是在与前一节描述的相同的语境中呈现的,特别是围绕业务事件为利益相关者非正式地连接和协作。然而,更直白的解释是将流程扩展到移动接触点的能力,这是iBPM的一个关键原则。

事实上,工作已经是流动的了。iPad、iPhone以及其他智能手机和平板电脑的结合正达到饱和。移动正迅速连接到云端并释放出大量的大数据,成为个人计算的主要手段,这为吸引客户和增强知识工作者的能力提供了新的机遇,但他们也可以迅速压倒现有的IT资源。您是否能够以足够快的速度发展企业应用程序以跟上发展步伐?大概不会。然而,这正是iBPMS提供跨应用程序管理事件生命周期的能力的地方,它既提供了到现有应用程序的连接点,也提供了独立管理业务事件的能力。

3.7.8 结论

在业务环境中,变化仍然是一个可靠的常量,但是现在的变化速度可能比以前任何时候都要快。过去的挑战可能只是为了跟上步伐,如今,社交、移动、云和大数据的巨大趋势似乎一夜之间共同重新定义了IT领域。有效利用这些资源,为吸引客户和增强知识工作者提供了新的机遇。然而,目前很少有企业系统能够做到这一点。相反,我们认为iBPM和iBPM系统能够支持动态和自适应的工作模式,正是这种能力能够使这种类型的协作工作在数据驱动的工作中因适应性而蓬勃发展。

最后,我们将遵循五条基本经验法则以培养适应性,使移动和云技术在今天为您工作,并在未来证明您的IT投资。第一,从考虑企业内的目标驱动流程开始,并考虑当前架构如何阻碍和支持这些流程(例如,确定障碍,以评估您的适应性需求)。第二,评估您作为业务事件集合管理流程的能力,以及如何使业务事件管理

how to enable a business event management framework to quickly and effectively manage the ability to read and act on signals of change.

The third rule, a critical one, is the leverage of standards. Consider how you will take advantage of certain key standards, notably Content Management Interoperability Services (CMIS), XML Process Definition Language (XPDL), and Business Process Model and Notation (BPMN). You need to know that the assets that you're creating and managing today can be accessible and follow the same evolutionary curve as the systems that you're using to access them. Fourth, you must have the ability to understand social technologies and the means for tagging and adding value to business events and to understand the business event half-life, and why responding to an event in a timely fashion is so critical.

Finally, look at the infrastructure within your organization that is increasingly commoditized, and prioritize where you can leverage cloud and consumerization to offset future upgrades and maintenance costs. Similarly, bring your own device (BYOD) access is making it easy to get mobile devices in the hands of workers—because they're putting them in their own hands. Take advantage of this, and look to leverage Intelligent BPM as a framework for creating, delivering, and managing new capabilities to a larger network of stakeholders and process participants.

End Notes

1. A set of architectural principles by Web services that focus on a system's resources, including how resource states are addressed and transferred over HTTP by a wide range of clients written in different languages.
2. The Twitter feed of the Associated Press reported that Barack Obama had been injured in an explosion at the White House (April 23, 2013).
3. Martin Reeves and Mike Deimler, "Adaptability: The New Competitive Advantage," HBR (July 2011).

框架能够快速、有效地管理读取和根据变化信号采取行动的能力。第三,一个关键的规则是标准的杠杆作用。考虑如何利用某些关键标准,特别是CMIS、XML流程定义语言(XPDL)和业务流程模型和符号(BPMN)。您需要知道,正在创建和管理的资产是可以访问的,并且遵循与访问它们的系统相同的进化曲线。第四,必须有能力理解社会技术和为商业事件标记和增加价值的手段,理解商业活动的半衰期,以及为什么及时对活动作出反应是如此重要。第五,看看您的组织中日益商品化的基础设施,并确定您可以利用云和消费来抵消未来升级和维护成本的优先顺序。类似地,可穿戴设备(bring your own device,BYOD)访问使移动设备更容易落入员工手中,因为他们将移动设备置于自己的手中。利用这一点,并将iBPM作为一个框架,用于为更大的利益相关者和流程参与者创建网络、交付和管理新功能。

参考文献

[1] A set of architectural principles by Web services that focus on a system's resources, including how resource states are addressed and transferred over HTTP by a wide range of clients written in different languages.

[2] The Twitter feed of the Associated Press reported that Barack Obama had been injured in an explosion at the White House (April 23, 2013).

[3] Martin Reeves and Mike Deimler, "Adaptability: The New Competitive Advantage," HBR (July 2011).

Evidence-Based Business Process Management

Marlon Dumas, Fabrizio Maria Maggi

INTRODUCTION

Traditional business process management (BPM) practice has largely relied on rough estimations and manual data gathering techniques when discovering, analyzing, and redesigning business processes. For example, traditional approaches to discovering "as-is" business processes rely on interviews or workshops with managers or process stakeholders, or in situ observation of process work by business process analysts. These techniques are effective when it comes to capturing and conceptualizing the "happy paths" of a process, but they fall short of providing a fully detailed picture thereof, including the numerous exceptions and deviations that generally characterize the performance of the process on a day-to-day basis. Similarly, when performing quantitative analysis, business process analysts generally rely on rough estimations of execution times, error rates, branching probabilities, and other relevant parameters of the process. This may lead analysts to misestimate the actual bottlenecks and sources of defects of the process.

Evidence-based BPM aims at addressing the shortcomings of traditional intuitionistic BPM practices. Evidence-based BPM is the practice of systematically using data produced during the execution of business processes to discover, analyze, and continuously monitor and improve business processes. Evidence-based BPM has garnered significant momentum in recent years, thanks to the widespread adoption of enterprise systems that store detailed business process execution data, as well as advances in data mining techniques and tools.

EVIDENCE-BASED BPM: WHAT FOR?

Evidence-based BPM enables business process analysts and managers (including process owners) to answer a range of questions across the entire spectrum of the BPM lifecycle, but more specifically in the following:

- Process design phase: for example, what will be the impact of a given process redesign decision given what we have observed in the past?
- Process monitoring phase: for example, what is the likelihood that a given case of the process will end up in a negative outcome, or what is the likelihood that we will violate our service-level agreement given the current state of the process and past observations?
- Process analysis: for example, why do certain cases of the process take too long to complete? What are the root causes of deviations with respect to our service-level objectives?

The Complete Business Process Handbook. http://dx.doi.org/10.1016/B978-0-12-799959-3.00017-3

3.8　基于证据的BPM

Marlon Dumas, Fabrizio Maria Maggi

3.8.1　介绍

在发现、分析和重新设计业务流程时,传统BPM实践在很大程度上依赖于粗略估计和手动数据收集技术。例如,发现"as-is"业务流程的传统方法依赖于与经理或流程利益相关者的面谈或研讨会,或者业务流程分析师对流程工作的现场观察。这些技术在捕捉和概念化流程的快乐路径(happy paths)时是有效的,但是它们不能提供一个完整的详细画面,包括大量的异常和偏差,这些异常和偏差通常是流程日常绩效的特征。同样,在执行定量分析时,业务流程分析师通常依赖于对流程执行时间、错误率、分支概率和其他相关参数的粗略估计。这可能会导致分析人员错误估计流程的实际瓶颈和缺陷来源。

3.8.2　基于证据的BPM：为何？

基于证据的BPM使业务流程分析师和经理(包括流程所有人)能够回答整个BPM生命周期范围内的一系列问题,但更具体地说,这些问题包括以下几点。

- 流程设计阶段：例如,考虑到我们过去观察到的情况,给定流程重新设计决策的影响是什么?
- 流程监控阶段：例如,流程的某个特定案例最终会产生负面结果的可能性有多大,或者考虑到流程的当前状态和过去的观察结果,我们违反服务级别协议的可能性有多大?
- 流程分析：例如,为什么流程的某些案例需要太长时间才能完成？与我们的服务水平目标相关的偏差的根本原因是什么?

Evidence-based BPM relates directly to the operational model of the BPM life-cycle, as it helps analysis to build accurate pictures of the current state of the process (including process models). It is also directly related to the performance model, because it allows analysis to answer key questions related to the performance of the process at present, in the past, and into the future.

For example, in the context of a business process for stock replenishment, specific examples of questions that evidence-based BPM is intended to address include the following:

1. What distinguishes the executions where the stock is replenished on time from those where the stock is replenished too early (overstocks) or too late (out-of-stocks)? In particular, what patterns can be used for early detection of deviant cases leading to overstock or understock?
2. Why do some stock replenishment cases lead to incorrect purchase orders or invoice payments that do not comply with the company policy? Which patterns can be used for early detection of noncompliant cases?

The first of these questions relates to the performance model, while the second one relates to the operational model (focusing on compliance aspects). In this respect, evidence-based BPM is a powerful tool to ensure on a continuous basis the integrity of the process with respect to its performance objectives and compliance rules. This is a highly relevant problem in large organizations where integrity and compliance assurance currently requires significant amounts of manual effort, thus leading to high costs. For example, according to the independent analyst Ponemon Institute, large organizations spend on average USD 3.5 million/year on compliance checking, while the costs of dealing with noncompliance issues (including fines and rectification costs) are much higher, at nearly USD 9.4 million/year.[1]

Evidence-based BPM is supported by a range of techniques to extract knowledge from business process execution logs. These techniques fall under the umbrella of *process mining*.[2] In the rest of this chapter, we provide an overview of process mining techniques, and we discuss by means of case studies how these techniques enable evidence-based BPM in practice.

THE ANSWER: PROCESS MINING

Process mining is concerned with the analysis of collections of event records produced during the execution of a business process. Such event records represent events signaling the start, end, abortion, or other relevant state change of a process or an activity therein, or any other event of relevance to the execution of a process, such as the allocation or deallocation of a process worker to a task, the receipt or dispatching of a message to a process participant, and so on.

In general, the main input of a process mining technique is a business process event log, which is a collection of event records relevant to a given business process. An event log is generally structured (but not necessarily) as a set of traces, where each trace consists of the sequence of events produced by one execution of the process

基于证据的BPM直接与BPM生命周期的操作模型相关,因为它有助于分析构建流程当前状态的准确图像(包括流程模型)。它还与绩效模型直接相关,因为它允许分析人员回答与流程当前、过去和将来的绩效相关的关键问题。

例如,在库存补充业务流程的背景下,基于证据的BPM要解决的特定问题示例包括以下几个。

(1)是什么使按时补货的执行与那些补货太早(积压)或太迟(缺货)的执行有区别? 特别是,哪些模式可用于早期检测导致库存积压或库存不足的异常情况?

(2)为什么有些库存补货案例会导致不正确的采购订单或发票付款不符合公司政策? 哪些模式可用于早期检测不符合情况?

这些问题中的第一个与绩效模型相关,而第二个与操作模型相关(侧重于遵从性方面)。在这方面,基于证据的BPM是一个强有力的工具,可以持续确保流程在其绩效目标和合规性规则方面的完整性。在大型组织中,这是一个高度相关的问题,在这些组织中,完整性和合规性保证目前需要大量的人工工作,从而导致高昂的成本。例如,根据独立分析机构波耐蒙研究所(Ponemon Institute)的数据,大型组织在合规性检查方面的平均支出为350万美元/年,而处理不合规问题(包括罚款和整改成本)的成本则要高得多,接近于940万美元/年[1]。

基于证据的BPM由一系列从业务流程执行日志中提取知识的技术支持。这些技术属于流程挖掘的范畴[2]。在本节的其余部分中,我们概述流程挖掘技术,并通过案例研究讨论这些技术如何在实践中实现基于证据的BPM。

3.8.3　答案是：流程挖掘

流程挖掘涉及对业务流程执行期间生成的事件记录集合进行分析。此类事件记录表示发出流程或其中活动的开始、结束、中止或其他相关状态更改信号的事件,或与流程执行相关的任何其他事件,如流程工作人员对任务的分配或解除分配、接收或向流程参与者发送消息等。

通常,流程挖掘技术的主要输入是业务流程事件日志,它是与给定业务流程相关的事件记录的集合。事件日志通常被构造为一组追踪(但不一定),其中每个追踪都由流程的一次执行(即案例)生成的事件序列组成。作为最低要求,事件记录至少包含事件引用的流程案例的标识符、时间戳,以及可能的一些附加属性。贷款

Table 1 *Extract of a Loan Application*			
Customer Identifier	**Task**	**Timestamp**	**...**
13219	Enter loan application	2007-11-09 11:20:10	–
13219	Retrieve applicant data	2007-11-09 11:22:15	–
13220	Enter loan application	2007-11-09 11:22:40	–
13219	Compute installments	2007-11-09 11:22:45	–
13219	Notify eligibility	2007-11-09 11:23:00	–
13219	Approve simple application	2007-11-09 11:24:30	–
13220	Compute installments	2007-11-09 11:24:35	–
...	...	...	...

(i.e., a *case*). As a minimum, an event record contains an identifier of the case of the process to which the event refers, a time stamp, and possibly a number of additional attributes. An example log of a loan application process is sketched in Table 1.

Based on such event logs, process mining techniques extract useful information that allows analysts to gain insights into the process and to formulate and validate hypotheses about the process. The output of process mining can be manifold, ranging from a complete model of the process, to a description of the main (frequent) paths of the process or its deviations, or a diagnosis of the reasons for deviations in a process.

Process mining techniques can be broadly classified into *descriptive techniques* and *predictive techniques*. Descriptive techniques aim at providing insights about the process as it is or as it has been observed in the logs. Predictive techniques aim at predicting how a process as a whole or a specific case of the process will behave in the future under certain hypotheses or conditions.

DESCRIPTIVE APPROACHES

Descriptive process mining approaches can be further classified into the following:

1. *Process performance analytics*, which seek to help analysts to analyze the performance, such as by extracting key performance indicators and uncovering bottlenecks in a business process.
2. *Automated process (model) discovery*, which seeks to automate the work involved in discovering a process that is in use so as to produce a model of the process for documentation and analysis purposes.
3. *Model enhancement*, which seeks to enhance or revise existing process models based on information extracted from the event log.
4. *Deviance mining*, which seeks to help analysts identify and analyze deviations of the business process with respect to normative or desired behavior.
5. *Process variant and outlier identification*, which seeks to help analysts identify and analyze the different variants of a process and to detect isolated cases that stand out with respect to others.

表1 贷款申请摘要

客户标识码	任 务	时 间 戳	……
13219	输入贷款申请	2007−11−09 11:20:10	—
13219	检索申请人数据	2007−11−09 11:22:15	—
13220	输入贷款申请	2007−11−09 11:22:40	—
13219	计算分期付款	2007−11−09 11:22:45	—
13219	通知合格性	2007−11−09 11:23:00	—
13219	批准简单申请	2007−11−09 11:24:30	—
13220	计算分期付款	2007−11−09 11:24:35	—
……	……	……	……

申请流程的日志示例如表1所示。

基于这些事件日志,流程挖掘技术提取有用的信息,使分析师能够深入了解流程,并制定和验证有关流程的假设。流程挖掘的输出可以是多方面的,从流程的完整模型到流程的主要(频繁)路径或其偏差的描述,或流程偏差原因的诊断。

流程挖掘技术可以大致分为描述技术和预测技术。描述性技术的目的是提供关于这个流程的真实情况,正如它是或已经在日志中观察到的那样。预测技术的目的是预测一个流程作为一个整体或一个流程的具体情况在未来的某些假设或条件下将如何表现。

3.8.4　描述性方法

描述性流程挖掘方法可进一步分为以下几类。

(1)流程绩效分析,旨在帮助分析师分析绩效,如提取KPI并发现业务流程中的瓶颈;

(2)自动化流程(模型)发现,旨在自动化发现正在使用的流程所涉及的工作,以便生成用于文档和分析目的的流程模型;

(3)模型增强,旨在根据从事件日志中提取的信息增强或修改现有流程模型;

(4)偏差挖掘,旨在帮助分析师识别和分析与规范或期望行为相关的业务流程偏差;

(5)流程变量和异常值识别,旨在帮助分析人员识别和分析流程的不同变量,并检测出与其他人不同的独立案例。

PROCESS PERFORMANCE ANALYTICS

Business process performance analytics refers to a family of techniques that take as input an event log and produce a performance assessment of the process as recorded in the log. Techniques for process performance analytics are widespread and well understood. They are an integral part of mainstream BPM methods and are supported by most commercial BPM tools.

The performance assessments produced via process performance analytics generally include mean values and other descriptive statistics of process metrics (e.g., cycle time, resource utilization), as well as associated graphs, charts, and other visualizations of process performance.[3] More sophisticated tools are able to detect and quantify the impact of bottlenecks and sources of defects in the process and to overlay this information on top of a process model.

Some tools also support *online process performance analytics* (also known as *real-time process performance analytics*), meaning that they continuously recalculate summary statistics of process metrics at runtime. These latter techniques take as input not a full event log, but rather an event stream generated by the information system supporting the business processes under observation. Such online process analytics tools provide dashboards that process managers or analysts can use to continuously monitor the state of the process. Such tools may also provide, in addition to real-time statistics, the ability to inspect specific cases of the process in order to analyze issues as they arise.

AUTOMATED PROCESS DISCOVERY

Automated process (model) discovery techniques take as input an event log and produce as output a process model. These techniques are generally geared towards producing models that capture the *frequent behavior* recorded in the input event log. They can be further classified into approaches that produce procedural models, such as business process model and notation (BPMN)[4] process models and Petri nets, and approaches that produce declarative models, such as a Declare model.[5] The former approaches are detail oriented in that they try to describe the full flow of control between tasks and events in the business process. These techniques produce closed process models—that is, process models in which everything that is not explicitly specified is assumed to be forbidden. Figure 1 shows an example of a business process model in the BPMN notation that could be automatically discovered from the log sketched in Table 1.

On the other hand, declarative approaches are *overview oriented* because they merely try to give a general overview on the process behavior. These approaches describe processes by means of open models—that is, rules that indicate what common behaviors are observed in the event log, without presupposing that other behaviors are not allowed.

As business processes evolve over time, it may be useful to discover not only a model of a snapshot of a business process at a particular point in time, but also to discover how the business process evolves over time. For instance, a business process

3.8.5　流程绩效分析

业务流程绩效分析是指将事件日志作为输入,并生成记录在日志中的流程绩效评估的一系列技术。流程绩效分析的技术是广泛的,并且被很好地理解。它们是主流BPM方法不可分割的一部分,并且受到大多数商业BPM工具的支持。

通过流程绩效分析生成的绩效评估通常包括流程度量的平均值和其他描述性统计(如周期时间、资源利用率),以及相关的图、图表和流程绩效的其他可视化[3]。更复杂的工具能够检测和量化流程中瓶颈和缺陷源的影响,并将这些信息叠加到流程模型之上。

一些工具还支持在线流程绩效分析(也称为实时流程绩效分析),这意味着它们在运行时持续重新计算流程度量的汇总统计信息。后一种技术采用的不是完整的事件日志,而是由支持观察中的业务流程的信息系统生成的事件流作为输入。此类在线流程分析工具提供了仪表盘,流程经理或分析师可以使用仪表盘持续监控流程的状态。除实时统计数据外,这些工具还可以提供检查流程特定案例的能力,以便在问题出现时进行分析。

3.8.6　自动化流程发现

自动化流程(模型)发现技术以事件日志为输入,以流程模型为输出。这些技术通常用于生成捕获输入事件日志中记录的频繁行为的模型。它们可以进一步分为生成流程模型的方法,如BPMN[4]流程模型和Petri网(20世纪60年代由C. A. Petri发明的,适合于描述异步的、并发的计算机系统模型),以及生成声明性模型的方法,如声明模型[5](declare model)。前一种方法是面向细节的,因为它们试图描述业务流程中任务和事件之间的完全控制流。这些技术产生封闭的流程模型,也就是流程模型,其中所有未明确指定的东西都被假定为禁止。图1显示了一个使用BPMN表示法的业务流程模型的示例,它可以从表1中绘制的日志中自动发现。

另一方面,声明性方法是面向概述的,因为它们只试图给出流程行为的一般概述。这些方法通过开放的模型来描述流程,也就是说,规则指示在事件日志中观察到的常见行为,而不预先假定不允许其他行为。

随着业务流程的不断发展,不仅可以在特定时间点发现业务流程快照的模型,而且还可以发现业务流程是如何随着时间发展的。例如,处理保险索赔的业务流

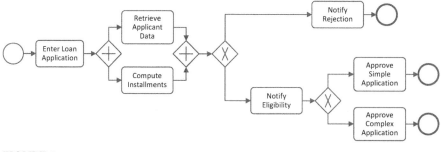

FIGURE 1

Automatically discovered process model in business process model and notation.

for handling insurance claims may be performed in one way when the workload is normal and in a completely different way when the number of claims spikes, such as in case of a natural disaster affecting a large number of customers. A family of process mining techniques known as *business process concept drift discovery* copes with the situation where the process is changing due to periodic or seasonal effects or due to changing environmental conditions. These techniques detect points in time when the observed behavior of the process has changed and provide a description of the changes that the process has undergone.

MODEL ENHANCEMENT

In process mining, model enhancement refers to a group of techniques that take as input an event log and a model and produce as output an enhanced model. Model enhancement techniques can be further classified into *model repair* techniques and *model extension* techniques.

Model repair uses information derived from a log to change an existing (procedural or declarative) process model to better reflect reality. Model repair typically involves two steps. In a first step, called *conformance checking*, the given process model is compared against the event log. In other words, the expected behavior captured in the process model is pitched against the actual behavior observed in the logs. Conformance checking produces as output a set of identified mismatches, where a mismatch is either an event observed in the log that cannot be explained by the process model or (conversely) a step that should occur according to the process model but is not observed in the log. Given the output of conformance checking, the second step in model repair is to add, remove, or modify elements in the process model, so that the resulting (repaired) model has fewer mismatches than the original model.

Model extension, on the other hand, is used to enrich an existing process model with information extracted from a log. A typical example of a model extension technique is *decision mining*, which seeks to automatically discover the branching conditions in a process model from the event log. In other words, given an event log and

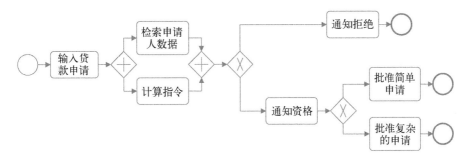

图1　在BPMN中自动发现流程模型

程可以在工作量正常时以一种方式执行,在索赔数量激增时以完全不同的方式执行,如在自然灾害影响大量客户的情况下。一系列被称为业务流程概念流动发现的流程挖掘技术可以处理流程因周期性或季节性影响或环境条件变化而发生变化的情况。当观察到的流程行为发生变化时,这些技术会及时检测出变化点,并提供流程所经历的变化的描述。

3.8.7　模型增强

在流程挖掘中,模型增强是指将事件日志和模型作为输入,并生成增强模型作为输出的一组技术。模型增强技术可以进一步分为模型修复技术和模型扩展技术。

模型修复使用从日志派生的信息来更改现有(流程性或声明性)流程模型,以更好地反映实际情况。模型修复通常包括两个步骤。在第一步(称为一致性检查)中,将给定的流程模型与事件日志进行比较。换言之,流程模型中捕获的预期行为与日志中观察到的实际行为相反。一致性检查作为输出生成一组已识别的不匹配事件,其中不匹配事件是在日志中观察到的无法由流程模型解释的事件,或者(相反)是根据流程模型应发生但在日志中未观察到的步骤。考虑到一致性检查的输出,模型修复中的第二步是在流程模型中添加、删除或修改元素,以使生成的(修复的)模型与原始模型的不匹配因素更少。

另一方面,模型扩展用于从日志中提取的信息充实现有的流程模型。模型扩展技术的一个典型例子是决策挖掘,它寻求从事件日志中自动发现流程模型中的分支条件。换句话说,给定一个事件日志流程模型和决策点(BPMN中的专用决策

a process model with decision points (e.g., exclusive decision gateways in BPMN), decision mining techniques discover the conditional expressions that determine when is a given branch of a given decision point taken—for example, that a given branch of a loan application process is taken when the amount of the requested loan exceeds $1000, while the other branch is taken for lower loan amounts.

Other model enhancement techniques are concerned with discovering resource allocation rules, such as that a given approval task in a loan application process is always allocated to the same loan officer who performed an earlier verification task in the same process.

DEVIANCE MINING

Business process deviance mining is a group of process mining techniques that take as input an event log and a specification of expected behavior in the form of rules that the process is expected to fulfill (e.g., compliance rules or process performance objectives). Deviance mining techniques produce as output a diagnostic of deviations in the business process with respect to the given specification. In other words, these techniques aim at identifying (positive or negative) deviations that occur in the input log with respect to the expected behavior described in the input specification. In addition to identifying occurrence of deviations, deviance mining techniques provide a diagnostic explaining why certain cases deviate from the given specification—for example, why certain cases outperform or underperform the given performance objective, or fail to fulfill a given compliance rule. The diagnostic can take different forms, but in any case its purpose is to enable the analyst to identify process improvement opportunities. For example, a common policy in the case of large purchases is that a purchase request and the corresponding payment should be handled by two different people (the "four-eyes principle"). Early detection of deviations with respect to this policy reduces the risk of fraud in such processes.

Starting from the deviations identified via deviance mining, it is possible to discover discriminative rules—that is, rules that discriminate between positive cases (cases that abide to the specification) and negative cases (cases that violate the specification). For example, given a specification stating that every case of a loan application process should be completed in less than 10 days, deviance mining techniques may extract a rule stating that loan applications where the applicant provides incorrect or inaccurate employment data are likely to take more than 10 days to complete. Such insight can help analysts uncover potential causes of undesirable deviations. Deviance mining can also be used to uncover and explain positive deviance, such as explaining what characterizes cases where the customer is served much faster than the norm.

Deviance mining is typically applied in an offline manner to understand the reason for observed deviance. However, deviance mining can also be applied in an online manner, such as in the context of *compliance monitoring*. Several approaches have been proposed to check compliance of an event stream with respect to a compliance model consisting of rules. In this context, compliance rules can be, for

网关),决策挖掘技术发现条件表达式,该表达式确定合适采用给定决策点的给定分支,例如,当申请贷款的金额超过1 000美元时,将采用贷款申请流程的给定分支机构,而另一个分支机构则采用较低的贷款金额。

其他模型增强技术关注于发现资源分配规则,例如,贷款申请流程中的给定批准任务总是分配给在相同流程中执行早期验证任务的相同信贷员。

3.8.8　偏差挖掘

业务流程偏差挖掘是一组流程挖掘技术,将事件日志和预期行为的规范作为输入,其形式为流程预期实现的规则(如符合性规则或流程绩效目标)。偏差挖掘技术根据给定的规范对业务流程中的偏差进行诊断,作为输出。换言之,这些技术旨在识别输入日志中出现的与输入规范中描述的预期行为相关的(正或负)偏差。除了识别偏差的发生,偏差挖掘技术还提供了一种诊断,解释为什么某些情况偏离了给定的规范,例如,为什么某些情况优于或低于给定的绩效目标,或未能满足给定的符合性规则。诊断可以采取不同的形式,但在任何情况下,其目的都是使分析人员能够识别流程改进的机会。例如,在大采购的情况下,一个共同的政策是,一个采购请求和相应的付款应该由两个不同的人处理(四眼原则,four-eyes principle)。及早发现与本政策相关的偏差可降低此类流程中的欺诈风险。

从通过偏差挖掘确定的偏差开始,可以发现区分正面和负面案例的规则,即区分正面案例(遵守规范的案例)和负面案例(违反规范的案例)的规则。例如,给定一个说明贷款申请过程的每一个案例都应在10天内完成的规范,偏差挖掘技术可以提取一个规则,说明申请人提供不正确或不准确的就业数据时,贷款申请可能需要10天以上才能完成。这种见解有助于分析人员发现不良偏差的潜在原因。偏差挖掘还可以用来发现和解释积极的偏差,如解释客户服务速度比标准快得多的情况的特征。

偏差挖掘通常以离线方式应用,以了解观察到的偏差的原因。但是,偏差挖掘也可以在线方式应用,如在合规性监控环境中。已经提出了多种方法来检查事件流是否符合由规则组成的符合性模型。在这种情况下,法规遵从性规则可以(如使用一些时态逻辑来表示)并转换为可用于在运行时有效监控规则的自动机。然

example, expressed using some temporal logic and translated into automata that can be used to efficiently monitor the rules at runtime. However, these monitoring approaches are reactive: They allow process stakeholders to identify a violation only after it has occurred rather than supporting them in preventing such violations in the first place. This feature is a key differentiator between these compliance monitoring techniques and predictive monitoring techniques introduced later.

PROCESS VARIANT AND OUTLIER IDENTIFICATION

Techniques for business process variant identification take as input an event log and produce as output a set of variants of the process recorded in the log. For example, given an event log of a loan application process, the output may be three variants of the process: "small loan applications," "large loan applications with mortgage," and "large loan applications without mortgage." These variants correspond to distinct versions of the process that differ in significant ways, such as in terms of the typical sequences of events observed in each variant.

Techniques for process variant identification and analysis usually rely on *trace clustering*, which is a data mining technique in which a set of objects (in this case, the set of traces forming the log) are partitioned into multiple subsets such that each cluster contains similar objects (traces). Each trace cluster produced in this way is a sublog on its own and can therefore be given as input to an automated process discovery technique in order to extract a model of a process variant.

Techniques for *outlier analysis* aim at identifying cases in the process that stand on their own, meaning that they have peculiarities that make them very different from other more common cases. Process mining techniques for outlier analysis also rely on trace clustering. In this context, traces belonging to clusters with a small number of elements are considered to be the outliers. Indeed, such small clusters represent rare behaviors that are sufficiently different from more frequent behavior (represented by larger clusters) to stand apart.

PREDICTIVE APPROACHES

Predictive process mining approaches are geared towards making predictions on the future state or performance of a process. A family of techniques falling in this category is *predictive deviance monitoring*, which is concerned with the early prediction (at runtime) of future deviations in the currently running cases of the process.

Predictive deviance monitoring techniques take as input a stream of events and produce recommendations for process workers during the execution of a case. These recommendations refer to a specific (uncompleted) case of the process and tell the user the impact of a given action on the probability that the case at hand will fail to fulfill the performance objectives or compliance rules. By setting thresholds on such probabilities, users are able to produce streams of warnings and alerts to which they can subscribe in order to monitor the process in real time.

而，这些监控方法是被动的：它们允许流程涉众仅在违规发生后才识别违规行为，而不是首先支持他们防止此类违规行为。此功能是这些合规性监控技术和稍后介绍的预测性监控技术之间的关键区别。

3.8.9　流程变量和异常值识别

业务流程变量标识技术将事件日志作为输入，并将记录在日志中的流程变量集作为输出生成。例如，给定一个贷款申请流程的事件日志，输出可能是流程的三个变体："小额贷款申请""有抵押的大额贷款申请"和"无抵押的大额贷款申请"。这些变体对应于流程的不同版本，这些版本在很大程度上有所不同，如在每个变量中观察到的典型事件序列方面。

流程变量识别和分析的技术通常依赖于追踪聚类，这是一种数据挖掘技术，在这种技术中，一组对象（在本例中，是构成日志的一组追踪）被划分为多个子集，以便每个集群包含类似的对象（追踪）。以这种方式生成的每个追踪集群本身就是一个子日志，因此可以作为自动流程发现技术的输入，以提取流程变量的模型。

异常值分析技术的目的是识别流程中独立存在的案例，这意味着它们具有使它们与其他更常见的案例非常不同的特性。异常值分析的流程挖掘技术也依赖于追踪聚类。在这种情况下，属于具有少量元素的集群的追踪被认为是异常值。事实上，这样的小集群代表了罕见的行为，这些行为与更频繁的行为（由更大的集群表示）有着充分的不同，因而可以脱颖而出。

3.8.10　预测方法

预测流程挖掘方法旨在预测流程的未来状态或绩效。属于这一类的一系列技术是预测偏差监控，它与当前运行的流程案例中未来偏差的早期预测（运行时）有关。

预测偏差监测技术将一系列事件作为输入，并在案例执行期间为流程工作者提出建议。这些建议涉及流程的特定（未完成）案例，并告诉用户给定操作对现有案例无法实现绩效目标或合规性规则的可能性的影响。通过设置这些概率的阈值，用户可以生成警告和警报流，他们可以订阅这些警告和警报流，以便实时监控流程。

A second family of predictive process mining approaches is *data-driven process simulation*. Data-driven process simulation refers to the practice of simulating a process model by replaying the events in an event log. In other words, rather than manually setting the simulation parameters based on rough estimations or guesswork, the simulation is driven entirely by the log, leading to simulation results that more accurately reflect reality. In the extreme case, the actual process model used for simulation may itself be extracted from the log by means of automated process discovery techniques, in which case both the process model and the stream of events generated during process simulation are designed to closely reflect reality.

Data-driven process simulation allows process analysts to perform "what-if" analysis of the process, in order to determine the potential impact or benefits of a given change to the process on key process metrics (e.g., cycle time). In particular, this technique allows analysts to understand the future performance of a "to-be" process, while taking into consideration historical observations of the process recorded in the log.

CASE STUDIES AND LESSONS IN EVIDENCE-BASED BPM

Several case studies of evidenced-based business process management based on process mining techniques have been reported in the past few years. A compendium of such techniques is maintained in the Website of the Institute of Electrical and Electronics Engineers task force on process mining.[6] In this section, we review a subset of these and other case studies related to automated process discovery, deviance mining, and predictive (deviance) monitoring.

CASE STUDIES IN AUTOMATED PROCESS AND VARIANT DISCOVERY

In *Application of Process Mining in Healthcare: A Case Study in a Dutch Hospital*,[7] analysts sought to obtain information about typical execution paths (i.e., careflows) followed by specific groups of patients in a gynecological oncology process. For these patients, all diagnostic and treatment activities were recorded by a billing system for financial purposes. Filtering and clustering techniques in combination with automated process discovery were then applied to analyze data coming from a group of 627 gynecological oncology patients treated in 2005 and 2006. The outcome of the study confirmed that by using process mining techniques, it is possible to produce understandable models for large groups of patients. The baseline for this analysis was a manually created flowchart for the diagnostic trajectory of the gynecological oncology healthcare process. The results produced using evidence-based analysis were comparable to the flowchart representation, but they provided further details not initially foreseen.

Another related case study has been reported at the hospital of São Sebastião in Santa Maria da Feira, Portugal.[8] The hospital has 300 beds and an in-house IT

预测流程挖掘方法的第二个系列是数据驱动流程模拟。数据驱动流程模拟是指通过在事件日志中重放事件来模拟流程模型的实践。换言之,模拟完全由日志驱动,而不是基于粗略估计或猜测手动设置模拟参数,从而产生更准确反映现实的模拟结果。在极端情况下,模拟所用的实际流程模型本身可以通过自动化流程发现技术从日志中提取,在这种情况下,流程模型和流程模拟过程中生成的事件流都被设计为紧密反映现实。

数据驱动的流程模拟允许流程分析师对流程执行"假设"分析,以确定给定流程更改对关键流程度量的潜在影响或好处(如周期时间)。特别是,这种技术允许分析师了解未来(to-be)流程的未来绩效,同时考虑日志中记录的流程历史观察结果。

3.8.11　基于证据的BPM的案例研究与教训

在过去的几年中,已经报告了基于流程挖掘技术的基于证据的BPM的几个案例研究。这类技术的概要保存在IEEE流程挖掘工作组的网站上[6]。在本节中,我们回顾了这些技术和其他与自动化流程发现、偏差挖掘和预测(偏差)监测相关的案例研究的子集。

3.8.12　流程自动化和变量发现中的案例研究

在医疗保健应用中的流程挖掘:荷兰一家医院的一项案例研究中[7],分析人员试图获得关于典型执行路径(即Careflows)的信息,然后是妇科肿瘤过程中特定患者组的信息。对于这些患者,所有的诊断和治疗活动都由计费系统记录下来,以便用于财务目的。然后,将过滤和聚类技术与自动化流程发现相结合,对在2005年和2006年接受治疗的627名妇科肿瘤患者的数据进行分析,研究结果证实,通过使用流程挖掘技术,可以为大量患者生成可理解的模型。该分析的基线是一个手动创建的妇科肿瘤医疗保健过程诊断轨迹流程图。使用基于证据的分析得出的结果与流程图表示法相当,但它们提供了最初无法预见的更多细节。

另一项相关案例研究已在葡萄牙Santa Maria da Feira的São Sebastião医院展示[8]。医院有300张床位和一个跨部门使用的内部IT系统,在事件日志中记录有关流程执行的数据。在案例研究中,对急诊患者的执行路径进行分析,以确定与

system used across different departments, recording data about process executions in event logs. In the case study, the execution paths of emergency patients were analyzed for activities related to triage, treatments, diagnosis, medical exams, and forwarding of patients. In this case study, several process mining techniques were adopted, such as the following:

1. Variants and infrequent behavior discovery
2. Performance analysis
3. Deviation mining (to discover discrepancies from medical guidelines)

In *Process Mining to Improve a Service Refund Process,*[9] a service refund process of an electronics manufacturer was analyzed using automated discovery techniques as well as performance and conformance analyses. The study aims at identifying the cause of some inefficiencies and too-long throughput times in the process, starting from concrete questions and problems coming from inspections and customer complaints and using the data collected in the event logs recorded by a service platform. Valuable insights have been derived from the discussion of the results of the data analysis with process managers.

In *Process Mining to Compare Procure-to-Pay Processes in Different Countries,*[10] a case study is discussed conducted in collaboration with AkzoNobel, the largest global paint and coatings company and a major producer of specialty chemicals headquartered in Amsterdam, the Netherlands, with operations in more than 80 countries. The analysis with process mining techniques revealed what was really happening in the different local procure-to-pay processes and has allowed process managers to get actionable insights on how to improve the process. Specifically, this study allowed management to

1. Obtain insights about cases in which the "first time right" principle was not realized
2. Compare the processes executed in different countries to identify process variants and best practices that can be adopted on the corporate level
3. Realize a compliance control to execute in accordance with corporate guidelines

CASE STUDIES IN DEVIANCE MINING

An increasing number of deviance mining case studies have been reported in recent years. One such case study in a large Australian insurance company was reported.[11] In this case, a team of analysts sought to find the reasons why certain simple claims, which should normally be handled within a few days, were taking substantially longer to be resolved. In other words, they needed to understand the difference between "simple quick claims" that were handled in less than X days and "simple slow claims" that took longer to be handled. They used a technique known as delta analysis, which consists of discovering one process model from each partition of an event log and comparing these models. In this case, one model was discovered for "simple slow claims" and another for "simple quick claims," and the resulting

分流、治疗、诊断、医学检查和转移患者相关的活动。在本案例研究中,采用了几种流程挖掘技术,如:

(1)变异和罕见的行为发现;

(2)绩效分析;

(3)偏差挖掘(发现与医疗指南的差异)。

在流程挖掘中,为了改进服务退款流程[9],使用自动发现技术以及绩效和一致性分析对电子产品制造商的服务退款流程进行了分析。本研究旨在找出流程中一些效率低下和吞吐量时间过长的原因,从具体的问题和来自检查和客户投诉的问题开始,并使用服务平台记录的事件日志中收集的数据。通过与流程经理讨论数据分析结果,得出了有价值的见解。

为了比较不同国家的采购与支付流程[10],我们与全球最大的油漆和涂料公司Akzonobel合作进行了一项案例研究,Akzonobel是一家总部位于荷兰阿姆斯特丹的专业化学品的主要生产商,在80多个国家开展业务。使用流程挖掘技术进行的分析揭示了不同的本地采购到支付流程中实际发生的情况,并使流程经理能够对如何改进流程获得切实可行的见解。具体来说,本研究允许管理层:

(1)了解“第一次权利”(“first time right”)原则未实现的情况;

(2)比较不同国家执行的流程,以确定可在公司层面采用的流程变体和最佳实践;

(3)实现合规控制以按照公司指导方针执行。

3.8.13　偏差挖掘案例研究

近年来,偏差挖掘的案例研究越来越多。据报道,一家大型澳大利亚保险公司进行了这样的案例研究[11]。在这种情况下,一组分析人员试图找出某些简单索赔(通常应在几天内处理)需要更长时间才能解决的原因。换句话说,他们需要了解在X天内处理的“简单快速索赔”和处理时间较长的“简单缓慢索赔”之间的区别。他们使用了一种称为Delta分析(delta analysis)的技术,该技术包括从事件日志的每个分区中发现一个流程模型并比较这些模型。在这种情况下,发现了一个模型用于“简单–缓慢索赔”,另一个用于“简单–快速索赔”,并手动比较结果模

models were manually compared. It was found that certain paths and cycles were more frequent for slow claims than for quick claims, and that two activity metrics distinguished slow versus quick claims, namely "an average number of occurrences of a given activity X (per case)" and "percentage of cases where a given activity X appears at least once." By calculating these metrics for each activity, the team traced the sources of delays to specific activities.

A similar idea was applied by Sun et al.[12] in the context of software defect handling processes in a large commercial bank in China. The authors took a log of more than 2600 defect reports of four large software development projects and examined the differences between defect reports that had led to a correct resolution (normal cases) versus those defect reports that had led to complaints by users (anomalous cases). The team defined a number of features to distinguish between normal and anomalous complaints, including the "number of occurrences of a given activity X in a case" (for each possible activity X) and the "number of occurrences of activity B after an activity A." Because there are many such combinations (A,B) and to avoid having a too-large number of features, the authors employed a discriminative item-set mining technique to identify the most relevant such pairs (A,B). Based on the resulting features, the authors constructed a decision tree that classified cases into "normal" and "anomalous." Finally, from the decision tree they extracted a set of seven rules that explained the majority of the anomalous cases, thus leading to potential improvement ideas.

In *A Process Deviation Analysis: A Case Study*,[13] instead of using item-set mining and a decision tree for discriminating between normal and anomalous cases, the authors proposed a methodology based on association rule mining. The proposed methodology is applied to a real-life case study pertaining to a procurement process in a European financial institution.

Another case study showing the potential of deviance mining, this time in the healthcare domain, was reported by Lakshmanan and Wang.[14] Here, the team applied deviance mining techniques to understand the differences between cases leading to positive clinical outcomes versus those leading to negative outcomes in the process of treatment of congestive heart failure at a large US-based healthcare provider. In this case, the team employed a combination of delta analysis (as in the Australian insurer case study mentioned above) with sequence mining techniques. Specifically, the authors used sequence mining to detect typical sequences of activities (e.g., activity B occurring sometime after activity A) that were common for positive outcomes but not common for negative ones, or vice versa. The observations made using sequence mining were complemented with additional observations obtained by comparing a process model discovered from cases with positive outcomes with the model obtained for cases with negative outcomes. In this way, the authors extracted a number of pathways and patterns that discriminate between positive and negative cases.

Another case study reported by Bose and vad der Aslst,[15] applied a technique for extracting patterns that discriminate between event traces associated with malfunctions (versus normal traces) in components of remotely monitored X-ray

型。研究发现，与快速索赔相比，缓慢索赔的某些路径和周期更频繁，并且两个活动指标区分了缓慢索赔和快速索赔，即"给定活动X的平均发生次数（每种情况）"和"给定活动X至少出现一次的情况百分比"。通过计算每个活动的这些指标，团队跟踪特定活动的延迟源。

Sun等[12]也在中国某大型商业银行软件缺陷处理过程中应用了类似的想法。作者记录了四个大型软件开发项目的2 600多个缺陷报告，并检查了导致正确解决的缺陷报告（正常情况）与导致用户投诉的缺陷报告（异常情况）之间的差异。该小组定义了一些特征来区分正常和异常投诉，包括"特定活动X在一个案例中的发生次数"（对于每个可能的活动X）和"活动A之后的活动B的发生次数"。因为有许多这样的组合（A、B），并为了避免出现太多特征值。作者采用了一种判别项目集的挖掘技术来识别最相关的组合（A、B）。基于所得到的特征，作者构建了一个将案例分为正常和异常的决策树，最后从决策树中提取了一套解释大多数异常案例的七条规则，从而引出了潜在的改进思路。

在一个流程偏差分析的案例研究中[13]，作者提出了一种基于关联规则挖掘的方法，而不是使用项目集挖掘和决策树来区分正常和异常情况。所提出的方法适用于与欧洲金融机构采购流程相关的实际案例研究。

Lakshmanan和Wang报道了另一个显示偏离采矿潜力的案例研究，这次是在医疗领域[14]。在这里，研究小组运用偏离挖掘技术来了解在美国大型医疗保健机构治疗充血性心力衰竭的过程中，导致积极临床结果的病例与导致消极结果的病例之间的差异。在这种情况下，团队采用了Delta分析（如上述澳大利亚保险公司案例研究中所述）与序列挖掘技术的结合。具体地说，作者使用序列挖掘来检测典型的活动序列（例如，活动A之后某个时间发生的活动B），这些活动序列对积极结果很常见，但对消极结果则不常见，反之亦然。使用序列挖掘进行的观察与通过比较从有积极结果的案例中发现的流程模型和从有消极结果的案例中获得的模型得到的额外观察结果相补充。通过这种方式，作者提取了一些鉴别阳性和阴性病例的途径和模式。

Bose和van der Aalst[15]报道的另一个案例研究应用了一种模式提取的技术，该技术可以区分与远程监控X射线机部件故障（与正常记录道）相关的事件记录道。他们所采用的技术属于一个更广泛的技术家族，称为识别序列挖掘技术[16,17]，

machines. The techniques they employed fall under a wider family of techniques known as discriminative sequence mining techniques,[16,17] which, in a nutshell, allow one to extract sequential patterns that discriminate between multiple classes of sequences (e.g., sequences leading to normal outcome vs. sequences containing deviations).

CASE STUDIES IN PREDICTIVE MONITORING

Case studies of predictive process monitoring in the field of transportation and logistics have been reported.[18,19] These case studies showed how predictive process monitoring can be used to explain and predict "late show" events in a transportation process. Here, a "late show" refers to a delay between expected and actual time of delivering the goods to a carrier (e.g., airline). In this case study, standard statistical techniques are used to find correlations between "late show" events and external variables such as weather conditions or road traffic. The uncovered correlations are then used to define complex event processing rules that detect situations where "late show" events are likely to occur.

A challenge for predictive process monitoring in this setting is that transportation processes are generally not "case based" because goods emanating from different customers are often aggregated and unaggregated at different points in the process. In other words, multiple "cases" of a transportation process will typically merge and split at runtime; thus, delays affecting one delivery might end up affecting others.

CONCLUSION

Evidence-based business process management based on process mining has gained significant momentum in recent years, as evidenced by numerous reported case studies in various fields. Also, over the past few years, an increasing number of tools supporting process mining have emerged and reached various levels of sophistication, including Disco by Fluxicon,[20] QPR Process Analyzer,[21] Perceptive Process Mining,[22] and the open-source ProM toolkit.[23] In parallel, methods for applying process mining have emerged, such as van der Aalst's L* method.[24]

Moving forward, we foresee the emergence of more sophisticated process mining methods capable of handling larger, more heterogeneous, and noisy datasets with high levels of accuracy. Equally or more importantly, though, we foresee the emergence of evidence-based business process governance methods, allowing managers to effectively set up and steer evidence-based BPM initiatives in large organizations, so that evidence-based BPM becomes part of the organization's management culture. This will bring BPM to the level of modern marketing approaches, which are often data driven. In an ideal evidence-based organization, every business process redesign decision will be made with data, backed by data, and continuously put into question based on data.

简而言之,它允许提取区分多类序列(如导致正常结果的序列与包含偏差的序列)的序列模式。

3.8.14 预测监测中的案例研究

在运输和物流领域进行预测过程监测的案例研究早已被报道过[18,19]。这些案例研究表明了预测性流程监控如何用于解释和预测运输过程中的"延迟展示"事件。在这里,"延迟展示"指的是将货物交付给承运人(如航空公司)的预期时间和实际时间之间的延迟。在本案例研究中,采用标准统计技术,找出"延迟展示"事件与外部变量(如天气条件或道路交通)之间的相关性。然后使用未发现的相关性定义复杂的事件处理规则,以检测可能发生"延迟展示"事件的情况。

在这种情况下,预测性流程监控的一个挑战是运输流程通常不是"基于案例"的,因为来自不同客户的货物通常在流程的不同点进行聚合和未聚合。换句话说,运输过程的多个案例通常会在运行时合并和拆分,因此,影响一个交付的延迟可能最终会影响其他交付。

3.8.15 结论

基于流程挖掘的基于证据的BPM近年来获得了显著的发展势头,许多不同领域的案例研究也证明了这一点。此外,在过去几年中,越来越多的支持流程挖掘的工具出现并达到了各种复杂程度,包括FluxIcon的Disco[20]、QPR流程分析器[21]、感知流程挖掘[22]和开源PROM工具包[23]。同时,也出现了应用流程挖掘的方法,如van der Aalst's L* 方法[24]。

展望未来,我们预计会出现更复杂的流程挖掘方法,能够以高精度处理更大、更异构、更嘈杂的数据集。同样或者更重要的是,我们预见到了基于证据的业务流程治理方法的出现,使管理者在大型组织中有效地建立和指导基于证据的BPM计划,从而使基于证据的BPM成为组织管理文化的一部分。这将使BPM达到现代营销方法的水平,而现代营销方法通常是数据驱动的。在一个理想的基于证据的组织中,每一个业务流程重新设计决策都将以数据为基础,以数据为后盾,并根据数据不断地提出问题。

End Notes

1. Ponemon Institute, "The true cost of compliance," *Benchmark Study of Multinational Organizations*, (January 2011). http://www.tripwire.com/tripwire/assets/File/ponemon/True_Cost_of_Compliance_Report.pdf.

2. W.M.P. van der Aalst, *Process Mining - Discovery, Conformance and Enhancement of Business Processes* I-XVI (Springer, 2011), 1–352.

3. M. zur Mühlen and R. Shapiro, "Business process analytics," in *Handbook on Business Process Management 2*, (Springer, 2010), 137–157.

4. BPMN (Business Process Model and Notation) is a standard process modeling notation defined by the Object Management Group (OMG)–http://bpmn.org/.

5. W.M.P. van der Aalst, Maja Pesic, and Helen Schonenberg, "Declarative workflows: balancing between flexibility and support", *Computer Science - R&D* 23 no. 2 (2009): 99–113.

6. http://www.win.tue.nl/ieeetfpm/doku.php?id=shared:process_mining_case_studies.

7. R.S. Mans, H. Schonenberg, M. Song, W.M.P. van der Aalst and Piet J. M. Bakker, "Application of process mining in healthcare - A case Study in a Dutch hospital," in *BIOSTEC (Selected Papers)*, (2008), 425–438.

8. A. Rebuge and D.R. Ferreira, "Business process analysis in healthcare environments: a methodology based on process mining," *Information Systems* 37, no. 2 (2012): 99–116.

9. Fluxicon, "Process mining to improve a service refund process," http://fluxicon.com/blog/2012/11/case-study-process-mining-to-improve-a-service-refund-process/, (retrieved on 28 05 2014).

10. Fluxicon, "Process mining to compare Procure-to-Pay processes in different countries," http://fluxicon.com/blog/2012/11/case-study-process-mining-to-compare-procure-to-pay-processes-in-different-countries/, (retrieved on 28 05 2014).

11.] S. Suriadi, M.T. Wynn, C. Ouyang, A.H.M. ter Hofstede, and N.J. van Dijk, "Understanding process behaviors in a large insurance company in Australia: a case study," in *Proc. of the International Conference on Advanced Information Systems Engineering (CAiSE)*, (Springer, 2013), 449–464.

12. C. Sun, J. Du, N. Chen, S.-C. Khoo and Y. Yang, "Mining explicit rules for software process evaluation," in *Proc. of the International Conference on Software and System Process (ICSSP)*, (ACM, 2013), 118–125.

13. J. Swinnen, B. Depaire, M.J. Jans and K. Vanhoof, "A process deviation analysis - a case study," in *Proc. of Business Process Management Workshops*, (Springer, 2011), 87–98.

14. G.T. Lakshmanan, S. Rozsnyai, F. Wang, F., "Investigating clinical care pathways correlated with outcomes," in *Proc. of the International Conference on Business Process Management*, (Springer, 2013) 323–338.

15. R.P.J.C. Bose, W.M.P. van der Aalst, "Discovering signature patterns from event logs", in *Proceedings of the IEEE Symposium on Computational Intelligence and Data Mining (CIDM)*, (IEEE, 2013) 111–118.

16. D. Lo, H. Cheng, J. Han, S.-C. Khoo and C. Chengnian, "Classification of software behaviors for failure detection: a discriminative pattern mining approach," in *Proc. of the International Conference on Knowledge Discovery in Databases (KDD)*, (2009) 557–566.

17. X. Xing, J. Pei and E.J. Keogh, "A brief survey on sequence classification," *SIGKDD Explorations* 12 no. 1 (2010) 40–48.

18. A. Metzger, R. Franklin and Y. Engel," Predictive monitoring of heterogeneous service-oriented business networks: the transport and logistics case," in *Proc. of the SRII Global Conference*, (2012), 313–322.

参考文献

［ 1 ］ Ponemon Institute, "The true cost of compliance," Benchmark Study of Multinational Organizations, (January 2011). http://www.tripwire.com/tripwire/assets/File/ponemon/True_Cost_of_Compliance_Report. pdf.

［ 2 ］ W.M.P. van der Aalst, Process Mining—Discovery, Conformance and Enhancement of Business Processes I–XVI (Springer, 2011), 1–352.

［ 3 ］ M. zur Mühlen and R. Shapiro, "Business process analytics," in Handbook on Business Process Management 2, (Springer, 2010), 137–157.

［ 4 ］ BPMN (Business Process Model and Notation) is a standard process modeling notation defined by the Object Management Group (OMG)— HYPERLINK http://bpmn.org/.

［ 5 ］ W.M.P. van der Aalst, Maja Pesic, and Helen Schonenberg, "Declarative workflows: balancing between flexibility and support", Computer Science—R&D 23 no. 2 (2009): 99–113.

［ 6 ］ http://www.win.tue.nl/ieeetfpm/doku.php?id=shared:process_mining_case_studies.

［ 7 ］ R.S. Mans, H. Schonenberg, M. Song, W.M.P. van der Aalst and Piet J. M. Bakker, "Application of process mining in healthcare—A Case Study in a Dutch Hospital," in BIOSTEC (Selected Papers), (2008), 425–438.

［ 8 ］ A. Rebuge and D.R. Ferreira, "Business process analysis in healthcare environments: a methodology based on process mining," Information Systems 37, no. 2 (2012): 99–116.

［ 9 ］ Fluxicon, "Process mining to improve a service refund process," http://fluxicon.com/blog/2012/11/case-study-process-mining-to-improve-a-service-refund-process/, (retrieved on 28 05 2014).

［ 10 ］ Fluxicon, "Process mining to compare Procure-to-Pay processes in different countries," http://fluxicon. com/blog/2012/11/case-study-process-mining-to-compare-procure-to-pay-processes-in-different-countries/, (retrieved on 28 05 2014).

［ 11 ］ S. Suriadi, M.T. Wynn, C. Ouyang, A.H.M. ter Hofstede, and N.J. van Dijk, "Understanding process behaviors in a large insurance company in Australia: a case study," in Proc. of the International Conference on Advanced Information Systems Engineering (CAiSE), (Springer, 2013), 449–464.

［ 12 ］ C. Sun, J. Du, N. Chen, S.-C. Khoo and Y. Yang, "Mining explicit rules for software process evaluation," in Proc. of the International Conference on Software and System Process (ICSSP), (ACM, 2013), 118–125.

［ 13 ］ J. Swinnen, B. Depaire, M.J. Jans and K. Vanhoof, "A process deviation analysis—a case study," in Proc. of Business Process Management Workshops, (Springer, 2011), 87–98.

［ 14 ］ G.T. Lakshmanan, S. Rozsnyai, F. Wang, F., "Investigating clinical care pathways correlated with outcomes," in Proc. of the International Conference on Business Process Management, (Springer, 2013) 323–338.

［ 15 ］ R.P.J.C. Bose, W.M.P. van der Aalst, "Discovering signature patterns from event logs", in Proceedings of the IEEE Symposium on Computational Intelligence and Data Mining (CIDM), (IEEE, 2013) 111–118.

［ 16 ］ D. Lo, H. Cheng, J. Han, S.-C. Khoo and C. Chengnian, "Classification of software behaviors for failure detection: a discriminative pattern mining approach," in Proc. of the International Conference on Knowledge Discovery in Databases (KDD), (2009) 557–566.

［ 17 ］ X. Xing, J. Pei and E.J. Keogh, "A brief survey on sequence classification," SIGKDD Explorations 12 no. 1 (2010) 40–48.

［ 18 ］ A. Metzger, R. Franklin and Y. Engel," Predictive monitoring of heterogeneous service-oriented business networks: the transport and logistics case," in Proc. of the SRII Global Conference, (2012), 313–322.

19. Z. Feldman, F. Fournier, R. Franklin and A. Metzger," Proactive event processing in action: a case study on the proactive management of transport processes," in *Proc. of ACM International Conference on Distributed Event-Based Systems (DEBS)*, (ACM, 2013) 97–106.

20. http://www.fluxicon.com.

21. http://www.qpr.com.

22. http://www.perceptivesoftware.com/products/perceptive-process/process-mining.

23. http://promtools.org.

24. W. M. P.van der Aalst, Ibid.

［19］ Z. Feldman, F. Fournier, R. Franklin and A. Metzger," Proactive event processing in action: a case study on the proactive management of transport processes," in Proc. of ACM International Conference on Distributed Event-Based Systems (DEBS), (ACM, 2013) 97–106.

［20］ http://www.fluxicon.com.

［21］ http://www.qpr.com.

［22］ http://www.perceptivesoftware.com/products/perceptive-process/process-mining.

［23］ http://promtools.org.

［24］ W. M. P. van der Aalst, Ibid.

Social Media and Business Process Management

Henrik von Scheel, Zakaria Maamar, Mona von Rosing

INTRODUCTION

As with the internet itself, social media has come to stay and is fundamentally affecting the way organizations are interacting, both internally and externally. However, when organizations adopt social media initiatives, they will not achieve its full potential unless the changes are integrated with both its operational processes and its approach to continuous improvement. Exploiting social media by making the requisite connections to the Business Process Management (BPM) Center of Excellence (CoE) interaction and feedback loop enables the creation of customer-centric process design, as well as providing access to stakeholder input about the maturity of BPM processes related to customer services, which again drives process-improvement decisions. This chapter will discuss the changing market around social media as well as how Socially Oriented Process Modeling can be used effectively to enable business model renewal, service improvements, customer performance focus, as well as address critical debates about whether to automate certain processes or keep them manual.

THE DIGITAL MIND-SET IS CHANGING

We live in a 'connected world' and organizations can no longer ignore this nor take it for granted. What many organizations are not realizing is that we are not in a shifting technical period, nor in a business or economic shift. We are experiencing a major cultural shift, and that shift affects technology, business, and economies. It is a revolution that is creating a 'digital mind-set' that is fundamentally transforming and affecting the way organizations are doing business, and it is vital to their survival that they understand these changes. A digital mind-set is different from the transactional mind-set that it is supplanting. A digital mind-set is open, connected, participating, selective, and controlling. It is fundamentally a very different way of thinking and working, in which the digital mind-set builds followers, from embracing one and becoming a 'friend,' whereas the transactional mind-set is looking for suspects who they can turn into prospects and will love you only if you become or have the potential to become a customer (Figure 1).

In the digital mind-set the person can be the user, consumer, community member, participant, producer, as well as customer, all at the same time. The important aspect is that such a mind-set operates in an open, connected, and participating

3.9　社交媒体与BPM

Henrik von Scheel, Zakaria Maamar, Mona von Rosing

3.9.1　介绍

与互联网本身一样,社交媒体已经存在,并且从根本上影响了组织内部和外部的互动方式。但是,当组织应用社交媒体时,除非将变更与"其运营流程及其持续改进方法"相结合,否则它们将无法充分发挥其潜力。通过与BPM CoE交互和反馈循环建立必要的连接来利用社交媒体,可以创建以客户为中心的流程设计,并提供有关BPM流程成熟度的利益相关者意见的访问权限与相关客户服务,这再次推动了流程改进决策。本节将讨论围绕社交媒体的不断变化的市场及如何有效地使用面向社交的流程建模来实现业务模型更新、服务改进、客户绩效关注,以及处理关于是否将某些流程自动化或将其保留为手动的关键辩论。

3.9.2　数字思维模式正在发生改变

我们生活在一个"相互联系的世界",组织不能再忽视这一点,也不能把它视为理所当然。许多组织没有意识到的是:我们并没有处于技术变革时期,也没有处于业务或经济变革时期。我们正在经历一场重大的文化变革,这种变革影响着技术、商业和经济。这是一场革命,它正在创造一种数字思维模式,从根本上改变和影响着企业的经营方式,理解这些变化对企业的生存至关重要。数字思维模式不同于它正在取代的交易思维模式。数字思维模式是开放的、连接的、参与的、选择性的和控制的。这根本上是一个非常不同的思考和工作方式,数字思维模式构建追随者,从拥抱他人进而成为朋友,而传统的思维模式是寻找嫌疑人,即他们可以变成潜在客户,同时只有您成为或有潜力成为客户时才会爱上您(图1)。

在数字思维中,人可以同时是用户、消费者、社区成员、参与者、生产者以及顾客。重要的方面是这种思维模式以开放、连接和参与的方式运作。虽然不是每个

Digital Mindsets

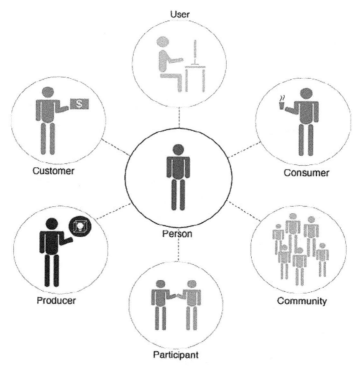

FIGURE 1

The digital mind-sets.

way. Although not everyone is open or even supportive to such a mind-set, as the internet in its very nature is open and connected such that it sees things that are hidden or obscured as "damaged," which it seeks to repair by bringing into the open the entities, for example, people/persons who successfully 'interact' in the internet naturally pursue the most openness, connectedness, and participation.

It can be argued that few of us treat others as we want to be treated; most of us look for short cuts and manipulatively clever tricks that can fool people into buying. So why is such a digital mind-set so good for business? Because, when organizations have a digital mind-set, they are more connected and they can more quickly solve problems.

As discussed in the earlier chapters, some of the Global University Alliance research has revealed a difference between outperforming and underperforming organizations. One of those differences is that outperforming organizations have empowered employees to communicate more effectively through many tools, including social media channels and technology.

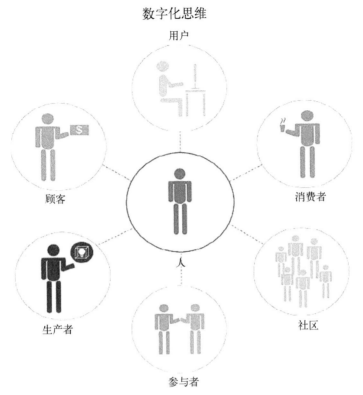

图1　数字思维模式

人都是开放的，甚至不支持这样的思维方式，因为互联网本质上是开放和连接的，所以它将隐藏或模糊的东西视为"受损"，它试图通过开放实体来修复，例如，在互联网上成功"互动"的人自然追求最开放、最连通和最参与。

可以说，我们很少有人像对待我们一样对待别人。我们大多数人都在寻找可以欺骗人们购买的捷径和操纵巧妙的技巧。那么，为什么这样的数字思维对商业如此有益呢？因为，当组织拥有数字思维模式时，他们就会更加紧密联系，他们可以更快地解决问题。

正如前面章节中所讨论的，全球大学联盟的一些研究揭示了表现优异和表现不佳的组织之间的差异。其中一个不同之处在于：表现优异的组织使员工能够通过许多工具（包括社交媒体渠道和技术）更有效地进行沟通。

SOCIAL MEDIA ARE RESHAPING BUSINESS

Traditional business models, particularly in large organizations, have had, as one common characteristic, careful limitation of direct contact between those within the organization and those outside. Only certain specific individuals (most frequently in roles such as sales, customer service, and field consulting) are designated as "customer-facing" personnel. Organizations further limited outside access to internal employees through filtering mechanisms such as publishing only a main switchboard number (whether routed through a live receptionist or an interactive voice response system), generic mailing, or email addresses such as "sales@" or "info@".

As discussed in the chapter on "outperformers and underperformers", success is based on doing what the outperformers know: being successful means constantly changing. And, therein lies the irony—to be successful you have got to change something…and something that must be changed must be something that has allowed you to be successful up to now ("we don't change a winning team"). However, where do you start? What needs to change? We all know that aspects of a process and activities need to be changed. Nevertheless, this is not the start of the journey. So the real question remains, what needs to change? Now, combine that reality with day-to-day business challenges such as: entry of new competitors, rising price of doing business, new technologies, evolving customer expectations, changing consumer demands, and your team's desire to better interact. Today, many organizations have "satisfied" customers and a sound business model, and they are confident that social media will have a nominal impact on their organizations. It is apparent, however, that social media have already had an impact on customers' daily lives and are here to stay, and, therefore, this confidence is both falsely held and ultimately likely dangerous for those that hold the view. We are still very early in the social media era, however, and some changes are already evident in the market. The question is not whether your organization is adapting any abilities around social media. The real question is to what extent are your customers or even your competitors already doing it. The only way to find out is to start watching your environment for the influence these changes will bring to your business. Moreover, here is the second irony, watching for the impact of social media requires that you employ social media.

We see that social media have the ability to change the Business Model in the following areas (Figure 2):

- From selling to connecting with customers
- To create a new era of brand management that recognizes the role and nature of bonding, advantage, performance, relevance, and presence
- From large campaigns to small and rapid actions
- From controlling the message to transparency
- From hard to reach to available everywhere

Social and mobile business is affecting most organizations in different ways, but some will be more impacted than others. Some organizations don't even realize it.

3.9.3　社交媒体正在重塑商业

传统的商业模式(特别是在大型组织中)有一个共同的特点,就是对组织内部和外部人员之间的直接联系有严格的限制。只有特定的个人(最常见的角色是销售、客户服务和现场咨询)被指定为"面向客户"的人员。组织通过制度等机制进一步限制外部人员与内部员工的沟通,例如,只允许通过主总机号码(无论是通过现场接待员还是交互式语音响应系统)、普通邮件或电子邮件(如"sales@"或"info@")的方式进行。

正如"卓越绩效者和表现不佳者"这一节所讨论的那样,表现出众的基础是:成功意味着不断变化。而且,具有讽刺意味的是,要想成功,您必须改变一些东西,而必须改变的东西必须是让您能够成功到现在的东西("我们不会改变一个胜利的团队")。但是,您从哪里开始?需要改变什么?我们都知道需要改变一个过程和活动的各个方面。然而,这不是旅程的开始。所以真正的问题仍然存在,需要改变什么?现在,将这一现实与日常业务挑战结合起来,例如:新竞争对手的进入、成本价格上涨、新技术、不断变化的客户期望、不断变化的消费者需求以及团队更好地互动等。如今,许多组织都为使客户"满意"而建立了良好的商业模式,他们相信社交媒体仅仅会对其组织产生非实质的影响。然而,显而易见的是:社交媒体已经对客户的日常生活产生了影响并且仍然存在,因此,这种信心被错误地保留,并且最终可能对那些持有这种观点的人带来危险。然而,我们还处于社交媒体时代的早期阶段,并且市场上已经出现了一些变化。问题不在于您的组织是否正在调整社交媒体的任何能力。真正的问题是您的客户甚至您的竞争对手已经在多大程度上做到了这一点。找出答案的唯一的办法就是开始观察您的环境,看看这些变化会给您的企业带来什么样的影响。另外,第二个讽刺是,观察社交媒体的影响需要您使用社交媒体。

我们发现,社交媒体能够在以下方面改变商业模式(图2):

- 从销售到与客户联系;
- 创造一个品牌管理的新时代,认识到联系、优势、绩效、相关性和存在的作用和性质;
- 从大型活动到小型快速行动;
- 从控制信息到透明;
- 从难以触及到无处不在。

社交和移动业务以不同的方式影响着大多数组织,但有些组织会比其他组织受到更大的影响。有些组织甚至没有意识到这一点。以下是五个明显的例子,其

The following are five clear examples in which we see the impact of social media on business models in various industries:

1. **Organizations that own assets and make them available to consumers for rent.** For example, hospitality and car rental companies face new competition from peer-to-peer models. In New York, hotels now compete against 10,000 rooms, apartments, and even spare couches offered by consumers in social media communities.

2. **For most consumers, the car is one of the most expensive assets owned**; yet the average consumer uses their car just 8% of the time. It is this low utilization that is leading some to offer their cars for rent. In addition, as consumers get access to the cars they need when they need them, ownership becomes less attractive. One study found that people who use car-sharing services were 72% less likely to buy or lease a car in the future.

3. **Companies that facilitate business between consumers.** If you are in the business of earning fees to take something from one customer and get it to another customer, social business models will challenge your business. We are not referring to eBay or Craigslist--they already are the standard for Peer-to-Peer (P2P) disintermediation and reintermediation, having taken a big piece of the newspapers' classified ad business; rather we are referring to applications that allow consumers to hail cabs, provide job search tools, and so on.

4. **Financial business model.** Look at banks, which take money from savers and lend it to borrowers. Today, savers get little, but this is not the case for folks lending money on Prosper and LendingClub. Although the risks are greater,

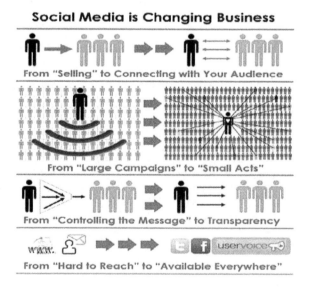

FIGURE 2

Social media are changing business.

中我们看到了社会媒体对不同行业的商业模式的影响。

（1）拥有资产并提供给消费者出租的组织。例如,酒店和租车公司面临着来自对等模式的新竞争。在纽约,酒店与社交媒体社区的10 000间客房、公寓,甚至是消费者提供的备用沙发竞争。

（2）对于大多数消费者来说,汽车是拥有的最昂贵的资产之一,然而消费者平均使用汽车的时间只有8%。正是这种低利用率导致一些人提供他们的汽车出租。此外,随着消费者在需要的时候能够获得他们需要的汽车,车主的吸引力就会降低。一项研究发现,使用汽车共享服务的人未来购买或租赁汽车的可能性降低了72%。

（3）促进消费者之间业务的公司。如果您从事的业务是从一个客户那里拿东西并把它卖给另一个客户,那么社会化的商业模式将挑战您的业务。我们不是指eBay或Craigslist,他们已经是对等(P2P)非中介化和再中介的标准,已经占据了报纸分类广告业务的很大一部分,而是指允许消费者呼叫出租车、提供求职工具等的应用程序。

（4）金融商业模式。看看银行,它们从储户那里取钱,然后把钱借给借贷者。今天,储蓄者得到的很少,但这不是为繁荣和借贷俱乐部贷款的人的情况。尽管风险更大,但回报也是如此。虽然作为银行的监管障碍很高,但企业能够避开监管,

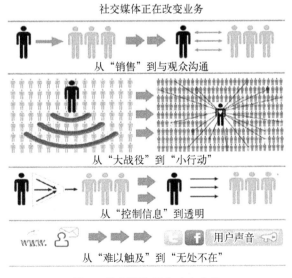

图2　社交媒体正在改变业务

the rewards are as well. Although the regulatory hurdles for being a "bank" are high, companies are able to skirt the regulations and bring down costs to consumers and provide new services with capabilities such as "mobile wallets", P2P money transfer, and P2P lending models.

5. **Business models that manufacture durable goods**. Younger consumers are less interested in obtaining drivers licenses and prefer to meet friends online and seek to decrease the miles they drive as a means to save the environment. It is clear that P2P and sharing business models will affect the auto business (and related industries such as auto parts and auto insurance).

For both hotels and cars, more supply means lower costs for consumers and less revenue for providers. In addition, the new social business competition has a vastly different cost structure from traditional providers. The social media rental rooms, couch surfing, rental sharing, or used CD, games, or book sales do not need to purchase, own, or maintain assets, resources, and locations the same way the traditional business models do.

All the aforementioned changes to the assumptions and conditions of the affected business models reflect how and why organizations should rely on their social networks referred to as customer, supplier, competitor, and partner. These four networks are established based on the interaction circles, which show the "social" or communications distance to be overcome when interacting, which define the environment of a company (Figure 3).

- Scenario 1. A company uses a customer social network to contact a customer's "friends" to disseminate information on products and services and, hence, boost sales.

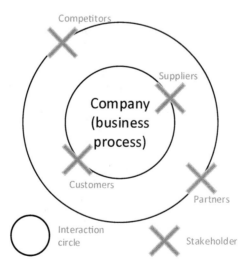

FIGURE 3

Interaction circles (From Badr & Maamar, 2009).

降低消费者的成本,并提供具有"移动钱包"、P2P转账和P2P借贷模式等功能的新服务。

(5)制造耐用商品的商业模式。年轻的消费者对获得驾驶执照不太感兴趣,他们更喜欢在网上结识朋友,并设法减少他们的行驶里程,以此来保护环境。很明显,P2P和共享业务模式将影响汽车业务(以及汽车零部件和汽车保险等相关行业)。

对于酒店和汽车来说,更多的供应意味着更低的消费者成本和更少的供应商收入。此外,新的社会商业竞争与传统供应商的成本结构有很大的不同。社交媒体租赁房间、沙发客、租赁共享或二手CD、游戏或书籍销售不需要像传统商业模式那样购买、拥有或维护资产、资源和位置。

上述对受影响商业模式假设和条件的所有更改都反映了组织应如何以及为什么依赖其社交网络(称为客户、供应商、竞争对手和合作伙伴)。这四个网络是建立在交互圈的基础上的,交互圈显示了交互时要克服的"社会"或通信距离,定义了公司的环境(图3)。

- 场景1。公司使用客户社交网络联系客户的朋友,传播有关产品和服务的信息,从而促进销售。

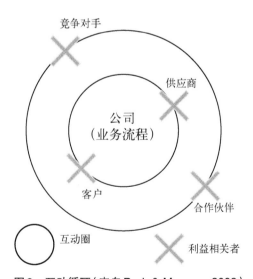

图3　互动循环(来自Badr & Maamar,2009)

- Scenario 2. A company uses a supplier social network to establish the reliability of a supplier, as perceived by other parties (e.g., customers and suppliers) that have dealt with this particular supplier in the past.
- Scenario 3. A company uses a competitor social network to work with competitors, and partners of competitors, to establish standards and common norms.
- Scenario 4. Finally, a company uses a partner social network to identify the individuals and groups with which it can join forces to tackle complex initiatives.[1]

Modern business strategy and models must expect and can exploit the fact that well-informed consumers will have more access to information in real time and can avail themselves of new social and mobile business models that save money. Social and mobile-centric business models have a way of empowering consumers so they can make better decisions. The introduction of these changes will mean that many organizations will have to scramble to keep up with lean new competitors and consumers' rapidly changing technology habits and sharing behaviors.

When changes occur in a business model, this means that, fundamentally, the organization must change their services and processes as well. A social media strategy that is supported by an organization that is able to rapidly alter business processes, listen to, identify, and manage risks early on, and generate sales opportunities, will be an effective one. Therefore, innovating and transforming one's process models is, in social media, the cornerstone for adapting the organization to a new service and value to the customer.

Consider how social media can impact the following operational and management processes. These provide good examples of how operational processes can be changed to take advantage of social media:

- **Purchase requisition to goods received**. Quality issues in goods received may immediately trigger an interaction between customer and supplier via a blog or other method to discuss how to best address the issues.
- **Order-to-delivery**. Use of online "Process communities" can help align all resources and people involved in a particular scenario and, through a focus on customer-centric metrics, such as low delivery variance, enhance delivery performance.
- **Human resources**. Recruitment processes can be enriched through social media. Consider the use of sites such as LinkedIn as part of the hiring process. Companies that develop software might look to recruit those most involved in forums or developing apps based on the company's application program interface. Blogs can highlight talent in particular areas as well.
- **Marketing**. By finding out what customers and prospects are saying about your products or those in your market space, you can enhance and target marketing efforts. Facebook, Pinterest, and other sites offer new ways to drive traffic and interest in products and services. Online focus groups can be conducted via online social media sites. Mini-surveys on a company's Facebook page often elicit responses that can rapidly close the feedback loop, making for effective marketing.
- **Customer service**. As never before, people are saying what they think in social media. Monitoring and engaging with customers using these tools can help drive customer centricity and customer loyalty. Engage your customers wherever they are speaking out, from Yelp to Tripadvisor to Facebook.

- 场景2。公司使用供应商社交网络来建立供应商的可靠性,就像其他方(如客户和供应商)过去与该特定供应商打交道时所感知到的那样。
- 场景3。公司使用竞争对手的社交网络与竞争对手以及竞争对手的合作伙伴合作,建立标准和共同规范。
- 场景4。最后,公司利用合作伙伴的社交网络来识别个人和团体,以便联合起来处理复杂的事务[1]。

现代商业战略和模式必须预期并能够利用这样一个事实,即消息灵通的消费者将有更多的机会实时获取信息,并能够利用新的社交和移动商业模式来节省成本。以社交和移动为中心的商业模式能够增强消费者的能力,让他们做出更好的决定。这些变化的引入意味着许多组织将不得不努力跟上精益的新竞争对手和消费者快速变化的技术习惯和共享行为。

当业务模型发生变化时,从根本上说,这意味着组织也必须更改其服务和流程。一个组织支持的能够快速改变业务流程、尽早倾听、识别和管理风险并产生销售机会的社交媒体策略将是一个有效的策略。因此,在社交媒体中,创新和转变一个流程模型是使组织适应新服务和为客户创造价值的基石。

考虑社交媒体如何影响以下运营和管理流程。这些提供了如何更改操作流程以利用社交媒体的良好示例。

- 对收到的货物进行采购申请。遇到货物质量问题可能会通过博客或其他方法立即触发客户与供应商之间的互动,以讨论如何更好地解决问题。
- 订单到交货。使用在线"流程社区"可以帮助调整特定方案中涉及的所有资源和人员,并通过关注以客户为中心的指标,如降低交付差异,提高交付绩效。
- 人力资源。可以通过社交媒体丰富招聘流程。考虑使用LinkedIn等网站作为招聘流程的一部分。开发软件的公司可能希望招募那些参与论坛或根据公司的应用程序界面开发应用程序的人,博客也可以突出显示特定领域的人才。
- 营销。通过了解客户和潜在客户对您的产品或您的市场空间的看法,您可以增强和定位营销工作。Facebook、Pinterest和其他网站提供了新的方式来提高产品和服务的流量和兴趣。在线服务小组可以通过在线社交媒体网站进行工作。公司Facebook页面上的小型调查通常可以实现工作闭环,从而实现有效的营销。
- 客户服务。人们从未像现在这样在社交媒体上说出他们的想法。使用这些工具监控和与客户互动有助于提高客户中心性和客户忠诚度,让客户无论身在何处,从Yelp到Tripadvisor再到Facebook。

- **Research and development**. Instead of formulating product direction in a vacuum, one electronics company that produces accessories started listening to early iPad owners speaking out on social media and let that drive their product strategy, with very strong results.
- **Operational processes** can be enhanced and supported by strategic engagement with social media, both inside and outside the company.

ENABLING CUSTOMER-CENTRICITY

The most important aspect of social media is understanding the social consumer decision life cycle journey. As discussed, the traditional consumer decision funnel is changing because it is outdated or not applicable in the world of social media. The graphical model in Figure 4 below describes the consumer decision life-cycle journey. This model was introduced by Henrik von Scheel and Prof. Mark von Rosing in November 2003, based on a "consumer influencer decision making cycle in e-Commerce" model developed with IKEA and Google, which uniquely combines specific defined online consumer persona behaviors with the traditional marketing purchase funnel to support a traditional customer journey and a distinct online purchase funnel. The model shows the states a consumer goes through within the cycle as they determine whether to purchase something, what an organization can do to act on the decision as an influencer, and how the two are connected.

This model contrasts with the pure version of the more traditional purchase funnel developed to map the consumer journey from the time they are initially aware of a brand or product to the point of action or purchase. This staged process is summarized as:

- **Awareness**. The customer is aware of the existence of a product or service
- **Interest**. Actively expressing an interest in a product group
- **Desire**. Aspiring to a particular brand or product
- **Action**. Taking the next step towards purchasing the chosen product

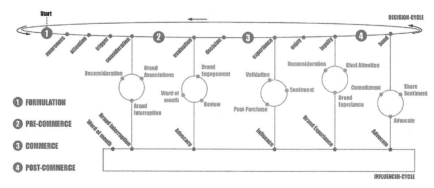

FIGURE 4

Consumer decision life cycle.

- 研究与开发。一家生产配件的电子公司没有凭空制定产品方向,而是开始聆听早期的iPad用户在社交媒体上发表意见,让他们推动他们的产品战略,并取得非常强劲的成果。
- 通过与公司内外的社交媒体进行战略性互动,可以增强和支持运营流程。

3.9.4 实现以客户为中心

社交媒体最重要的方面是了解社交消费者决策生命周期之旅。如上所述,传统的消费者决策漏斗正在发生变化,因为它已经过时或不适用于社交媒体领域。下面图4中的图形模型描述了消费者决策生命周期之旅。该模型由H. von Scheel和M. von Rosing教授基于宜家和谷歌开发的"电子商务中的消费者影响者决策周期"模型开发,于2003年11月推出,该模型将特定的在线消费者角色行为与传统的营销购买渠道相结合,以支持传统的客户旅程和独特的在线购买渠道。该模型显示了消费者在周期内经历的状态,他们决定是否购买某些东西、组织作为影响者采取什么行动来影响这些决定,以及两者如何相互联系。

这个模型与纯粹版本的传统购买漏斗形成了对比,后者用于绘制消费者从最初了解某个品牌或产品到行动或购买的过程。这一阶段的过程总结如下。

- 认知。客户知道产品或服务的存在。
- 兴趣。积极表达对产品组的兴趣。
- 欲望。对特定品牌或产品有抱负。
- 行动。购买所选产品的下一步。

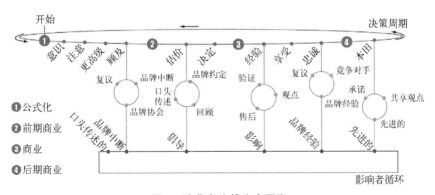

图4 消费者决策生命周期

Today's consumers take a much more complex and more iterative path that extends through and beyond purchase. The classic funnel shows an ever-narrowing array of decisions and choices until purchase, when in fact the social, connected, and channel-surfing customer today often is expanding the set of choices and decisions after their initial consideration.

Just as important, the consumer decision life cycle treats the post–purchase process with the same level of importance as the prepurchase journey. It's a simple concept, really, but it adds significantly to the insight that can be gained as to how to engage consumers and the implications for organization design, including business process design, because this visually highlights and isolates the most important aspects of the journey in this new paradigm:

- **Consider.** What brands/products do consumers have in mind as they contemplate a purchase?
- **Evaluate.** Consumers gather information to narrow down their choices.
- **Buy.** Consumers decide on a brand and buy it.
- **Post–purchase.** Consumers reflect on the buying experience, creating expectations/considerations that will inform a subsequent purchase.
- **Advocate.** Consumers tell others about the product or service they bought.

The funnel model also ignores and the consumer decision life cycle journey exposes the role that influencers play in modern consumer behavior, a reality that must be reflected in modern business design and managed through BPM capabilities. This visualization of the journey helps focus conversations on where to invest time, money, and attention, where the opportunities are, what sorts of people and processes you need to deliver on them, where you are weak and your competitors strong. The consumer decision journey approach helps clarify the issues that are undermining a brand, or where a brand has an opportunity to grow. What part of the consumer decision journey is critical to a brand? Social media provides organizations with a wealth of information on their customers. Although social media and BPM can definitely enable customer centricity, the long-term success for organizations will depend on the enterprise-wide social media monetization strategy and the agility of internal business processes.

Social media complemented by agile business processes can definitely aid in creating "customer-centric" organizations. These organizations will be able to successfully attract and retain customers, align and position their offerings in line with customer requirements, effectively optimize their marketing strategies, and will find ample opportunities for improved customer experience and enhanced customer relationship management.

LESSONS LEARNED AROUND SOCIAL-ORIENTED PROCESS MODELING

The customer acquisition process consists of identifying potential prospects, qualifying them, and converting them to live customers.

Organizations can observe customer patterns and behaviors on social media platforms by using the right tools and technologies, and identifying potential

今天的消费者采取了更加复杂和更加迭代的途径,这种途径延伸到购买之外。经典的渠道显示了一系列决策和选择,直到购买,实际上今天的社交、互联和渠道促使客户经常在最初的考虑之后扩大选择和决策的范围。

同样重要的是,消费者决策生命周期使购后过程与购前过程同等重要。这确实是一个简单的概念,但它极大地增强了我们对如何吸引消费者的洞察力,以及对组织设计(包括业务流程设计)的影响,因为这在视觉上突出并隔离了这个新范例中旅程中最重要的方面。

- 考虑:消费者在考虑购买什么品牌/产品时,心里想的是什么?
- 评估:消费者收集信息以缩小他们的选择范围。
- 购买:消费者决定购买一个品牌。
- 售后:消费者会对购买体验进行反思,从而产生预期/考虑因素,为后续的购买提供信息。
- 倡导者:消费者告诉其他人他们购买的产品或服务。

漏斗模型也忽略了消费者决策生命周期之旅揭示的影响因素在现代消费者行为中所扮演的角色,这一现实必须反映在现代商业设计中并通过BPM能力进行管理。这种旅程的可视化有助于讨论在哪儿投入时间、金钱和注意力、机会所在的位置,您需要提供的人员和流程类型、弱势群体以及强大的竞争对手。消费者决策之旅的方法有助于明确正在损害一个品牌的问题,或者一个品牌在哪里有机会成长。消费者决策过程的哪一部分对品牌至关重要?社交媒体为企业提供了大量关于客户的信息。尽管社交媒体和BPM肯定能够支持以客户为中心,但组织的长期成功将取决于企业范围的社交媒体货币化策略和内部业务流程的敏捷性。

由敏捷业务流程补充的社交媒体肯定可以帮助创建"以客户为中心"的组织。这些组织将能够成功地吸引和留住客户,使他们的产品符合客户的需求,有效地优化他们的营销策略,并为改善客户体验和加强客户关系管理找到充足的机会。

3.9.5 关于面向社交的流程建模的经验教训

客户获取过程包括识别潜在的客户、对他们进行资格认证,并将他们转化为活生生的客户。

组织可以通过使用正确的工具和技术观察客户在社交媒体平台上的模式和行为,并识别潜在的前景。结构化分析和聚合可以帮助客户分析和细分。随后,有针

prospects. Structured analysis and aggregation can help in customer profiling and segmentation. Subsequently, targeted marketing campaigns and customer dialog can help in customer qualification and acquisition. However, all of this can be done only if business processes are developed and deployed to provide these capabilities and only if they have been integrated into the other business processes to align with an overarching strategy, a fact that of necessity means the adoption of BPM.

An example of where this transition has been made and the necessary capabilities developed includes as an example, UBS Bank. This organization launched a new credit card with special features and a preapproved credit limit targeting students. To reach potential customers, it could then look at the basic profile information of students on Twitter. This is the first level of customer segmentation. At the next level, UBS can now examine the tweets posted by the target prospects. Let's explore what this means by assuming that a couple of these prospects are tweeting about their need for a credit card that offers a minimum level of interest. If the bank is able to detect this, a process can be triggered to facilitate customer conversion, by initiating a social media-centered marketing campaign, which sends customized information to the prospects about specific credit card products. Such a campaign is much cheaper than traditional campaigns to establish and develop a new credit card product. Based on the interest and response from target customers, subsequent steps of the business process can be orchestrated to facilitate customer acquisition through target marketing campaigns, by improving prospect qualification, through new approaches to customer relationship management, and by creating process flows that reflect and leverage the properties of social media. These approaches are discussed in the remainder of this chapter.

TARGET MARKETING CAMPAIGNS WITH SOCIAL MEDIA

Figure 5 shows the possible interaction between platforms in a targeted social media campaign strategy that:

- Changes the flow of process, information, and services to **reach** more people
- Keeps the followers **informed**
- Allows people to **sample/put** together their wishes and wants
- Offers product and/or services to be **sold** to the customer
- Has content/information that customers can pass along to **involve** their connections/friends

Business processes that use customer patterns and behaviors on social media platforms mixed with analytical data from previous purchases as an input to trigger the right marketing campaigns, can prove very effective. Using a portfolio of social media platforms with applicable processes can integrate these prospects into processes that can efficiently convert them into new customers.

对性的营销活动和客户对话可以帮助客户认证和获取资格。然而,只有当业务流程被开发和部署以提供这些功能,并且只有当它们被集成到其他业务流程中以与总体策略保持一致时,才能完成所有这些工作,这一事实必然意味着采用BPM。

例如,瑞银银行(UBS Bank)已经完成了这种转换,并开发了必要的功能。该组织针对学生推出了一种具有特殊功能和预先批准的信用额度的新信用卡。为了接触潜在客户,它可以查看学生在Twitter上的基本个人信息。这是客户细分的第一步。在下一步,瑞银银行现在可以检查目标客户发布的推文。让我们来探讨一下这意味着什么,假设这些潜在客户中的一些人正在推特上谈论他们对提供最低利率的信用卡的需求。如果银行能够检测到这一点,就可以启动一个流程,通过发起以社交媒体为中心的营销活动,将定制的信息发送到特定信用卡产品的前景,从而促进客户转换。这样的活动比建立和开发一种新的信用卡产品的传统活动要便宜得多。根据目标客户的兴趣和反应,业务流程的后续步骤可以通过目标策划促进顾客营销活动,通过改善前景资格,通过客户关系管理的新方法,通过创建流程流,反映和利用社交媒体的属性。本节的其余部分将讨论这些方法。

3.9.6　利用社交媒体进行营销活动

图5显示了目标社交媒体活动策略中各平台之间可能的交互:

- 改变流程、信息和服务流,以吸引更多人;
- 让关注者知情;
- 允许人们对他们的愿望和需求进行采样/整理;
- 提供可销售给客户的产品和/或服务;
- 具有客户可以传递给他们的关系人/朋友的内容/信息。

在社交媒体平台上使用客户模式和行为的业务流程与先前购买行为的分析数据相结合,作为触发正确营销活动的输入,可以证明是非常有效的。使用具有适用流程的社交媒体平台组合可以将这些潜在客户整合到可以有效地将其转换为新客户的流程中。

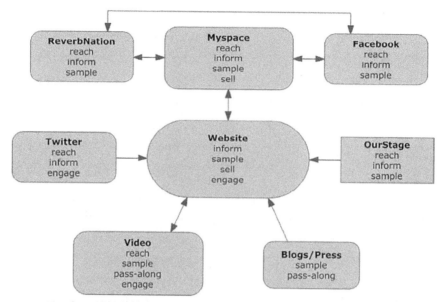

Reach: twitter, video, blogs/press, myspace, facebook, reverbnation, ourstage

Inform: twitter, myspace, facebook, reverbnation, ourstage, website, e-mail

Sample: myspace, facebook, reverbnation, ourstage, website, video, blogs/press

Sell: myspace, website

Pass-along: video, blogs/press

Engage: video, twitter, website/artist blog

FIGURE 5

Social media platforms enabling a marketing campaign.

IMPROVING THE PROSPECT QUALIFICATION PROCESS

A well-defined business process, the collection of data, and performance measures for prospect qualification are important aspects of customer acquisition. Partial/complete automation of the steps in this process can be achieved by leveraging the power of a process flow, steps, and activities with business rules defined in accordance with an organization's business objectives.

CUSTOMER PROFILE DATA FOR PROCESS (SIMPLIFICATION)

One of the key benefits of social media platforms is the availability of customer profile information in the public domain. Additional customer profile data can be collected from these platforms by using analytical tools. This customer profile data can

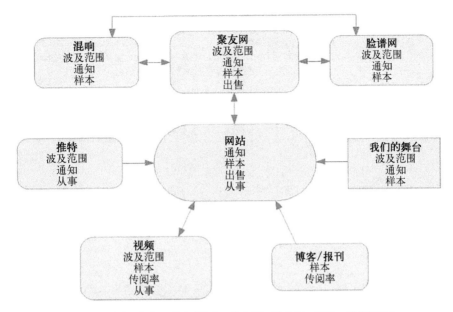

波及范围：推特，视频，博客/报刊，聚友网，脸谱网，混响，我们的舞台

通知：推特，聚友网，脸谱网，混响，我们的舞台，网站，邮件

样本：聚友网，脸谱网，混响，我们的舞台，网站，视频，博客/报刊

出售：聚友网，网站

传阅率：视频，博客/报刊

从事：视频，推特，网站/艺术家博客

图5　启用营销活动的社交媒体平台

3.9.7　改进潜在客户资格认证流程

定义良好的业务流程、数据收集和潜在客户资格的绩效衡量是客户获取的重要方面。通过利用流程流、步骤和活动的能力，并根据组织的业务目标定义业务规则，可以实现此流程中步骤的部分/完全自动化。

3.9.8　流程的客户配置文件数据（简化）

社交媒体平台的一个主要好处是在公共领域提供客户档案信息。通过使用分析工具，可以从这些平台收集其他客户概要数据。这些客户概况数据可以提供重

provide vital information for use in defining, optimizing, and simplifying the steps in the customer acquisition process, thus playing a significant role in identifying where the proactive or reactive process should be set up to influence the decision life cycle. Configuration and execution of process steps based on the kind of customer profile information can significantly reduce customer acquisition costs.

CUSTOMER NOTIFICATIONS FOR PROCESS (VISIBILITY)

One of the most important aspects of customer acquisition is continuous interaction and communication on the status and progress of the customer acquisition process. Process notification steps can be configured to provide customized customer communications based on the status of the requests, and sent to customers via channels like Twitter, blogs, direct messages, texts, and so on. These processes and the information they generate, however, need to be integrated into other aspects of the business operation.

ALTERNATIVE CHANNELS FOR SALES

Social media platforms are already being used as alternative sales channels for initiating specific customer requests, by extending the existing process and systems to include the social media platforms as input channels for initiating the sales process. A process-based solution for channel unification can also be considered.

SELECTION OF THE RIGHT OFFERING/SOLUTION

Customer patterns and behaviors on social media platforms can be effectively used as key decision variables in predefined business processes to aid in the selection of the right offering/solution for customers. In addition to specific data provided by customers, social media analytics can be leveraged to provide recommendations to customers on the best solutions for their requirements, provided the insight gained about patterns and behaviors can be communicated to, and exploited by, the design, production, distribution, service, and other arms of the enterprise.

SOCIAL MEDIA AND BPM FOR CUSTOMER SERVICING

The customer-servicing process consists of understanding and working on the customers' requirements/requests to address their specific needs. Social media platforms provide a wealth of information on customer expectations, preferences, and most importantly, real-time updates on changing customer needs. Customer expectations revealed on social media are an invaluable resource for organizations striving to be customer-centric. Processes can be orchestrated to connect with the customer, and to validate whether these changing customer expectations and preferences should be factored in to service their requests.

要信息,用于定义、优化和简化客户获取过程中的步骤,从而在确定应在何处设置主动或被动流程以影响决策生命周期方面发挥重要作用。基于客户概要信息的流程步骤的配置和执行可以显著降低客户获取成本。

3.9.9　流程的客户通知(可见性)

客户获取的一个最重要的方面是对客户获取流程的状态和进展进行持续的交互和沟通。流程通知步骤可以配置为根据请求的状态提供定制的客户通信,并通过Twitter、博客、直接消息、文本等渠道发送给客户。但是,这些流程及其生成的信息需要集成到业务操作的其他方面。

3.9.10　其他销售渠道

社交媒体平台已经被用作启动特定客户请求的替代销售渠道,通过扩展现有流程和系统,将社交媒体平台包括为启动销售流程的输入渠道。还可以考虑基于流程通道的统一解决方案。

3.9.11　选择合适的产品/解决方案

社交媒体平台上的客户模式和行为可以有效地用作预定义业务流程中的关键决策变量,以帮助为客户选择正确的产品/解决方案。除了客户提供的特定数据外,还可以利用社交媒体分析为客户提供有关其需求的最佳解决方案的建议,前提是所获得的有关模式和行为的见解,在设计、生产、分销、服务和其他企业目标中进行沟通和利用。

3.9.12　用于客户服务的社交媒体和BPM

客户服务流程包括:理解和处理客户的需求,以满足他们的特定需求。社交媒体平台提供了大量关于客户期望、偏好以及最重要的实时更新客户需求的信息。在社交媒体上公布的客户期望对于努力以客户为中心的组织来说是非常宝贵的资源。可以对流程进行编排,以便与客户进行连接,并验证是否应该将这些不断变化的客户期望和首选项考虑在内,以满足他们的请求。

In a case within Vodafone, for example, they found a situation in which a customer had subscribed to a prepaid plan, and, unfortunately, errors occurred in billing. The customer was unable to reach the Customer Care section of Vodafone and understandably was upset and frustrated that he was unable to get a satisfactory explanation for the erroneous billing, and tweeted his dissatisfaction. Vodafone has active listening agents who read the Twitter or Facebook messages, and contacted the dissatisfied customer to help resolve the billing issue. Because of this, Vodafone was able to intercept the problem and initiate a process to recover the situation. Once the process is initiated, internal systems and resources work collaboratively to ensure that either a response is provided in the form of an acknowledgment, a clarification is provided, or the issue is resolved. This integration of social media-based processes properly integrated with internal systems turned what could have been a difficult experience into a constructive customer experience.

Every interaction a customer has with an organization is a 'Moment of Truth' – the possibility of an experience which can be positive, negative, or neutral. On the one hand positive experiences usually translate into repeat business and customer loyalty, and referrals, recommendations or reviews on several channels. Hence the organization can initiate processes to further enhance and strengthen customer relationships through dialog, cross-selling/up-selling other products and services. On the other hand, negative experiences can result in negative reviews and feedback through these channels, or termination of the business relationship with an organization. Here the organization should track these experiences and then trigger the process of raising a customer service ticket to address the specific customer concern. Once triggered, internal systems, people, and resources are aligned to execute the processes.

There are however, many other ways in which BPM along with social media can be leveraged for customer servicing such as: change request initiation; enriched customer profile data; sustained focus on operational efficiency; service request initiation (complaint/query/feedback and customer notifications) for process and filtration and prioritization of requests.

CUSTOMER RELATIONSHIP MANAGEMENT

The process of Customer Relationship Management (CRM) involves listening to customers, providing satisfactory responses (as applicable), advice, recommendations (as required), and, most importantly, continuously engaging with customers. Social media provide several opportunities for organizations to observe, listen, and communicate with their customers, creating new relationships and nurturing existing ones through meaningful dialogs. Based on the kind of interaction, several organizational processes can be initiated, modeled, created, modified, or decommissioned as needed.

The social CRM process and environment is set out in Figure 6. In this figure, we see communities interacting through social media that are subject to ongoing and active monitoring using listening tools, which are then able to pass the results into the CRM environment.

例如,在沃达丰(Vodafone)内部的一个案例中,他们发现了这样一种情况:客户订阅了预付费计划,不幸的是,账单出现了错误。该客户无法联系到沃达丰的客户服务部门,他无法就错误的账单得到满意的解释,因此感到沮丧和失望,并在Twitter上表达了他的不满,这是可以理解的。沃达丰有积极的倾听代理,他们会阅读Twitter或Facebook上的信息,并联系不满意的客户,帮助解决账单问题。正因为如此,沃达丰才能够拦截问题,并启动一个程序来恢复局面。一旦流程启动,内部系统和资源就会协同工作,以确保以确认、澄清或解决问题的形式提供响应。这种基于社交媒体的流程与内部系统的适当集成,将原本困难的体验变成了建设性的客户体验。

客户与组织的每一次互动都是一个"关键时刻":体验可能是积极的、消极的或中性的。一方面,积极的体验通常转化为重复的业务和客户忠诚度,以及几个渠道的推荐或评论。因此,组织可以启动流程,通过对话、交叉销售/向上销售其他产品和服务,进一步加强和强化客户关系。另一方面,通过这些渠道,消极的经历可能会导致负面的评论和反馈,或终止与组织的业务关系。在这里,组织应该跟踪这些经历,然后启动提高客户服务票据的流程,以解决特定的客户关注点。一旦被触发,内部系统、人员和资源将进行协调以执行流程。

然而,还有许多其他方式可以利用BPM和社交媒体为客户服务,如更改请求发起、丰富客户资料、持续注重业务效率、服务请求发起(投诉/查询/反馈和客户通知),用于处理、筛选和确定请求的优先级。

3.9.13　客户关系管理

客户关系管理(CRM)的过程包括倾听客户的意见,提供满意的回应(尽可能适用),建议,推荐(必要时),以及最重要的,持续与客户接触。社交媒体为组织提供了几个观察、倾听和与客户沟通的机会,通过有意义的对话建立新的关系并培育现有的关系。根据交互的类型,可以根据需要启动、建模、创建、修改或退役几个组织流程。

社交CRM流程和环境如图6所示。在图6中,我们看到社区通过社交媒体进行交互,这些社交媒体使用监听工具进行持续和积极的监视,然后这些工具能够将结果传递到CRM环境中。

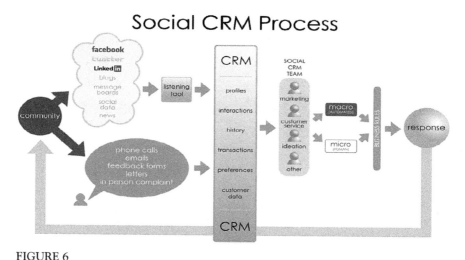

FIGURE 6

Components of social customer relationship management (CRM).

Although these processes augment the more traditional and passive channels, such as telephone, email, or letters, they also support new social media channels.

A CRM business process is among the multiple business processes in a company that could be designed from a social perspective. In "The Network-Based Business Process" by Ejub et al.,[2] the authors develop an approach to design social business processes. The three components that define a process are task, person, and machine. A task is a work unit (e.g., deliver report) that constitutes, with other tasks, a business process and that a person and/or machine execute. Execution is either manual (i.e., person only), automatic (i.e., machine only), or mixed (i.e., person and machine). Because the variety of interactions occur during the completion of business processes, it becomes possible to map some of these interactions onto specific social relations between these three components. Indeed, tasks are put together to form processes, persons collaborate on complex tasks, and machines replace each other in the case of failure. These examples offer a glimpse into the social relations that business process management systems exhibit and, hence, can be captured. Although Ejub et al. acknowledge that tasks and machines cannot "socialize" (in the strict sense), combining tasks and machines presents a lot of similarities with how people behave daily. Supporting the importance of socializing tasks and machines, Tan et al. state that "…*Currently, most social networks connect people or groups who expose similar interests or features. In the near future, we expect that such networks will connect other entities, such as software components, Web-based services, data resources, and workflows. More importantly, the interactions among people and nonhuman artifacts have significantly enhanced data scientists' productivity*".[3] Examples of social relations between business process components include coupling, interchange, delegation, and partnership. The different social relations are used for developing a configuration network of tasks, a social network of persons, and a support network of machines.

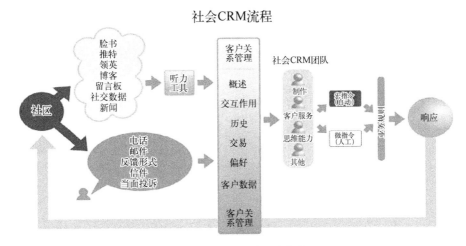

图6　社交CRM的组成部分

　　虽然这些流程增加了更传统和被动的渠道(如电话、电子邮件或信件),但它们也支持新的社交媒体渠道。

　　CRM业务流程是公司中可以从社交角度设计的多个业务流程之一。在Ejub等[2]的"基于网络的业务流程"中,作者开发了一种设计社交业务流程的方法。定义流程的三个组件是:任务、人员和机器。任务是工作单元(如递送报告),其与其他任务一起构成人和机器执行的业务流程。执行包括手动(即仅限人)、自动(即仅机器)或混合(即人和机器)方式。由于在业务流程完成期间发生了各种交互,可以将这些交互中的一些映射到这三个组件之间的特定社交关系中。实际上,任务被组合在一起形成流程,人们在复杂的任务上协作,并且在失败的情况下互相替换。这些示例提供了BPM系统展示的社交关系的一瞥,因此可以被我们总结到。虽然Ejub等承认任务和机器不能社交(严格意义上说),但任务和机器与人们每天的行为方式有很多相似之处。Tan等支持社交任务和机器的重要性,声明"……目前,大多数社交网络将那些暴露相似兴趣或特征的人或群体联系起来。在不久的将来,我们希望此类网络将连接其他实体,如软件组件、基于Web的服务、数据资源和工作流。更重要的是,人与非人工制品之间的相互作用大大提高了数据科学家的生产力[3]。"业务流程组件之间的社会关系示例包括耦合、交换、授权和伙伴关系。不同的社交关系用于开发任务的配置网络、人员的社交网络和机器的支持网络。

SOCIAL MEDIA PROCESS FLOW

When remodeling processes to align them to the chosen social media strategy, we see many organizations faced with challenges of understanding not only how to change manual processes, but also the importance of automated processes and the ability of automated learning and sophisticated decision rules within the process. The realignment of users sets up process flows, defines decision points within each flow, and connects to touch-point systems to capture events at those decision points. The **system** can then be used to automatically correlate event outcomes with the new social media channels, offers, customer attributes, and other factors. In addition to that, the social media process flow should include standard campaign measurements and reporting. Decision rules within the process can incorporate and link to multiple goals, each assigned a relative weight, and multiple business choices, each assigned a value toward reaching each goal. The system scores each choice by adding up the value it contributes to each goal, adjusted for the probability that the customer will accept that choice if offered. Users within the process of the customer interaction can also weigh goals differently for different customer segments: for example, retention might be more important for high-value customers, whereas cost reduction could be a priority for customers who are less profitable. The same goal definitions can apply to multiple decisions, reducing work, and ensuring consistency throughout the process.

We can, therefore, derive the following important aspects for success in social media process flow modeling:

1. Define specific goals for the social media process flow
2. Different flows involved, such as the response information and service flow
3. Roles involved in terms of "who does what"
4. The various services and their actual delivery of value to customers
5. The platforms that are used to enable parts of process automation
6. From a process-technology point of view, processes are automated with dedicated technology, which considers infrastructure aspects, for example, the process rule engine resides on infrastructure components and infrastructure services support the platform services

Figure 7 illustrates a basic view of such a Social Media Process Flow.

In addition to developing a social media process flow, do we recommend developing a posting interaction response? This can, among others, include some of the following considerations:

1. How can the posting response create customer value? Organizations realize that customer value creation is subject to the relationship between business processes and their resources, tasks, events, and the service they deliver.
2. Include interaction and response rules, including time frame and tone for response
   - response handling
   - tone of response
   - response content structure
   - Social media Frequently Asked Questions and expected issues and answers

3.9.14　社交媒体流程

当重构流程以使其与所选择的社交媒体策略保持一致时,我们看到许多组织不仅面临着如何更改手动流程的挑战,而且还面临着自动化流程的重要性以及流程中自动化学习和复杂决策规则的能力的挑战。用户的重新组合设置流程流,在每个流程中定义决策点,并连接到接触点系统以捕获这些决策点上的事件。然后,该系统可用于自动将事件结果与新的社交媒体渠道、报价、客户属性和其他因素关联起来。除此之外,社交媒体流程应该包括标准的活动度量和报告。流程中的决策规则可以合并并链接到多个目标(每个目标分配了相对权重)和多个业务选择(每个选择为实现每个目标分配了一个值)。系统通过将其对每个目标的贡献值相加来为每个选择打分,并根据客户接受该选择的可能性进行调整。客户交互过程中的用户还可以根据不同的客户细分来衡量目标:例如,顾客维系对高价值客户来说可能更重要,而降低成本则可能是利润较低的客户的优先考虑。相同的目标定义可以应用于多个决策,减少工作,并确保整个过程的一致性。

因此,我们可以在社交媒体流程流建模中获得以下成功的重要方面:

(1)为社交媒体流程流定义特定的目标;

(2)涉及不同的流,如响应信息和服务流;

(3)涉及"谁做什么"方面的角色;

(4)各种服务及其实际交付给客户的价值;

(5)用于支持流程自动化部分的平台;

(6)从流程技术的角度来看,流程是由专用技术自动化的,这些技术考虑基础设施方面,例如,流程规则引擎驻留在基础设施组件上,而基础设施服务支持平台服务。

图7展示了此类社交媒体流程流的基本视图。

社交媒体流程。除了开发一个社交媒体流程流,我们是否建议开发一个发布交互响应? 除其他外,这可以包括下列一些考虑。

(1)发布响应如何创造客户价值? 组织认识到,客户价值的创造取决于业务流程及其资源、任务、事件和它们提供的服务之间的关系。

(2)包括互动和回应规则,包括回应的时间框架和语气:

- 响应处理;
- 回应的语气;
- 响应内容结构;
- 社交媒体上经常出现的问题以及预期的问题和答案。

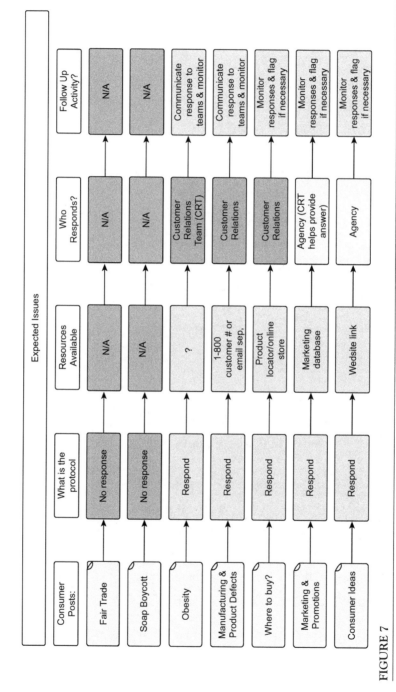

FIGURE 7

Overview of the social media process flow.

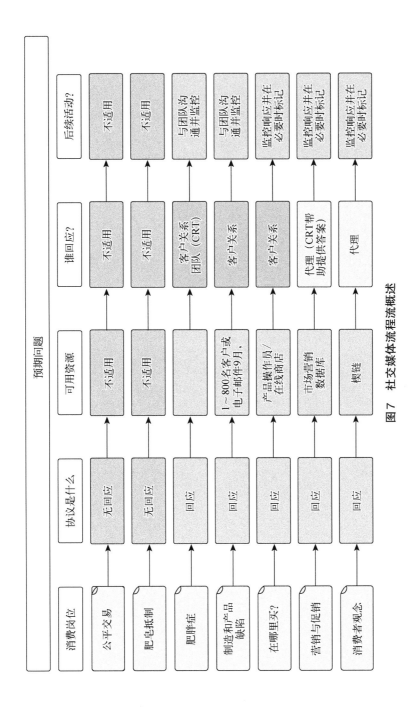

图 7 社交媒体流程流程概述

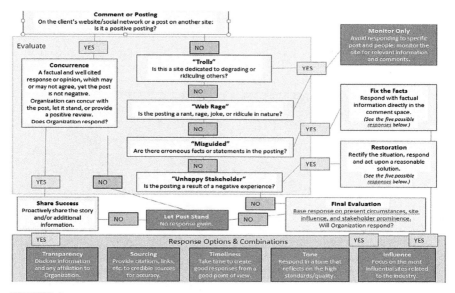

FIGURE 8

Overview of interaction response and combination.

3. Make sure the comment policy is integrated into the customer interaction processes
4. Identify which applications and data are involved in the various processes
5. Ensure process measurement and process monitoring capabilities
6. Enable the different technology media involved and used within the process
7. he various social media customer channels, for example, portal, business-to-customer, and business-to-business commerce need to be applying the same posting response

The illustration of such a posting interaction response can be seen in Figure 8:

CONCLUSION

The outperforming organizations investing in social media, both internally and externally, did this through mobile technology, cloud technology, data analytics, as well as new service constructs and knowledge workers. For most organizations, the initial promise of social media is therefore about customer interaction, customer/market data, brand, reach, profit, as well as the possibility of employee self-service and automation. The Global University Alliance analysis revealed that the promise of social media enables for most the following benefits:

- **Cost**: Content-rich social media are free, and can be accessed in an automated and repeatable way. This means that social media are less expensive than traditional customer research.

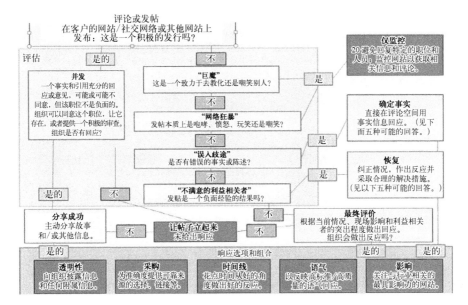

图8　交互响应和组合概述

（3）确保将评论策略集成到客户交互流程中。

（4）确定哪些应用程序和数据涉及各个流程。

（5）确保过程测量和过程监控能力。

（6）启用流程中涉及和使用的不同技术媒体。

（7）各种社交媒体客户通道，如门户、企业对客户，以及企业对企业的商业需要应用相同的发布响应。

这种发布交互响应的说明如图8所示。

3.9.15　结论

无论是内部还是外部，投资于社交媒体的优秀企业都是通过移动技术、云技术、数据分析，以及新的服务结构和知识工作者来实现这一点的。因此，对于大多数组织来说，社交媒体最初的承诺是关于客户互动、客户/市场数据、品牌、覆盖范围、利润，以及员工自助服务和自动化的可能性。全球大学联盟的分析显示，社交媒体的前景能够带来以下大部分好处。

- 成本：内容丰富的社交媒体是免费的，可以通过自动化和可重复的方式访问。这意味着社交媒体比传统的客户调查要便宜。

- **Accuracy**: Given the advances in analytics and intelligence technology, we should be able to accurately identify consumer's preferences, behaviors, and patterns, what customers are discussing, and the emotional disposition of those conversations. This means that social media will tell us more about online consumer persona behaviors, the online purchase funnel, and the customer journey, as well as create insight into the general purchase funnel, evolving social media into an automated proactive intelligence with more accurate information to enable informed decisions.
- **Relevance**: So much content is available on the web that one can find narratives on almost any topic or perspective. This means that social media enable specific views, trends, and content. Gathering and customizing data to make your information relevant and stand out is a unique opportunity.
- **Representativeness**: Given the census-like approach to collecting content from the web, any analysis should be a more accurate representation of market experiences than other methods—so extensive sampling or weighting will not be as necessary.
- **End-to-end flow**. Today, it is all about easy access and delivering on time. This requires a rethinking of how to identify, create, innovate, and manage an end-to-end services, information, data, and process flow.

Socially Oriented Process Modeling can be used effectively, internally and externally, to share solutions within an organization and to reach out to customers who talk about the business in their own social media interactions, facilitating new processes, and process changes. It can help solve critical debates about whether to automate certain processes or keep them manual. It is both a stand-alone tool in the BPM toolbox and an asset for improving existing BPM technology. Fundamentally, Socially Oriented Process Modeling is a powerful asset for a process-driven organization that will manifest itself in new and currently unforeseen ways. The organizations that embrace social media and BPM concepts will have some of the most developed process capabilities in their fields, and will be the most resilient in the face of change. However, Social Media will not achieve their full potential in an organization unless they are integrated both strategically and with operational process improvement. We have demonstrated how to apply social media and BPM to enable customer centricity. **Some of the critical success factors include:**

- Requirements needed to address the Social Media Strategy
- Processes that will be affected by the Requirements (these are likely to be new processes)
- Where new processes may affect workflow
- The preferred use of the Social Media Applications
- Information/Data that will be exchanged. This could be in the form of advertising content going to the Social Media Application or consumer feedback coming from a Social Media Application.
- Who is responsible for Social Media Application(s) that will be used and monitored to achieve the Social Media Strategy

- **准确性**：鉴于分析和智能技术的进步，我们应该能够准确识别消费者的偏好、行为和模式，消费者在讨论什么，以及这些对话的情感倾向。这意味着社交媒体将告诉我们更多关于在线消费者角色行为、在线购买漏斗和客户旅程的信息，并对一般的购买漏斗产生洞察，将社交媒体演变为自动主动智能，并提供更准确的信息以启用明智的决定。

- **相关性**：网络上有如此多的内容，您几乎可以找到任何主题或观点的叙述。这意味着社交媒体支持特定的视图、趋势和内容。收集和定制数据，使您的信息相关和脱颖而出是一个独特的机会。

- **代表性**：考虑到从网络上收集内容的普查式方法，任何分析都应该比其他方法更准确地反映市场经验，因此没有必要进行广泛的抽样或加权。

- **端到端流程**。今天，一切都是为了方便获取和按时交付。这需要重新考虑如何识别、创建、创新和管理端到端服务、信息、数据和流程流。

面向社会的流程建模可以有效地在内部和外部使用，以在组织内共享解决方案，并与在自己的社交媒体交互中讨论业务的客户联系，促进新流程和流程变更。它可以帮助解决关于是否自动执行某些流程或使其保持手动的关键性争论。它既是BPM工具箱中的独立工具，也是改进现有BPM技术的资产。从根本上说，面向社会的流程建模是流程驱动型组织的强大资产，它将以新的和目前无法预料的方式表现出来。拥抱社交媒体和BPM概念的组织将拥有其所在领域中一些最发达的流程功能，并且在面对变化时将是最具弹性的。但是，社交媒体不会在组织中充分发挥其潜力，除非它们在战略上和运营流程改进方面进行整合。我们已经演示了如何应用社交媒体和BPM来实现以客户为中心。一些关键的成功因素包括以下方面。

- 解决社交媒体战略所需的要求。
- 受需求影响的流程（这些流程很可能是新流程）。
- 新流程可能会影响工作流程。
- 社交媒体应用程序的首选用途。
- 将要交换的信息/数据。这可以是以社交媒体应用程序的广告内容或来自社交媒体应用程序的消费者反馈的形式展现。
- 谁负责社交媒体应用程序，这些应用程序将被使用和监控以实现社交媒体战略。

- Updates to the point-of-sale application to direct consumers to the organization's Social Media presence
- The types of Measurements used to determine success of any new Social Media Strategy

End Notes

1. Y. Badr and Z. Maamar, "Can Enterprises Capitalize on Their Social Networks? Cutter IT Journal," *Special Issue on Measuring the Success of Social Networks in the Enterprise*, 22 no. 10 (October 2009).
2. K. Ejub, F. Noura, M. Zakaria, L. Alfred, P. Aldina, and Z.S. Quan, "The Network-Based Business Process," *IEEE Internet Computing*, 18, no. 2 (March/April 2014).
3. W. Tan, M.B. Blake, I. Saleh, and S. Dustdsar, "Social-Network-Sourced Big Data Analytics," *IEEE Internet Computing*, 17, no. 5 (September/October 2013).

- 更新销售点应用程序，以引导消费者加入组织的社交媒体。
- 用于确定任何新社交媒体策略成功的度量类型。

参考文献

[1] Y. Badr and Z. Maamar, "Can Enterprises Capitalize on Their Social Networks? Cutter IT Journal," Special Issue on Measuring the Success of Social Networks in the Enterprise, 22 no. 10 (October 2009).

[2] K. Ejub, F. Noura, M. Zakaria, L. Alfred, P. Aldina, and Z.S. Quan, "The Network-Based Business Process," IEEE Internet Computing, 18, no. 2 (March/April 2014).

[3] W. Tan, M.B. Blake, I. Saleh, and S. Dustdsar, "Social-Network-Sourced Big Data Analytics," IEEE Internet Computing, 17, no. 5 (September/October 2013).

BPM and Maturity Models

Henrik von Scheel, Gabriella von Rosing, Krzysztof Skurzak, Maria Hove

INTRODUCTION

For organizations to perform well in a global competitive world, it is important to identify the competitive advantages from which they can benefit. Models to assess the status of one's capabilities and identify improvement opportunities, and, in particular, maturity models that can help organizations assess their current capabilities in a structured way to implement changes and improvements, have become essential. A maturity model can be described as a structured collection of elements that describe certain aspects of capability maturity in an organization. A maturity model may provide, for example:

- a situational analysis of one's capabilities
- a place to start
- the benefit of a community's prior experiences
- a common language and a shared vision
- a framework for prioritizing actions
- a way to define what improvement means for your organization.
- as a benchmark for comparison and as an aid to understanding.

In this section, we will focus on maturity models, what they are, their historic development, how they can be used, and when business process management (BPM) can use maturity concepts. This includes a detailed BPM maturity self-assessment, a benchmark among the various aspects that are related to the BPM maturity context as well as a BPM maturity development path.

The Complete Business Process Handbook. http://dx.doi.org/10.1016/B978-0-12-799959-3.00019-7

3.10 BPM和成熟度模型

Henrik von Scheel, Gabriella von Rosing, Krzysztof Skurzak, Maria Hove

3.10.1 介绍

要使组织在激烈的全球竞争中表现卓越,重要的是要确定他们的竞争优势。评估一个组织的能力状态并确定改进方面的模型,尤其是能够对帮助组织以结构化方式评估其当前能力以实施变更和改进的成熟度模型已变得至关重要。成熟度模型可以定义为描述组织中能力成熟度的特定方面的结构化集合。成熟度模型可以提供如下方面的信息:

- 对能力的情境分析;
- 一个开始的地方;
- 社区以往的先进经验;
- 共同语言和共同愿景;
- 优先行动的框架;
- 定义您组织的改进方法;
- 作为比较的基准和帮助理解。

在本节中,我们将关注成熟度模型是什么、它们的发展历史、如何使用它们以及BPM何时可以使用成熟度概念。这包括一个详细的BPM成熟度自我评估,一个与BPM成熟度语义以及BPM成熟度开发路径相关的各个方面的标准。

HISTORIC DEVELOPMENT OF MATURITY MODELS

Maturity models are used in multiple areas ranging from Software,[1] Organizational Project Management Maturity,[2] People Capability Maturity Model,[3] Portfolio, Programme and Project Management Maturity,[4] to concepts like E-learning Maturity.[5] Maturity Models have existed for close to 40 years and are therefore not a new way of evaluating the maturity level of a business. Although maturity models for the most part are attributed to Carnegie Mellon University Software Engineering Institute[6] this is not really correct. The first published maturity model was developed by Richard L. Nolan, who in 1973 published the Stages of Growth model for IT organizations. It didn't take more than 6 years for Philip B. Crosby to publish his 1979 book, Quality is Free,[7] the Quality Management Maturity Grid (QMMG), which is an organizational maturity matrix. The QMMG is used by a business or organization as a benchmark of how mature are their processes, and how well they are embedded in their culture, with respect to service or product quality management. The staged structure of the framework is based on total quality management (TQM) principles that have existed for nearly a century. The work of Frederick Taylor and Frank Gilbreth on "scientific management" and time andmotion studies in the early 1900s eventually led to the new discipline of industrial engineering.[8] In the 1930s, Walter Shewhart, a physicist at AT&T Bell Laboratories, established the principles of statistical quality control. These principles were further developed and successfully demonstrated in the work of such authorities as W. Edwards Deming[9] (1986) and Joseph M. Juran[10] (1988).

In recent years, the TQM concepts have been extended from manufacturing processes to service and engineering design processes. The software process[11] can be defined as a set of activities, methods, practices, and trans formations that people use to develop and maintain software and the associated products. As an organization matures, the software process[12] becomes better defined and more consistently implemented throughout the organization. This, in turn, leads to higher-quality software, increased productivity, less rework, and improved software project plans and management. Crosby describes five evolutionary stages in adopting quality practices. As seen in Table 1, the quality management maturity grid applies five stages to six measurement categories in subjectively rating an organization's quality operation.

The five stages of the QMMG are:

3.10.2　成熟度模型的历史发展

成熟度模型用于多个领域,从软件[1]、组织项目管理成熟度[2]、人员能力成熟度模型[3]、投资组合、项目和项目管理成熟度[4],到在线学习成熟度等概念[5]。成熟度模型已存在接近40年。因此,这不是评估企业成熟度的新方法。尽管成熟度模型大部分都归功于卡内基梅隆大学软件工程研究所(Software Engineering Institute, SEI)[6],但这并不正确。第一个公布的成熟度模型由R. L. Nolan开发,他于1973年发布了IT组织的增长阶段模型。P. B. Crosby花了不到6年的时间于1979年出版了他的著作《质量是免费的》[7],他论述了质量管理成熟度网格(quality management maturity grid, QMMG)是一个组织成熟度矩阵。在服务或产品质量管理方面,QMMG被企业或组织用作衡量其流程是否成熟,以及它们在企业文化中的嵌入程度的基准。该框架的阶段结构基于已存在近一个世纪的TQM原则。F. Taylor和F. B. Gilbreth在20世纪初对科学管理以及工时和动作的研究,最终催生了工业工程这一新的学科[8]。20世纪30年代,AT&T贝尔实验室的物理学家W. Shewhart建立了"统计质量控制"的原则。这些原则在诸如W. E. Deming[9](1986年)和J. M. Juran[10](1988年)等权威人士的工作中得到了进一步的发展和成功的证明。

近年来,TQM的概念已经从制造过程扩展到服务和工程设计过程。软件流程[11]可以定义为一组活动、方法、实践和转换,人们使用这些活动、方法、实践和转换来开发和维护软件和相关产品。随着组织的成熟,软件流程[12]在整个组织中得到更好的定义和更一致的实现。这进而导致更高质量的软件,提高生产力、减少返工,以及改进软件项目计划和管理。P. B. Crosby描述了采用质量实践的5个演进阶段。如表1所示,QMMG将5个阶段应用于6个度量类别,用于对组织的质量运营进行主观评级。

QMMG的5个阶段如下。

Table 1 *The Quality Management Maturity Grid (QMMG)*

	Stage 1: Uncertainty	Stage 2: Awakening	Stage 3: Enlightenment	Stage 4: Wisdom	Stage 5: Certainty
Management understanding and attitude	No comprehension of quality as a management tool. Tend to blame quality department for "quality problems."	Recognizing that quality management may be of value but not willing to provide money or time to make it all happen.	While going through quality improvement program, learning more about quality management; becoming supportive and helpful.	Participating. Understand absolutes of quality management. Recognize their personal role in continuing emphasis.	Consider quality management as an essential part of company system.
Quality organization status	Quality is hidden in manufacturing or engineering departments. Inspection probably not part of organization. Emphasis on appraisal and sorting.	A stronger quality leader is appointed but main emphasis is still on appraisal and moving the product. Still part of manufacturing or other.	Quality department reports to top management, all appraisals are incorporated and manager has role in management of company.	Quality manager is an officer of company; effective status reporting and preventive action. Involved with customer affairs and special assignments.	Quality manager on board of directors. Prevention is main concern. Quality is a thought leader.
Problem handling	Problems are fought as they occur; no resolution; inadequate definition; lots of yelling and accusations.	Teams are set up to attack major problems. Long-range solutions are not solicited.	Corrective action communication established. Problems are faced openly and resolved in an orderly way.	Problems are identified early in their development. All functions are open to suggestion and improvement.	Except in the most unusual cases, problems are prevented.
Cost of quality as % of sales	Reported: unknown Actual: 20%	Reported: 3% Actual: 18%	Reported: 8% Actual: 12%	Reported: 6.5% Actual: 8%	Reported: 2.5% Actual: 2.5%
Quality improvement actions	No organized activities. No understanding of such activities.	Trying obvious "motivational" short-range efforts.	Implementation of a multi-step program (e.g., Crosby's 14-step) with thorough understanding and establishment of each step.	Continuing the multi-step program and starting other pro-active/preventive product quality initiatives.	Quality improvement is a normal and continued activity.
Summary of company quality posture	"We don't know why we have problems with quality."	"Is it absolutely necessary to always have problems with quality?"	"Through management commitment and quality improvement we are identifying and resolving our problems."	"Defect prevention is a routine part of our operation."	"We know why we do not have problems with quality."

表1　QMMG

	阶段1：不确定性	阶段2：觉醒	阶段3：启蒙活动	阶段4：智慧	阶段5：确定
管理理解与态度	没有理解质量作为一种管理工具。倾向于将质量问题归咎于质量部门	认识到质量管理可能是有价值的，但不愿意提供金钱或时间来实现这一切	通过质量改进项目，学习了更多的质量管理知识，变得支持和乐于助人	参加。理解质量管理的绝对原则。确认他们在持续强调中的个人角色	把质量管理作为公司制度的重要组成部分
质量组织状态	质量隐藏在制造或工程部门。检查不是组织的一部分。重视评估和分类	任命了一位更高素质的领导者，但主要重点仍然是评估和推动制造业其他部分	质量部门向高层管理人员报告，所有评估人员都纳入，经理在公司管理中发挥作用	质量经理是公司的一名官员，有效的状态报告和预防措施。涉及客户事务和特殊任务	质量经理是董事会成员。预防是主要问题。质量是思想的领导者
问题处理	问题随时会发生；没有决议；定义不足；很多大喊大叫和指责	组建团队来主要解决问题。不征求长期解决办法	建立纠正措施沟通。公开地面对问题，有序地解决问题	在开发的早期就发现了问题。欢迎所有职能部门提出建议和改进	除了在最常见的情况下，问题是可以避免的
质量成本占销售额的百分比	报道：未知 实际：20%	报道：3% 实际：18%	报道：8% 实际：12%	报道：6.5% 实际：8%	报道：2.5% 实际：2.5%
质量改进措施	缺乏有组织的活动。不了解此类活动	尝试明显的激励式的短期努力	实施一个多步骤的计划（如 P. B. Crosby 的14步），全面了解和建立每个步骤	继续多步骤计划，并开始其他主动/预防产品质量倡议	质量改进是一项正常和持续的活动
公司质量态势总结	"我们不知道为什么质量有问题。"	"总有质量问题对出现的吗？"	"通过管理承诺和质量改进，我们正在识别和解决我们的问题。"	"缺陷预防是我们日常工作的一部分。"	"我们知道为什么我们在质量上没有问题。"

The QMMG is credited with being the precursor of all maturity models. In August 1986, the Software Engineering Institute (SEI) at Carnegie Mellon University, with assistance from the MITRE Corporation, began developing a process maturity framework that would help organizations improve their software processes. This effort was initiated in response to a request to provide the federal government with a method for assessing the capability of their software contractors. In June 1987, the SEI released a brief description of the software process maturity[13] framework and, in September 1987, a preliminary maturity questionnaire. Based on experience in using

Subject and Reference	Approach
Quality Management Maturity Grid (Crosby, 1979)	Grid, 6 issues, detailed description at each level
R&D Effectiveness Audit (Szakoryi, 1994)	Grid, 10 issues, detailed description at each level
Quality Management Process Maturity Grid (Crosby, 1996)	Grid, 5 issues, captions describing performance at each level
Technical Innovation Audit (Chiesa and others, 1996)	Grid, 8 areas, 23 issues, detailed descriptions at each level
Product & Cycle Time Excellence (McGrath, 1996)	Grid, 10 issues, detailed description at each level
Design Maturity Model (Fraser & Moultrie, 192001)	Grid, 5 areas, 21 issues, detailed descriptions and captions
Product & Cycle Time Excellence - Mark 2 (McGrath, 2002)	Grid, Revision of earlier model
Collaboration Maturity Model (Fraser & Gregory, 2002)	Grid, 7 issues, detailed descriptions and captions
Design Atlas - Design Capability (Design Council, 2002)	Grid, 5 areas, 15 issues, detailed descriptions at each level
Supplier Relationships (Macbeth & Ferguson, 1994)	Grid / Likert Hybrid, 9 issues, brief descriptions at 3 levels plus 7 point scale
Continuous Improvement in NPD (Caffyn, 1997)	Global levels defined, 6 core abilities, 10 key behaviors
ISO 9004 (EN ISO 9004, 2000)	Global levels defined, 5 questions, 11 issues
Project Management Maturity (Dooley and others, 2001)	Likert style questionnaire, 15 areas, 85 issues, no descriptions of performance
Software CMM - Staged Maturity Levels (Pauk and others, 1993)	CMM Style
Agility (change proficiency) Maturity Model (Dove, 1996)	CMM Style
Usability - Human Factors Maturity (Earthy, 1998)	CMM Style
CMMI - Continuous Capability Levels (Shrum, 2000)	CMM Style
Free (collaboration) Capability Assessment Framework (Wognum & Faber, 2000)	CMM Style

FIGURE 1

Overview of various maturity model concepts.[14]

　　QMMG被认为是所有成熟度模型的先驱。1986年8月,卡内基梅隆大学SEI在MITRE公司的帮助下,开始开发一个流程成熟度框架,帮助改进他们组织的软件流程。这项工作始于应要求向联邦政府提供一种评估其软件承包商能力的方法。1987年6月,SEI发布了软件流程成熟度[13]框架的简要方案描述,并于1987年9月发布了初步成熟度问卷。

主题和参考	方法
质量管理成熟度方格 (Crosby, 1979)	方格, 6个问题, 每个级别的 详细描述
研发有效性审计 (Szakory, 1994)	方格, 10个问题, 每个级别的 详细描述
质量管理流程成熟度方格 (Crosby, 1996)	方格, 5个问题, 描述每个 级别的绩效的说明
技术创新审计 (Chiesa等人, 1996)	方格, 8个区域, 23个问题, 每个级别的详细描述
产品和周期卓越 (McGrath, 1996)	方格, 10个问题, 每个级别的 详细描述
设计成熟度模型 (Fraser & Moultrie, 192001)	方格, 5个区域, 21个问题, 详细描述和说明
产品和周期时间卓越——Mark 2 (McGrath, 2002)	方格, 早期模型的修订
协作成熟度模型 (Fraser和Gregory, 2002)	方格, 7个问题, 详细描述 和说明
设计图集——设计能力 (设计委员会, 2002年)	方格, 5个区域, 15个问题, 每个级别的详细描述
供应商关系 (Macbeth和Ferguson, 1994)	方格/Likert Hybrid, 9个问题, 3级加7点比例的简要描述
NPD持续改善 (Caffyn, 1997)	全球级别定义, 6个核心能力, 10个关键行为
国际标准化组织9004 (EN ISO 9004, 2000)	全球级别定义, 5个问题, 11个议题
项目管理成熟度 (Dooley等人, 2001)	风格问卷, 15个领域, 85个 问题, 无绩效描述
软件CMM——阶段性成熟度水平 (Pauk等人, 1993年)	能力成熟度模型样式
敏捷(变更熟练程度)成熟度模型 (Dove, 1996)	能力成熟度模型样式
可用性——人为因素成熟度 (Earthy, 1998)	能力成熟度模型样式
CMMI-连续能力水平 (Shrum, 2000)	能力成熟度模型样式
免费(协作)能力评估框架 (Wognum & Faber, 2000)	能力成熟度模型样式

图1　各种成熟度模型概念的概述[14]

the software process maturity framework and the maturity questionnaire for diagnosing problems and improving processes, the SEI[15] formalized the concepts as the Capability Maturity Model for Software[16] (Software CMM[17]). Version 1.0[18] of the model was published in 1991.[19] Version 1.1[20] was released in 1993.[21] The Software CMM[22] was then retired in favor of the CMM Integration (CMMI[23]) model. CMMI was developed by the CMMI project, which aimed to improve the usability of maturity models by integrating three different models into one framework.

The project consisted of members of industry, government, and the Carnegie Mellon Software Engineering Institute[24] (SEI). The main sponsors included the Office of the Secretary of Defense (OSD) and the National Defense Industrial Association.[25] CMMI currently addresses three areas of process interest:

- Development—addresses product and service development
- Acquisition[26]—addresses supply chain management, acquisition, and outsourcing
- Services[27]—addresses guidance for delivering services.

However, as shown in Figure 1, whereas the CMM/CMMI evolved and matured, so did many of the other maturity model approaches; e.g., Agility, Usability of Human Factors, as well as Continuous Capability Levels and Free (collaboration) Capability Assessment maturity models emerged.

From the described Quality Management Maturity Grid from Crosby emerged not only the maturity models but numerable other Grid approaches, such as Research and Development, Product Cycle,[28] Continuous Improvement levels and approaches, as well as Project Management Maturity. As it many times happens one model and framework inspires the work and content of another standards and frameworks in related engineering and/or management areas and disciplines.

THE DIFFERENT STAGES OF MATURITY MODELS

In the software process maturity framework,[29] Humphrey identified five maturity levels that, even though they are based on the idea of Crosby, are claimed to describe successive foundations for process improvement and defined an ordinal scale for measuring the maturity of an organization's software processes. The descried concepts underlying maturity levels have remained stable through the evolution of the Software CMM. In discussions of this early work, Bill Curtis, Humphrey's[30] successor as director of the Process Program, identifies the focus on identifying and managing project commitments and managing to a plan as one of the few differences between maturity models and Crosby's maturity grid. It also reflects Beer, Eisenstat, and Spector's (1990) observation that senior managers create a climate for change in successful change programs, but this change needs to start at the grass roots level rather than top-down.

The general idea with the maturity or grids levels is to provide possible improvement priorities or define levels of possible development—guidance for selecting levels of improvement activities:

- At **Level 1**, the initial level, the stage is typically characterized as ad hoc, not recognized, informal, uncertainty, occasionally even chaotic, and no formal

基于使用软件流程成熟度框架的经验和用于诊断问题和改进流程的成熟度问卷,SEI[15]将概念形式化为software[16](Software CMM[17])的能力成熟度模型。该模型的1.0版[18]于1991年[19]发布,1.1版[20]于1993年[21]发布,软件CMM[22]随后退役,采用CMM集成(capability maturity model integration, CMMI[23])模型(软件能力成熟度集成模型)。CMMI是由CMMI项目开发的,该项目旨在通过将三个不同的模型集成到一个框架中来提高成熟度模型的可用性。

该项目由工业界、政府和卡内基梅隆大学SEI[24]的成员组成。主要赞助商包括国防部长办公室(Office of the Secretary of Defense, OSD)和国防工业协会(National Defense Industrial Association)[25]。CMMI目前涉及以下三个过程兴趣领域。

- 开发:解决产品和服务开发问题;
- 获得[26]:解决供应链管理、收购和外包问题;
- 服务[27]:提供服务提供指导。

然而,如图1所示,虽然CMM/CMMI发展和成熟,但许多其他成熟度模型方法也是如此,例如,敏捷性、人为因素的可用性,以及持续能力水平和自由(协作)能力评估成熟度模型的出现。

来自P. B. Crosby描述的QMMG不仅出现了成熟度模型,还出现了其他网格方法,如研究与开发、产品周期[28]、持续改进水平和方法,以及项目管理成熟度。因为很多时候,一个模型和框架激发了相关工程、管理领域和学科中另一个标准和框架的工作和内容。

3.10.3　成熟度模型的不同阶段

在软件过程成熟度框架[29]中,W. S. Humphrey确定了5个成熟度级别,即使它们基于P. B. Crosby的想法,也声称描述了流程改进的连续基础,并定义了用于衡量组织软件过程成熟度的序数量表。成熟度水平的隐含概念通过软件CMM的发展保持稳定。在讨论这一早期工作时,作为流程计划主管的Humphrey[30]的继任者B. Curtis将重点放在识别和管理项目承诺以及将计划作为成熟度模型与P. B. Crosby成熟度网格之间的少数差异之一。它还反映了Beer(比尔)、Eisenstat(艾森斯塔特)和Spector(斯佩克特)(1990年)的观察结果,即高级管理人员为成功的变革计划创造了变革的气氛,但这种变化需要从基层开始,而不是自上而下。

成熟度或网格级别的一般思想是提供优先级的改进,或定义可能的开发指导级别,以选择改进活动的级别。

- 在第1级,即初始级,阶段的典型特征是临时的、未被识别的、非正式的、不确定的,有时甚至是混乱的,没有正式的方法。很少有明确的活动,成功取

approach. Few activities are defined, and success depends on individual effort and heroics. The challenge with the first stage activities is that it is difficult to predict performance and value realization or learn from experience when everything is new and unique. In nearly all the maturity or grid approaches, the first level is more defined by the failure to satisfy the requirements for Level 2.

- At **Level 2**, which is more the repeatable level, basic, initial efforts, regression, and repeatable activities are established to track cost, schedule, and functionality. The necessary process discipline is in place to repeat earlier successes on projects with similar experience. The focus at Level 2 does not explicitly include operational activities, because the major problems Level 1 organizations face are for the most part managerial, not operational, problems. Operational activities are planned and tracked at Level 2, but they are not described in detail—or even listed in most versions of the different models.
- At **Level 3**, the awakening and defined level, both strategic (management) and operational activities are documented, standardized, and integrated into a set of standard competencies for the organization. Programs, portfolio, and projects use an approved, tailored version of the organization's set of standard approaches, methods, and processes. The operational processes are first explicitly addressed at Level 3, but they must be implemented at Level 1 if the organization is, for example, developing a product, creating quality management, or building software, even if those engineering processes are informal, ad hoc, and inconsistently performed. The emphasis of Level 3, however, is more centered on organizational learning via competency and process definition and improvement.
- At **Level 4**, the wisdom, enlightenment, excellence, improvement integrated, and/or managed level, detailed measures of the process, and product quality are collected. Both the competencies and activities, and thereby process and products, are quantitatively understood and controlled. This implies statistical thinking[31] and evidence-based management,[32] although these terms were not used in the early formulations of the different models. It also should be noted that measurement and analysis could occur at all levels of the models, although it comes to the forefront in Levels 4 and 5.
- At **Level 5**, the certain, collaborative, enterprise-wide integration, continuous improvement, culturally embedded, best in class, mastered as well as institutionalized or optimized level, should be enabled by feedback from the competencies, activities, and processes, and from piloting innovative ideas and technologies. Applying statistical and analytical thinking enables the organization to understand their competencies as well as their process and activities and confirm when measurably significant differences occur in performance.

As shown in Figure 2, the basic level approaches are all based upon and further developed from the one grid approach developed by Crosby. Most of them have five levels/stages, and the ones that have less or even more have split some of the stages or joined them, but the biggest difference is the focus of the subjects

决于个人的努力和英雄主义。第1级活动的挑战是：当一切都是新的和独特的时候，很难预测性能和价值实现或从经验中学习。在几乎所有的成熟度或网格方法中，第1级的定义更多的是由于未能满足第2级的需求。

- 在第2级，即可重复级别上，建立基本的、初始的工作、回归和可重复的活动来跟踪成本、进度和功能。必要的流程规程用于在具有类似经验的项目上重复早期的成功。第2级的重点没有明确包括业务活动，因为第1级组织面临的主要问题大部分是管理问题，而不是业务问题。业务活动在第2级进行规划和跟踪，但是它们没有详细描述，甚至没有在不同模型的大多数版本中列出。

- 在第3级，唤醒和定义级别，战略（管理）和业务活动都被文档化、标准化，并集成到组织的一组标准能力中。程序、投资组合和项目使用组织的标准方法、方法和过程集的经过批准的、定制的版本。操作过程首先在第3级显式地处理，但是如果组织正在开发产品、创建质量管理或构建软件，则必须在第1级实现它们，即使这些工程过程是非正式的、临时的和执行不一致的。然而，第3级的重点更集中于通过能力和流程定义和改进的组织学习。

- 在第4级，收集智慧、启蒙、卓越、改进集成或管理级别、流程的详细度量和产品质量。能力和活动以及流程和产品都是定量理解和控制的，这意味着统计思维[31]和基于证据的管理[32]（尽管这些术语并未用于不同模型的早期表述中）。还应该注意的是：测量和分析可能发生在所有级别的模型中，尽管它在第4级和第5级中处于最前沿。

- 在第5级，特定的、协作的、企业范围的集成、持续改进、文化嵌入、最佳级别、精通的、制度化的或优化的级别，应该通过能力、活动和流程的反馈，以及创新理念和技术的试点来实现。组织应用统计思维[31]和分析思维[32]能够了解他们的能力以及他们的流程和活动，并确认在绩效方面出现明显差异时的情况。

如图2所示，基本级方法都是基于P. B. Crosby开发的一种网格方法并进一步开发的。他们中的大多数都有5个级别/阶段，而那些拥有更少甚至更多的阶段已经分裂了一些阶段或加入了它们，但最大的区别是主题和领域的重点，如质量管

and areas, e.g., Quality Management, R&D Effectiveness, Technical Innovation, Excellence, Design, Change, Project Management, and/or Relationship Management.

Subject and Reference	Maturity Levels						Approach
Quality Management Maturity Grid (Crosby, 1979)	Level 1 Uncertainty	Level 2 Awakening	Level 3 Enlightenment	Level 4 Wisdom	Level5 Certainty		Grid 6 issues, detailed description at each level
R&D Effectiveness Audit (Szakonyi, 1994)	Level A Not recognised	Level B Initial efforts	Level C Skills	Level D Methods	Level E Responsibilities	Level F Continuous Improvement	Grid 10 issues, detailed description at each level
Quality Management Process Maturity Grid (Crosby, 1996)	Level 1 Uncertainty	Level 2 Regression	Level 3 Awakening	Level 4 Enlightenment	Level 5 Certainty		Grid 5 issues, captions describing performance at each level
Technical Innovation Audit (Chiesa and others, 1996)	1	2	3	4			Grid 8 areas, 23 issues, detailed descriptions at each level
Product & Cycle Time Excellence (McGrath, 1996)	Stage 0 Informal	Stage 1 Functionally focused project managed	Stage 2 Cross functional project management	Stage 3 Enterprise wide integration of prod. dev.			Grid 10 issues, detailed description at each level
Design Maturity Model (Fraser & Moultrie, 192001)	Level 1 None	Level 2 Partial	Level 3 Formal	Level 4 Culturally embedded			Grid 5 areas, 21 issues, detailed descriptions and captions
Product & Cycle Time Excellence - Mark 2 (McGrath, 2002)	Stage 0 Informal Management	Stage 1 Functional Excellence	Stage 2 Project Excellence	Stage 3 Portfolio Excellence	Stage 4 Collaborative		Grid Revision of earlier model
Collaboration Maturity Model (Fraser & Gregory, 2002)	Level 1 None	Level 2 Partial	Level 3 Formal	Level 4 Culturally embedded			Grid 7 issues, detailed descriptions and captions
Design Atlas - Design Capability (Design Council, 2002)	Level 1	Level 3	Level 4	Level 5			Grid 5 areas, 15 issues, detailed descriptions at each level
Supplier Relationships (Macbeth & Ferguson, 1994)	Level 1 Adversarial	Level 2 Transitional	Level 3 Partnership				Grid / Likert Hybrid 9 issues, brief descriptions at 3 levels plus 7 point scale
Continuous Improvement in NPD (Caffyn, 1997)	Level 1 Natural or background CI	Level 2 Structured CI	Level 3 Goal oriented CI	Level 4 Proactive, autonomous CI	Level 5 Full CI capability		Global levels defined 6 core abilities 10 key behaviors
ISO 9004 (EN ISO 9004, 2000)	Level 1 No formal approach	Level 2 Reactive approach	Level 3 Stable formal system approach	Level 4 Cont. improvement emphasized	Level 5 Best in class performance		Global levels defined 5 questions, 11 issues
Project Management Maturity (Dooley and others, 2001)	1	2	3	4	5		Likert style questionnaire 15 areas, 85 issues, no descriptions of performance
Software CMM - Staged Maturity Levels (Pauk and others, 1993)	Level 1 Initial	Level 2 Repeatable	Level 3 Defined	Level 4 Managed	Level5 Optimizing		CMM Style
Agility (change proficiency) Maturity Model (Dove, 1996)	Level 1 Accidental	Level 2 Repeatable	Level 3 Defined	Level 4 Managed	Level5 Mastered		CMM Style
Usability - Human Factors Maturity (Earthy, 1998)	Level X Unrecognised	Level A Recognised	Level B Considered	Level C Implemented	Level D Integrated	Level E Institutionalized	CMM Style
CMMI - Continuous Capability Levels (Shrum, 2000)	Level 0 Not performed	Level 1 Performed	Level 2 Managed	Level 3 Defined	Level 4 Qualitatively Managed	Level 5 Optimizing	CMM Style
Free (collaboration) Capability Assessment Framework (Wognum & Faber, 2000)	Level 2 Repeatable	Level 3 Defined	Level 4 Managed	Level 5 Optimizing			CMM Style

FIGURE 2

Historic development of the maturity grids and models.[33]

THE MISSING PARTS OF THE MATURITY MODELS

Although the adoption rate of the mentioned models is high, the one that is most developed and adopted is the discussed CMM and then CMMI The last years, however, have seen the CMMI approach/models heavily criticized both in theory[34] as well as in practice. In the following, we summarize the criticism, which is in six main areas:

1. The CMM/CMMI model is based on the experiences of large government contractors and of Watts Humprey's own experience in the mainframe world. It does not represent the successful experiences of many software companies

理、研发效率、技术创新、卓越、设计、变革、项目管理和/或关系管理。

主题和参考	成熟度等级						方法
质量管理成熟度方格 (Crosby, 1979)	1级 不确定性	2级 唤醒	3级 环境	4级 智慧	5级 确定性		方格，6个问题，每个级别的详细描述
研发有效性审计 (Szakory, 1994)	A级 未识别	B级 初始效应	C级 技能	D级 方法	E级 责任	F级 持续改进	方格，10个问题，每个级别的详细描述
质量管理流程成熟度方格 (Crosby, 1996)	1级 不确定性	2级 回归	3级 唤醒	4级 启示	5级 确定性		方格，5个问题，描述每个级别的绩效的说明
技术创新审计 (Chiesa等人, 1996)	1	2	3	4			方格，8个区域，23个问题，每个级别的详细描述
产品和周期卓越 (McGrath, 1996)	阶段0 非正式	阶段1以功能为中心的项目管理	阶段2跨职能项目管理	阶段3生产开发的企业范围一体化			方格，10个问题，每个级别的详细描述
设计成熟度模型 (Fraser & Moultrie, 192001)	1级无	2级 部分	3级 正式	4级 文化嵌入			方格，5个区域，21个问题，详细描述和说明
产品和周期时间卓越 Mark-2 (McGrath, 2002)	0阶段 非正式管理	1阶段 功能卓越	2阶段 项目卓越	3阶段产品 组合卓越	4阶段协作		方格，早期模型的修订
协作成熟度模型 (Fraser和Gregory, 2002)	1级无	2级 部分	3级 正式	4级 文化嵌入			方格，7个问题，详细描述和说明
设计图集——设计能力 (设计委员会, 2002年)	1级	3级	4级	5级			方格，5个区域，15个问题，每个级别的详细描述
供应商关系 (Macbeth和Ferguson, 1994)	1级敌对	2级过渡	3级合作关系				方格/Likert Hybrid，9个问题，3级加9点比例的简要描述
NPD持续改善 (Caffyn, 1997)	1级自然或 背景CI	2级结构 化CI	3级目标 导向CI	4级主动、 自主的CI	5级完整 CI能力		全球级定义，6个核心能力，10个关键行为
国际标准化组织9004 (EN ISO 9004, 2000)	1级无 正式方法	2级反 应式方法	3级稳定形 式系统方法	4级强调 持续改进	5级 最佳绩效		全球级定义，5个问题，11个议题
项目管理成熟度 (Dooley等人, 2001)	1	2	3	4	5		Likert风格问卷，15个领域，85个问题，无绩效描述
软件CMM——阶段性成熟度水平 (Paulk等人, 1993年)	1级 初始	2级 可重复	3级 定义	4级 管理	5级 优化		能力成熟度模型样式
敏捷(变更熟练程度)成熟度模型(Dove, 1996)	1级 意外	2级 可重复	3级 定义	4级 管理	5级 掌握		能力成熟度模型样式
可用性——人为因素成熟度(Earthy, 1998)	X级 未识别	A级 认可	B级 考虑	C级 实施	D级 综合	E级 制度化	能力成熟度模型样式
CMMI-连续能力水平 (Shrum, 2000)	0级 未执行	1级 执行	2级 管理	3级 定义	4级 质量管理	5级 优化	能力成熟度模型样式
免费(协作)能力评估框架(Wognum&Faber, 2000)	2级 可重复	3级 定义	4级 管理	5级 优化			能力成熟度模型样式

图2   成熟度网格和模型的历史发展[33]

3.10.4   成熟度模型的缺失部分

尽管所提到的模型的采用率很高，但是最早开发和采用的模型是讨论的 CMM，然后是CMMI。然而，最近几年，CMMI方法/模型在理论[34]和实践中都受到严厉批评。在下文中，我们总结了6个主要领域的批评：

（1）CMM/CMMI模型基于大型政府承包商的经验以及 W. Humprey（以"软件质量之父"闻名）在大型机领域的经验，它并不代表许多软件公司的成功经验。

that, as a matter of fact, would be judged to be a "Level 1" organization by the CMM/CMMI levels. For example, the CMM or CMMI for software development[35] was arguably irrelevant to successful software development and therefore criticized for the applicability of the narrow capability view. For some of the most successful software companies like Microsoft, IBM, Apple, Oracle, Google, Softbank, SAP, CSC, Yahoo, Software AG, and Symantec. Though these companies may have successfully developed their software, they would not necessarily have considered or defined or managed their processes as the CMM/CMMI described as level three or above, and so would have fit Level 1 or 2 of the model.[36] This did not change the successful development of their software. As CMM/CMMI is not built on empirical research, but rather is built on experience, the experience/best practice would somehow have to build on the industry leaders to be a foundation of best practice standardization, which the CMM/CMMI is not.

2. CMMI ignores the importance of people involved with the process by assuming that processes can somehow render individual excellence less important. For this to be the case, people/team tasks would somehow have to be included in the process itself, which the CMMI does not address.

3. CMMI does not effectively describe any information on process dynamics, which confuses the study of the relationships between practices and levels within the CMMI. The CMMI does not perceive or adapt to the conditions of the combined capabilities of an organization. Arguably, most, and perhaps all, of the key practices of the CMMI at its various levels could be performed usefully at Level 1, depending on the particular dynamics of an organization. Instead of modeling these process capability dynamics, the CMMI merely satisfies them.

4. CMMI's focus is only on process capability, which is only one side of the coin, for a company can not separate one's capabilities from the relationship of another related capability that are connected. Therefore, a company should not only look at its capability maturity model of one area, but rather look at its related Enterprise Maturity. However, CMMI does not address this.

5. CMMI reveres the institutionalization of process for its own sake. This guarantees nothing, and in some cases, the institutionalization of processes may lead to oversimplified public processes, ignoring the actual successful practice of the organization. For one cannot look at a process in itself, without taking into consideration which other capabilities are attached to the process and activity. To consider which other capabilities are attached to the process and its activity, other capability maturity models would have to be interlinked with and measure the process capabilities, which the CMMI does not address. Therefore, a process maturity model would have to consider the related aspects to the process, which gives it context. This includes the purpose and goal, the organizational context (competencies

事实上，它们将被CMM/CMMI级别判定为"一级"组织。例如，用于软件开发的CMM或CMMI[35]可以说与成功的软件开发无关，因此批评了狭隘的功能视图的适用性。对于一些最成功的软件公司，如微软、IBM、Apple、甲骨文、谷歌、软银、SAP、CSC、雅虎、AG软件公司和赛门铁克，虽然这些公司可能已经成功开发了他们的软件，但他们不一定会考虑或定义或管理他们的流程，因为CMM/CMMI被描述为第3级或更高级别，所以适合模型的第1级或第2级，这并没有改变他们的软件的成功开发。由于CMM/CMMI不是建立在实证研究之上的，而是建立在经验之上的，经验/最佳实践必须以某种方式建立在行业领导者之上，才能成为最佳实践标准化的基础，而不是CMM/CMMI。

（2）CMMI忽略了参与这个流程的人的重要性，因为它假定流程可以以某种方式使个人的卓越性变得不那么重要。要做到这一点，人员/团队任务必须以某种方式包含在流程本身中，而CMMI并没有解决这个问题。

（3）CMMI没有有效的描述任何有关流程动态的信息，这混淆了对CMMI中实践和级别之间关系的研究。CMMI不能感知或适应组织综合能力的条件。可以说，CMMI在其不同级别的大部分（也许是全部）关键实践都可以在第1级上有效地执行，这取决于组织的特定动态。CMMI并没有对这些过程能力动态进行建模，只是满足了它们。

（4）CMMI的重点仅仅是流程能力，这只是硬币的一面，因为一个公司不能将一个人的能力与另一个相关能力的关系分开。因此，一个公司不仅应该着眼于其一个领域的能力成熟度模型，而且应该着眼于其相关的企业成熟度。然而，CMMI并没有解决这个问题。

（5）CMMI尊重流程本身的制度化。这不能保证任何东西，而且在某些情况下，流程的制度化可能导致公共流程过于简化而忽略了组织的实际成功实践。因为如果不考虑附加到流程和活动的其他哪些功能，就无法查看流程本身。要考虑哪些其他功能附加到流程及其活动，其他功能成熟度模型必须与流程功能相关联，并度量流程功能，而CMMI没有处理这些功能。因此，流程成熟度模型必须考虑流程的相关方面，这就为它提供了相应的环境，这包括目的和目标、组织背景（能力和业务功能）、角色、所有者、流程、规则、遵从性方面、自动化部分（应用程序）、度

and business function), the roles, owners, flows, rules, compliance aspects, automated pieces (applications), measures, channels, media, platform, infrastructure, and the services delivered.

6. CMMI encourages the achievement of a higher maturity level with all aspects, in some cases by displacing the true mission, which is improving the process and overall competency in lowering the cost and increasing the revenue. In most cases, the cost to achieve a higher maturity level would be far greater than the possible gain. This may effectively "blind" an organization to the most effective use of its capabilities and resources.

This narrow focus makes CMMI limited in real essential improvement that a BPM maturity model would need.

BPM MATURITY MODEL

From the above discussion, we will in this section illustrate of which components a BPM Maturity Model should consist. These components include Business Process levels, their description, and the areas that give context to BPM maturity. We will then exemplify a BPM maturity benchmark and a BPM maturity development path.

MATURITY LEVELS
Business Process Maturity: Level 1

The organization's process portfolio and initiatives are functionally oriented and exist in multiple instances. The process initiatives are typically characterized as ad hoc in terms of specific for only one or a few business units/departments, thereby organizationally siloed, not fully recognized, or adaptable by others. The process solutions are thereby more department or business-unit centric, occasionally coordinated with others and sometimes even jointly developed. Such coordination and joint development are initial and, therefore, from an enterprise perspective the process strategy is unorganized and partly chaotic in having no formal process approach. Few cross-enterprise process strategies, developments, and improvements are defined, and success of these solutions depends on few individual heroic departments coordinating or a process Center of Excellence (CoE) effort. The challenge with the first stage is that, with multiple process solutions/instances, it is difficult to predict joint value creation or performance. At this stage, it is more difficult to learn from experience when everything is done initially in silos and, if done jointly, it is for the most part new for each business unit/department. In nearly all the maturity or grid approaches, the first level is therefore defined by the failure to satisfy the requirements for Level 2.

量、渠道、媒体、平台、基础设施和交付的服务。

（6）CMMI鼓励在各个方面实现更高的成熟度级别，在某些情况下，通过取代真正的任务（即改进流程和整体能力）以降低成本和增加收入。在大多数情况下，达到更高成熟度级别的成本将远远大于可能的收益。这可能会有效地"蒙蔽"组织，使其无法最有效地使用其能力和资源。

这种狭隘的关注使得CMMI在BPM成熟度模型所需的真正本质改进方面受到限制。

3.10.5　BPM成熟度模型

从上面的讨论中，我们将在本节演示BPM成熟度模型应该包含哪些组件。这些组件包括业务流程级别、它们的描述以及为BPM成熟度提供背景的区域。然后，我们将举例说明BPM成熟度基准和BPM成熟度开发路径。

3.10.6　成熟度级别

1. 业务流程成熟度：第1级

组织的流程组合和计划是面向功能的，并且存在于多个实例中。流程计划的典型特征是被临时性的针对一个或几个特定的业务单位/部门，从而在组织上是孤立的，未被其他人完全认可或适应。因此，流程解决方案更多地以部门或业务单元为中心，有时与他人协调，有时甚至是联合开发。这种协调和联合开发是初始的，因此，从企业的角度来看，流程策略是无组织的，而且在没有正式的过程方法时会部分混乱。很少有跨企业流程战略、开发和改进的定义，这些解决方案的成功取决于少数个人英雄部门协调或流程CoE的努力。第一阶段的挑战是：对于多个流程解决方案/实例，很难预测联合价值创造或绩效。在这个阶段，从经验中学习是非常困难的，因为最初所有的事情都是单独完成的，当初如果是共同完成的，那么对于每个业务部门/部门来说，它大部分都是新的。在几乎所有成熟度或网格方法中，第1级被定义未能满足第2级的要求。

Business Process Maturity: Level 2

The repeatable level is the level in which basic process standardization efforts and repeatable joint process development initiatives (workflow, programming, upgrades, blueprints, etc.) are established to track process development cost, schedule, and functionality. The necessary process CoE disciplines are in place to repeat earlier successes in areas/projects with similar experience. The focus at Level 2 does not explicitly include operational process system merger activities, because the major problems Level 1 organizations face are multiple process managerial problems (e.g., process solution development definition, development planning, value identification, performance measurements, initiatives, and joint reporting in process solutions), and not operational system problems. Joint operational process solution initiatives are planned and tracked at Level 2, but they are not described or executed in detail.

Business Process Maturity: Level 3

The defined and awakening level, around both management (strategic and tactical level) and operations, have a common documented process. The Level 2 standardizations around process are thereby documented and integrated into a set of standard joint process developments and joint competencies for the organization. Process joint development programs, portfolio, and projects use an approved, tailored version of the organization's set of process/solution framework, method, and approaches. The operational multi-instance challenges and possible process single-instance strategies and or initiatives are first explicitly defined and thereby addressed at Level 3. Nevertheless, they must be implemented at Level 1 if the organization is, for example, developing a process single-instance product, creating quality management, or building tools, even if those initiatives are informal, ad hoc, and inconsistently executed. The emphasis of Level 3, however, is centered more around organizational learning of their pain points, challenges, goals, competencies, process definition, and improvements of the standardized processes.

Business Process Maturity: Level 4

At the management level, process/solution excellence is managed across the organizational boundaries. The detailed system measures of the processes are collected in joint cockpits, dashboards, and scorecards, and are optimized and managed. Both the process strategy and competencies, and thereby the process initiatives, are quantitatively understood, monitored, controlled, and managed. This implies statistical thinking[31] and evidence-based management[32] about the process initiatives. It also should be noted that process measurement and analytical abilities could occur at earlier phases (Levels 1, 2 or 3), although it comes as full crossdiscipline to the forefront at Levels 4 and 5, when the process solution is optimized for joint enterprise performance and value creation.

2. 业务流程成熟度：第2级

可重复级别是建立基本流程标准化工作和可重复联合流程开发计划（工作流程、编程、升级、蓝图等）以跟踪流程开发成本、进度和功能的级别。必要的流程CoE学科已经到位，以便在具有类似经验的领域／项目中重复早期的成功。第2级的重点并未明确包括运营流程系统合并活动，因为处于第1级的组织面临的主要问题是多流程管理问题（例如，流程解决方案开发定义、开发计划、价值识别、绩效评估、主动性和在流程解决方案中的联合报告），而不是操作系统问题。联合运营流程解决方案计划在第2级进行规划和跟踪，但不会对其进行详细描述或执行。

3. 业务流程成熟度：第3级

围绕管理（战略和战术级别）、操作定义和唤醒级别有一个共同的文档化过程。因此，需要将围绕流程的第2级标准化记录下来，并集成到组织的一组标准的联合流程开发和联合能力中。流程联合开发计划、投资组合和项目使用组织的流程／解决方案框架、方法和方法集的经过批准的、定制的版本。可操作的多实例挑战和可能的流程单实例策略或计划首先被明确定义，因此在第3级处理。然而，如果组织正在开发流程单实例产品、创建质量管理或构建工具（即使这些计划是非正式的、特别的和不一致执行的），则必须在第1级实现这些计划。然而，第3级的重点更多地集中在组织学习他们的痛点、挑战、目标、能力、流程定义和标准化流程的改进。

4. 业务流程成熟度：第4级

在管理层，卓越的流程／解决方案是跨组织边界进行管理的。流程的详细系统度量收集在联合工作区、仪表盘和记分卡中，并进行优化和管理。流程战略和能力，以及由此产生的流程计划，都是定量理解、监控、控制和管理的。这意味着对流程计划进行统计思考[31]和基于证据的管理[32]。还应注意的是，当流程解决方案针对联合企业绩效和价值创造进行优化时，流程度量和分析能力可能出现在早期阶段（第1级、第2级或第3级），尽管它作为全面的跨学科出现在第4级和第5级。

Business Process Maturity: Level 5

The organization becomes process centric in terms of collaborative developments, enterprise-wide integration and, most important, continuous improvement becomes culturally embedded in the organization. The continuous process improvements support the business differentiation which the organizations are pursuing. On this maturity level, the continuous improvement of the process portfolio is enabled by feedback from the business competencies and their functions, tasks, and services. Applying strategic and analytical thinking enables the organization to understand its expert competencies as well as its processes and the activities that enable their processes. The organization optimizes and develops its processes when and where measurable significant differences in performance and value creation occur.

In Figure 3, we see how the levels are put together with an example of a maturity journey and the statistical Ease of Adoption curve, together with the Return on Investment (ROI) curve:

As we described earlier, CMMI reveres the institutionalization of process for its own sake. This guarantees nothing, and in some cases, the institutionalization of processes may lead to an oversimplified view, ignoring the successful practice of the organization and its process context. For one cannot look at a process itself, without taking into consideration other capabilities that are attached to the process and activity. As the organization progresses and ascends through each phase of maturity, the achievement of its critical success factors must also evolve. Leading organizations take a balanced approach to managing their different critical success factors and what makes them unique. Managed together, they represent the framework from which BPM competencies are built. This includes multiple factors.[37]

To consider what other capabilities are attached to the process and activity, other capability maturity models would have to be interlinked and used to measure the process capabilities. Therefore, a BPM maturity model would have to consider the related aspects of the process that give it context. This includes the purpose and goal, the organizational context (competencies and business function), the roles, owners, flows, rules, compliance aspects, automated pieces (applications), measures, channels, media, platform, infrastructure, and the services delivered. Therefore, the BPM maturity model would have to include the context of the maturity benchmark question that enables one to place it into a maturity level. In Figure 4 is an example of a BPM Maturity Model with related context for BPM maturity assessment.

In the following, we have listed the various aspects relevant for BPM maturity. The questions and the list do not claim to be complete, but more illustrative and representative for how such a BPM maturity benchmark works, and the questions that lead to placement of the organization into a maturity level (Table 2):

5. 业务流程成熟度: 第5级

在协作开发、企业范围的集成方面,组织变得以流程为中心,最重要的是,持续改进成为组织文化的一部分,持续的流程改进支持组织所追求的业务差异化。在这个成熟度级别上,流程组合的持续改进是通过业务能力及其功能、任务和服务的反馈实现的。运用战略和分析思维,使组织能够了解其专家能力,以及其流程和实现其流程的活动。当绩效和价值创造出现可测量的显著差异时,组织会优化和开发其流程。

在图3中,我们看到了级别应用的一个例子,即统计易用性曲线与ROI曲线相结合的成熟度的应用。

正如我们之前所描述的那样,CMMI本身就是为了使流程制度化。但这并不保证什么,在某些情况下,流程的制度化可能导致过于简化的观点,进而忽视了组织的成功实践和流程语义。如果没有考虑附加到流程和活动的其他功能,则无法查看流程本身。随着组织在成熟度的每个阶段的进步和提升,其关键成功因素的实现也必须发展。领先的组织采取平衡的方法来管理他们不同的关键成功因素以及使他们独特的因素。它们共同管理、构建代表BPM能力的框架。这包括多种因素[37]。

要考虑流程和活动附加了哪些其他功能,其他功能成熟度模型必须相互关联并用于衡量流程功能。因此,一个BPM成熟度模型将不得不考虑为其提供流程语义的相关方面。这包括目的和目标、组织背景(能力和业务职能)、角色、所有者、流程、规则、合规性方面、自动化部件(应用程序)、度量、渠道、媒体、平台、基础设施和提供的服务。因此,BPM成熟度模型必须包含成熟度基准问题的语义,使人能够将其放入成熟度级别。图4是一个BPM成熟度模型的例子,它具有用于BPM成熟度评估的相关语义。

在下面,我们列出了与BPM成熟度相关的各个方面。这些问题和列表并不声称是完整的,但更具说明性和代表性的是此类BPM成熟度基准是如何工作的,以及导致将组织置于成熟度级别的问题(表2)。

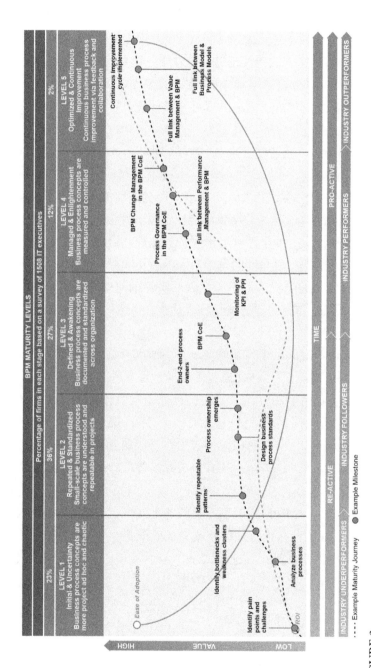

FIGURE 3

Example maturity journey and the statistical Ease of Adoption and Return on Investment (ROI) curve.[38]

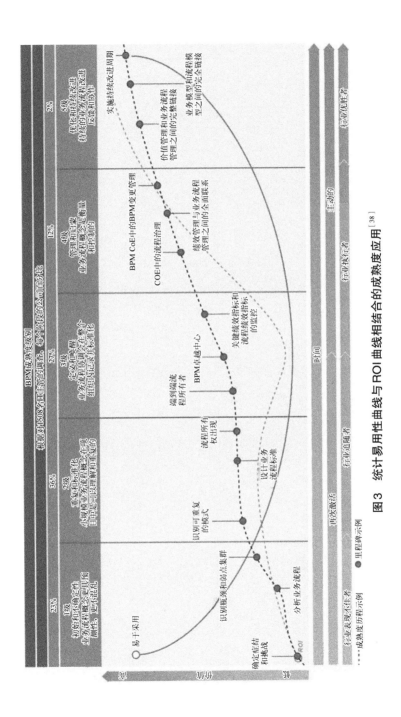

图3 统计易用性曲线与ROI曲线相结合的成熟度应用[38]

FIGURE 4

An example of context for a business process management (BPM) maturity assessment.[39]

BPM成熟度级别

流程的调查，每个阶段的级别百分比

23%	36%	27%	3% 12%	2%
1级 初始和不确定性 业务流程概念上是临时的，缺乏预测性，反应混乱	**2级** 重复和标准化 小规模业务流程概念上，目标是可以理解和重复的	**3级** 定义和觉醒 业务流程协同调查会整合组织机构内记录支持标准化	**4级** 管理和量需 业务流程概念是衡量和控制的	**5级** 优化和持续改进 持续的业务流程改进反馈和协作

哪些业务是业务能力
需求和目标
流程的需求和目标是什么
使用
流程中使用的对象类型是什么
所有者
使用流程的所有者
优先权
哪些流程能够被保护着
角色
哪些业务角色会与流程一起运行
规则和组织
哪些规则控制着流程
中
什么类型的电话播音作自动化流程作
改进
流程发展与哪种能作业
测量
哪些通用用文件衡量进行操作
数据
哪些数据被收集及和发生什么
整合
完成流程与哪些媒体交互
基础设施
哪些应用与基础支持业务流程执行
服务
哪些服务流程是为何的

时间

行业表现不佳者　行业追随者　行业执有者　行业优胜者

再次激活　主动的

图 4　BPM成熟度评估语义的示例[39]

Table 2 *The Business Process Management (BPM) Maturity Self-Assessment*

	Process Maturity Levels Over Time				
	Level 1 Initial and Uncertain	Level 2 Repeated and Standardize	Level 3 Defined and Awakening	Level 4 Managed and Enlightenment	Level 5 Continuous Improvement
Process	The organization's process portfolio and initiatives are functionally oriented and exist in multiple instances. The process initiatives are typically characterized as ad hoc in terms of specific for only one or a few business units/departments, thereby organizationally siloed, not fully recognized or adaptable by others. The process solutions are thereby more department- or business-unit centric, occasionally coordinated with others and sometimes even jointly developed. Such a coordination and joint development are initial and therefore from an enterprise perspective the process strategy is unorganized and partly chaotic in having no formal process approach. Few cross-enterprise process strategies, development, and improvements are defined, and success of these solutions depends on few individual heroic departments coordinating or a process center of excellence (CoE) effort. The	The repeatable level is the level at which basic process standardization efforts, and repeatable joint process development initiatives (workflow, programming, upgrades, blueprints etc.) are established to track process development cost, schedule, and functionality. The necessary process CoE disciplines are in place to repeat earlier successes in areas/projects with similar experience. The focus at Level 2, does not explicitly include operational process system merger activities, because the major problems Level 1 organizations face are multiple-process managerial problems (e.g., process solution development	The defined and awakening level, both around management (strategic and tactical level) and operations, has a common documented process. The Level 2 standardizations around process are thereby documented and integrated into a set of standard joint process developments and joint competencies for the organization. Process joint development programs, portfolio, and projects use an approved, tailored version of the organization's set of process/solution framework, method, and approaches. The operational multi-instance challenges and possible process single-instance strategies and/or initiatives are first explicitly defined and thereby addressed at Level 3. Nevertheless, they must be implemented at Level 1 if the organization is, for example, developing	At the management level, process/solution excellence is managed across organizational boundaries. The detailed system measures of the processes are collected in joint cockpits, dashboards, and scorecards and are optimized and managed. Both the process strategy and competencies, and thereby the process initiatives, are quantitatively understood, monitored, controlled, and managed. This implies statistical thinking[429] and evidence-based management[430] about the process initiatives. It also should be noted that process measurement and analytical abilities could occur	The organization becomes process centric in terms of collaborative developments, enterprise-wide integration and, most important, continuous improvement becomes culturally embedded in the organization. The continuous process improvement supports the business differentiation that the organizations are pursuing. On this maturity level, the continuous improvement of the process portfolio is enabled by feedback from the business competencies and their functions, tasks, and services. Applying strategic and analytical thinking enables the organization to understand its

表2　BPM成熟度自我评估

随时间变化的流程成熟度

	第1级 初始和不确定	第2级 重复和标准化	第3级 定义和觉醒	第4级 管理与启蒙	第5级 连续的提高
流程	组织的流程组合和计划以功能为导向,并存在于多个实例中。流程计划的典型特征是临时性的,针对一个或几个业务单位/部门特定的,从而在组织上是孤立的,未被其他组织充分认可或适应。因此,流程解决方案更多地以部门或业务为中心,有偶尔与其他人协调,甚至联合开发这种协调和联合开发是最初的,因此从企业角度来看,流程策略是无组织的,并且在没有正式的过程定义时部分混乱。很少有跨企业流程战略,开发和改进的定义,这些成功取决于很少的个人英雄部门协调或流程CoE的努力	可重复级别是建立基本流程标准化工作以及可重复联合流程开发计划(工作流程,蓝图等),以跟踪流程开发成本,进度和功能的级别。必要的流程CoE学科已经到位,以便在类似经验的领域/项目中重复早期的成功。第2级的重点并未明确包括运营流程系统合并活动,因为以1级组织面临的主要问题是多流程管理问题(例如,流程解决方案开发	围绕管理(战略和战术层面)和运营,定义和觉醒和水平具有过程。共同的第2级标准流程被记录了下来,并将其整合到一组标准化的联合流程开发和组织的联合能力中。项目组合和项目使用的批是一组经过组织的一组流程/解决方案框架/方法。首先明确定义的多实例战略和的流程单实例和可能处理,然后任在第3级正在开发	在管理层面,跨组织边界管理卓越流程/解决方案。这些流程的详细系统收集和联合着流程的第2级组合驾驶舱,仪表板和和进行优化和管理。流程策略和能力以及流程计划得到了定量的理解,监控,控制和管理。这意味着关于流程举措的统计思考[429]和基于据的管理。还应注意,过程测量和分析能力可能发生	在协作开发,企业范围的集成方面,组织变得以流程以为中心,最重要的是,流程持续改进在组织中成为文化的一部分。持续改进流程改进正在支持组织正在追求的业务差异。在这个业务和服务实现的成熟度级别上,流程组合的持续改进是通过对业务自业务能力以及其功能,应用战略性和分析的反思维使组织能够理解其

challenge with the first stage is that with multiple process solutions/instances, it is difficult to predict joint value creation and/or performance. At this stage it is furthermore difficult to learn from experience when everything is done initially in silos and if done jointly, it is for the most part new (for each business unit/department). In nearly all the maturity or grid approaches the first level is therefore more defined by the failure to satisfy the requirements for Level 2.	definition, development planning, value identification, performance measurements, initiatives, and joint reporting in process solutions), and not operational system problems. Joint operational process solution initiatives are planned and tracked at Level 2, but they are not described or executed in detail.	a process single-instance product, creating quality management, or building tools, even if those initiatives are informal, ad hoc, and inconsistently executed. The emphasis of Level 3, however, is centered more around organizational learning of their pain points, challenges, goals, competencies, process definition, and improvements of the standardized processes.	at earlier phases (Levels 1, 2, or 3), although it comes as full cross-discipline to the forefront at Levels 4 and 5, when the process solution is optimized for joint enterprise performance and value creation.	expert competencies as well as its processes and activities that enable their processes. The organization optimizes and develops its processes when and where measurable significant differences in performance and value creation occur.
Business competency Business competencies share only sporadic and ad hoc relation to business processes, and knowledge of this relation is not shared across business units, making the relation entirely silo-based. Furthermore, business competency potential is neither fully recognized nor adaptable to changing business requirements.	Business competencies and their connection to the business processes across business units is now part of basic standardization projects and initiatives. Successes from earlier initiatives have now become repeatable in future process-oriented projects and development initiatives.	Business competencies are now being defined and documented in detail. This knowledge is shared across organizational boundaries at the strategic, tactical, and operational management levels, and is centered around the learning of how competencies can be used within the existing process portfolio in future process-oriented projects and development initiatives.	All business competencies of the enterprise now share a direct relationship to all of the business processes across the organizational boundaries. Business competencies and their relation to business processes are now fully understood, managed, controlled, and monitored during process-oriented projects and development initiatives.	Business competencies play a significant role within the collaborative feedback loop of continuous business process improvement and optimization projects and initiatives. The relation between value-creating business processes, and business competencies of the organization help build business differentiation on the market.

Continued

（续表）

随时间变化的流程成熟度

	第 1 级 初始和不确定	第 2 级 重复和标准化	第 3 级 定义和觉醒	第 4 级 管理与启蒙	第 5 级 连续的提高
流程	第一个阶段的挑战是,对于多个流程解决方案的价值关联。很难预测性能或整合的价值,从经验中学习这个阶段是非常困难的,因为所有的事情都是在最初的孤岛上完成的,如果是在很大程度上是联合完成的,那么它在很少业务单元/部门)。因此,在几乎所有的成熟度方法中,第 1 级是由于 1 级所不平格式定义更多地是由于未能满足每个第 2 级的需求	定义、开发规划、价值识别、性能度量、计划和流程解决方案中的联合报告,而不是操作系统综合。联合流程跟踪在第 2 级,但没有详细描述或执行	流程单实例产品、创建质量管理或构建工具,即使这些计划是非正式的、临时的和执行不一致的,则必须在第 1 级实施它们。然而,第 3 级实施的重点集中在组织学习他们的痛点、挑战、目标、能力,流程定义和标准化流程的改进上	在早期阶段(第 1 级、第 2 级或第 3 级),尽管对关联在第 4 级和第 5 级,当流程解决方案针对创造全企业绩效时,它作为一个全面的跨学科走在了最前面	专家能力以实现其流程和价值活动。当性能和价值创造方面存在可衡量的显著差异时,组织会优化和发展其流程
业务能力	业务能力只与业务流程共享零星的和临时的关系,并且不会在业务部门之间共享关系的知识,使得这种关系完全基于知识点。此外,业务能力潜力既未得到充分认可,也未能适应不断变化的业务需求	业务能力及其与业务部门业务流程的关联现在已成为基本标准化项目和计划的一部分。现在在未来面向流程的项目和发展计划中变得可重复	现在正在详细定义和记录业务能力。这些知识在战略、战术和运营管理级别的业务上共享,并围绕流程的界上共享,并围绕流程的界面上在未来面向流程的项目和开发计划中如何在现有流程组合中使用能力	企业的所有业务能力现在都与跨组织边界的所有业务流程有直接关系。在面向流程的项目和开发活动期间,业务能力及其现在业务流程的关系现在已被完全理解、管理、控制和监视	业务能力在持续的业务流程改进和优化项目计划及设计的协作反馈循环中发挥着重要的作用。价值创造业务流程和组织的业务能力之间的关系有助于建立业务在市场上建立业务差异

Table 2 The Business Process Management (BPM) Maturity Self-Assessment—Cont'd

	Process Maturity Levels Over Time				
	Level 1 Initial and Uncertain	Level 2 Repeated and Standardize	Level 3 Defined and Awakening	Level 4 Managed and Enlightenment	Level 5 Continuous Improvement
Purpose and goal	Few to no value-centric aspects of what can, or does, give the business processes purposes and goals—in terms of both forces (external and internal), drivers (external and internal), value indicators, value propositions, performance indicators, strategy, goals, objectives, and quality thereof—exist as concepts in the organization. Value-centric aspects such as these often remain locked in silos, and are rarely shared across organizational boundaries. Furthermore, the aforementioned value aspects are rarely, if ever, linked to processes, and, if so, only applied in an ad hoc manner during process-oriented projects and initiatives within each individual business unit of the organization.	Some value-centric aspects around that which gives the business processes a purpose and a goal—in terms of forces (external and internal), drivers (external and internal), value indicators, value propositions, performance indicators, strategy, goals, objectives, and quality thereof—have become repeatable due to successful integration in previous process-oriented project initiatives that have been performed within individual business units. These successes are shared across organizational boundaries, and some of these value-centric aspects around the business processes of each business unit are now standardized and allow for repeatable development initiatives during process-oriented projects and initiatives.	Most value-centric aspects around the purpose and goals of business processes—such as forces (external and internal), drivers (external and internal), value indicators, value propositions, performance indicators, strategy, goals, objectives, and quality thereof—are explicitly defined across the organizational boundaries of all the business units, and share a common documentation point of reference at the strategic, tactical and operational levels within the organization.	All value-centric aspects around the purpose and goals of business processes—such as forces (external and internal), drivers (external and internal), value indicators, value propositions, performance indicators, strategy, goals, objectives, and quality thereof—are quantitatively understood, controlled, monitored, measured, and managed across organizational boundaries of the enterprise.	The enterprise-wide organization has become purpose- and goal-centric—in terms of forces (external and internal), drivers (external and internal), value indicators, value propositions, performance indicators, strategy, goals, objectives, and quality thereof—around the continuous improvement of business processes. The development, improvement, and optimization of business processes around value-adding aspects is now the central focus in process-oriented project initiatives across the enterprise. This development is supported by a collaborative feedback loop of the organization at the strategic, tactical, and operational levels.

（续表）

随时间变化的流程成熟度

	第 1 级 初始和不确定	第 2 级 重复和标准化	第 3 级 定义和觉醒	第 4 级 管理与启蒙	第 5 级 连续的提高
目的与目标	就力量（外部和内部）、驱动程序（外部和内部）、价值指标、价值主张、性能指标、策略、目标、目标和质量等作为而组织中存在有以价值的概念而言，很少或没有以价值为中心或能够为业务流程提供的以价值目的和目标。诸如此类的方面常常被锁定在类的以价值为中心的方面孤岛中，很少跨组织边界共享。此外，前面提到的方面很少（如果有的话）与流程相关联，仅在面向流程的每个业务单元中进行的主动性工作期间以特定的方式应用	围绕这一点的一些以价值为中心的方面以价值和目标为业务流程提供了目的和目标——力量（外部和内部）、驱动因素（外部和内部）、价值指标、价值主张、价值指标、战略、目标，由于成功建合以前在各业务部门内进行的面向流程的项目计划，其质量已经变得可重复。这些成功是跨组织边界共享的，围绕每个业务部门为中心的这些价值已经标准化，并允许在面向流程的项目和目的计划中进行可重复的开发计划	围绕业务流程的目的和目标的大多数以价值为中心的方面——如价值（外部和内部）、驱动因素（外部和内部）、价值指标、价值主张、战略、目标是明确定义又所有业务部门的组织边界，并在组织内的战略、战术和运营层面共享一个共同的文档参考点	围绕业务流程的目的和目标的所有方面以价值为中心的方面——如力量（外部和内部）、驱动因素（外部和内部）、价值指标、价值主张、绩效指标、战略、目标和质量——都是跨企业范围内的组织边界业进行定量理解、控制、监控、测量和管理	在力量（外部和内部）、驱动力（外部和内部）、价值指标、价值主张、策略、目标、性能指标、目标和质量方面，围绕着业务流程的持续改进、企业范围内目的和流程成为向的组织已经成为了组织和围绕业务增值面的开发、改进是整个企业面向流程的中个焦点。这一发展得到了这组织在目标、战略、战术和运营层面的协作反馈循环的支持

Objects (business, information and data)	Business information and data objects exist only in silos and—apart from data objects—share little to no coherency nor affiliation with the business processes of the organization. In the case that business and/or information objects are part of a business process, its participation is largely initial and sporadic in context.	Business information and data objects are gradually being implemented across business processes in the organization. This allows for repeating earlier development successes and makes room for basic standardization upon implementation across business units.	Business information and data objects are now being addressed and explicitly defined for use in process-oriented projects and initiatives. Furthermore, the objects share a common set of documentation standards across the strategic, tactical, and operational management levels across organizational boundaries.	All business information and data objects are now being individually mapped and related to all process-centric meta objects during process-oriented projects and development initiatives. This allows for a much higher degree of object management and control as well as continuous object governance and monitoring.	Business information and data objects are a central part of business process development and optimization during process-oriented projects and initiatives across organizational boundaries. Furthermore, the collaborative environment across all business units becomes "object"-centric during process modeling in a continuous effort to improve organizational business processes.

Continued

（续表）

随时间变化的流程成熟度

对象（业务、信息和数据）	第1级 初始和不确定	第2级 重复和标准化	第3级 定义和觉醒	第4级 管理与启蒙	第5级 连续的提高
	业务信息和数据对象仅存在于孤岛中，除了没有与组织数据岛之外几乎没有一致性，也没有与组织的业务流程相关联。在业务和信息对象是业务流程的一部分的情况下，其参与且在背景中初始的并且零星是零星	业务信息和数据对象正逐步跨组织的业务流程实现。这允许重复早期的开发成功，并作为跨业务单元的实现为跨业务单元标准化的基本标准提供空间	业务信息和数据对象现在正在被处理和明确定义，以便在面向流程的项目和计划中使用。此外，对象在跨组织边界的战略、战术和运营管理级别上共享一组管理通用的文档标准	在面向流程的项目和开发计划期间，所有业务信息和数据对象现在都被单独映射与中心。这允许所有项目和计划以流程为中心的元对象相关。这允许更高程度的对象管理和控制，以及持续的对象治理和监视	业务信息和数据对象是跨组织边界面向流程的项目和计划中的核心的核心部分。此外，在流程建模过程中，跨所有业务单元的协作环境将成为以对象为中心，从而持续改进组织业务流程

Table 2 The Business Process Management (BPM) Maturity Self-Assessment—Cont'd

	Process Maturity Levels Over Time				
	Level 1 Initial and Uncertain	Level 2 Repeated and Standardize	Level 3 Defined and Awakening	Level 4 Managed and Enlightenment	Level 5 Continuous Improvement
Process owner	Few to no process owners exist around the current process portfolio within each individual business unit. For the process owners that do exist, their responsibilities and accountability remains largely ad hoc and their roles are neither fully understood nor recognized during process-oriented projects and initiatives.	Process ownership is being established across the organizational boundaries, which allows for the repetition of earlier successes from previous process-oriented projects and initiatives. This allows for a basic standardized incorporation of process ownership in both new and existing projects and initiatives.	Process ownership has become explicitly defined and is fully standardized and incorporated in process-oriented projects and initiatives across the organizational boundaries. The role and purpose of process ownership is fully documented and shared across the strategic, tactical, and operational levels of the enterprise.	Process ownership and their relation to the other owners—such as the business, service, application, data, platform, and infrastructure owners—is being managed across the enterprise. Their role and responsibility is fully understood and controlled and their results are monitored and measured during process-oriented projects and initiatives.	Process ownership—including their collaboration with the other business services, application, data, platform, and infrastructure owners of the organization—play fundamental roles within the enterprise when it comes to the development and optimization of both new as well as existing business processes across the organizational boundaries. They play a key role in the collaborative feedback loop around the continuous improvement of business processes.

（续表）

随时间变化的流程成熟度

	第 1 级 初始和不确定	第 2 级 重复和标准化	第 3 级 定义和觉醒	第 4 级 管理与启蒙	第 5 级 连续的提高
流程 所有人	在每个单独的业务单元中,当前流程组合周儿乎没有流程所有人。对于确实存在的流程所有人,他们的职责和问责制在很大程度上仍然是临时的,在面向流程的项目计划中,他们的角色既没有得到充分地理解,也没有得到认可	流程所有权正在跨组织边界建立,这允许重复以前面向流程导向的项目和计划中基本的项目和计划中新的和现有的项目的标准化地合并流程所有权	流程所有权已明确定义并完全标准化,并在跨组织边界所定义的流程导向项目计划中纳入人。流程所有权的角色和计划在企业运营战略术和记录面得到分层面共享	流程所有权及其其他所有者(如业务、数据、服务、应用程序、平台和架构所有者)的关系正在整个企业中进行管理。他们的角色和责任得到充分理解和控制,他们的结果在面向流程的项目和计划中得到监控和衡量	流程所有人——包括它们与组织的其他业务服务、数据、应用程序、平台和基础设施所有者的协作——在整个组织边界开发和发现新业务流程及优化企业流程时,在有业务流程的持续改进的持续循环中扮演着关键角色

Process flow	Process flows are the natural part of all business processes in the respective business units, although they exist only in silos. Thus, knowledge of them is not shared across organizational boundaries (i.e., between business units).	Process flows have become an increasingly larger part of basic business process standardization across organizational units, and are now used in repeatable process-oriented projects and development initiatives.	Process flows are now being defined and documented in detail. This information is shared across all organizational units and on the strategic, tactical, and operational management levels of each business unit.	Process flows have become quantitatively understood, monitored, controlled, and are being managed across organizational boundaries. They also share a direct relationship to all other flows of the organization, such as the business work flows, service flows, other process flows, application flows, as well as data flows.	During process-oriented projects and development initiatives, process flows play a significant role in the collaborative feedback environment because of their direct relation to all other flows of the organization, such as the business work flows, service flows, other process flows, application flows, as well as data flows. They also represent an important aspect of continuous business process improvement initiatives across organizational boundaries.

Continued

（续表）

随时间变化的流程成熟度

	第1级 初始和不确定	第2级 重复和标准化	第3级 定义和觉醒	第4级 管理与启蒙	第5级 连续的提高
流程活动	流程是各个业务单元中所有业务流程的自然组成部分。尽管它们只存在于孤岛中。因此，关于它们的知识并不是跨组织边界共享的(如在业务单元之间)	流程流已经成为跨组织单元所有的基本业务流程标准化中越来越大的一部分。现在面向流程的项目和重复的项目和开发计划	现在正在详细定义和记录流程流。此信息流在所有组织单元之间共享，并在每个业务单元级别和操作共享。战略、战术级别上共享管理级别共享	流程流已经被定量地理解、监视、控制，并且正在跨组织边界进行管理。它们还与组织的所有其他流程共享工作的所有接关系，如业务流、服务流、应用程序流、数据流	在面向流程的项目和开发活动中，流程流在协作反馈环境中扮演重要角色。因为它们与组织的所有其他流程(如业务工作流、服务流、其他流程流、应用程序流以及数据流)直接相关。它们还代表了跨组织边界的持续业务流程改进活动的一个重要方面

Table 2 *The Business Process Management (BPM) Maturity Self-Assessment—Cont'd*

	Process Maturity Levels Over Time				
	Level 1 Initial and Uncertain	Level 2 Repeated and Standardize	Level 3 Defined and Awakening	Level 4 Managed and Enlightenment	Level 5 Continuous Improvement
Process roles	Process roles largely work ad hoc on an initial basis during all process-oriented projects and initiatives. Their role, responsibilities, and overall purpose within the organization are not fully recognized nor understood. Their work is also only carried out in silos—and the result is knowledge that is retained and static, and not shared across business units nor across organizational boundaries.	Process roles are now being established across all business units, as earlier successes from process-oriented projects and development initiatives allow for being repeated in new projects and initiatives. Process roles have also become part of basic business process standardization initiatives across business units.	Process roles have become explicitly defined and are fully standardized and incorporated in process-oriented projects and initiatives across the organizational boundaries. The role and purpose of process ownership is fully documented and shared across the strategic, tactical, and operational levels of the enterprise.	Process roles and their relation to the role owners—such as the business services, application, data, platform, and infrastructure roles—are being managed across the enterprise. Their role and responsibility is fully understood and controlled and their results are monitored and measured during process-oriented projects and initiatives.	Process roles—including their collaboration with the other business services, application, data, platform, and infrastructure roles of the organization—plays fundamental role within the enterprise when it comes to the development and optimization of both new as well as existing business processes across organizational boundaries. They play key roles in the collaborative feedback loop around the continuous improvement of business processes.

（续表）

随时间变化的流程成熟度

	第 1 级 初始和不确定	第 2 级 重复和标准化	第 3 级 定义和觉醒	第 4 级 管理与启蒙	第 5 级 连续的提高
流程角色	在所有面向流程的项目和计划中，流程角色基本上都是临时工作的。他们在初始工作中的角色、职责和总体目标的认识没有得到充分的认识和理解。他们的工作也只是在竖井中进行的，其结果是知识被保留并且是静态的，而不是跨业务单元或跨组织边界共享的	现在正在所有业务单元之间建立流程角色，因为面向流程的项目和开发计划的早期成功允许许多在新项目计划中重复。流程业务角色也成为基本业务流程标准化活动的一部分	流程角色已经被明确定义，并且完全标准化，并被纳入跨组织边界的面向流程的项目和计划中。流程所有在企权的角色，战术和操作业的战略，级别上得到充分的文档化和共享	流程角色及其与角色所有者的关系（如业务服务，应用程序、数据、平台和基础设施角色）正在整个企业中进行管理。在面向过程的项目和计划中，他们的角色和职责被充分理解和控制，他们的结果被监视和测量	流程角色（包括它们与其他业务服务，应用程序、数据、平台和组织的基础设施角色的协作）在跨组织边界开发和优化新业务流程及发现有业务流程时，在企业中扮演基本角色。它们在围绕持续改进业务流程的协作反馈循环中扮演着关键角色

Process rules	The process rules of each individual business unit share no connection or connection to the other rules of organization (i.e., business, service, application, data, platform, and infrastructure rules). This makes the process rules function only in silos, and their influence upon other business units is not fully understood.	Process rules share little to no connection to the other rules of the organization (i.e., business, service, application, data, platform, and infrastructure rules), but previous successful business process implementation initiatives allows for repetition. Process rules are now a natural part of business process standardization across organizational boundaries.	Process rules and their connection to business, service, application, data, platform, and infrastructure rules have become explicitly defined and documented for future development and implementation projects. Their definitions and the documentation thereof allow for knowledge sharing and use across organizational boundaries in all process-oriented projects and initiatives.	Process rules and their relation to business, service, application, data, platform, and infrastructure rules are now being efficiently managed, controlled, and monitored across organizational boundaries.	Process rules have become a central part of all process-oriented development and continuous improvement initiatives across all organizational business units. They also represent an important aspect of the enterprise-wide collaborative feedback loop.
Process compliance	Process compliance is entirely silo-based and shares no relation to any other form of compliance, regardless of which business unit of which the associated business process is a part.	The successful implementation of process compliance during earlier process-oriented projects and initiatives allows for basic process compliance standardization across organizational boundaries.	Process compliance is now standardized, defined, and fully documented. Process compliance is furthermore being addressed during all process-oriented projects and development initiatives across organizational boundaries.	Process compliance now shares a direct connection to business, application, data, platform, and infrastructure compliance. These connections are now fully understood, and their usability is managed, controlled, and monitored during all process-oriented projects and development initiatives.	Process compliance, along with process rules, have become a central part of all process-oriented development and continuous improvement initiatives across all organizational business units. They also represent an important aspect of the enterprise-wide collaborative feedback loop.

Continued

（续表）

随时间变化的流程成熟度

	第1级 初始和不确定	第2级 重复和标准化	第3级 定义和觉醒	第4级 管理与启蒙	第5级 连续的提高
流程规则	每个单独业务单元的流程规则（即业务、服务、数据、应用程序、平台和基础架构规则）没有任何关联关系或关联，仅在孤岛中起流程规则作用。这使得流程规则在孤岛中起作用，并且它们对其他业务单位的影响尚未完全理解	流程规则与组织的其他规则（即业务、服务、数据、应用程序和基础架构规则）几乎没有任何关系，但以前成功的业务流程实施计划允许许多重复。流程规则现在是跨组织流程标准化的自然组成部分	流程规则及其与业务、服务、应用程序、数据、平台和基础架构的连接已明确定义并记录在案，以供将来的开发和实施项目使用。他们的定义和文档允许在所有面向过程的项目和计划中跨组织边界进行知识共享和使用	流程规则及其与业务、服务、应用程序、数据、平台和基础架构的关系现在可以在跨组织边进行有效管理、控制和监控	流程规则已成为所有组织业务部门中所有面向流程的开发和持续改进计划的核心部分。它们还代表了企业反馈循环的一个重要方面
流程合规性	流程合规性完全基于任何其他规则，并且与任何其他业务流程属于哪个业务单元无关。形式上相关合规性无关，无论相关业务流程属于哪个业务单元	在早期面向流程的项目和计划中成功实施跨组织的流程合规性，可以跨组织边界实现基本流程合规性标准化	现在，流程合规性已经过标准化、定义和完整记录。在跨组织边界的所有面向流程的项目开发计划中，进一步解决了流程合规性问题	流程合规性现在与业务、应用程序、数据、平台合规性直接相关。现在可以完全理解这些连接，并在所有面向流程的项目开发计划中管理、控制和监视它们的可用性	流程合规性已成为所有组织业务部门中所有面向流程的开发和持续改进计划的核心部分。它们还代表了企业反馈循环的一个重要方面

Table 2 *The Business Process Management (BPM) Maturity Self-Assessment—Cont'd*

	Process Maturity Levels Over Time				
	Level 1 Initial and Uncertain	Level 2 Repeated and Standardize	Level 3 Defined and Awakening	Level 4 Managed and Enlightenment	Level 5 Continuous Improvement
Application	Logical and physical application components as well as application modules, features, functions, tasks, and system reports support the execution of business processes, although only in silos, and performance and implementation knowledge is not shared across organizational boundaries.	Using logical and physical application components as well as application modules, features, functions, tasks, and system reports to successfully implement and run business processes is not repeatable across organizational boundaries and allows for basic business process standardization.	Logical and physical application components as well as application modules, features, functions, tasks, and system reports are now being clearly defined and documented across organizational boundaries. Knowledge sharing happens across the strategic, tactical, and operational management levels across business units.	Logical and physical application components as well as application modules, features, functions, tasks, and system reports share a direct connection to all business processes, and are managed, controlled, and monitored during all process-oriented projects and development initiatives.	Logical and physical application components as well as application modules, features, functions, tasks, and system reports and their connection to business processes have become a part in collaborative feedback loop of continuous business process improvement and optimization across organizational boundaries.

（续表）

随时间变化的流程成熟度

	第1级 初始和不确定	第2级 重复和标准化	第3级 定义和觉醒	第4级 管理与启蒙	第5级 连续的提高
应用程序	逻辑和物理应用程序组件以及应用程序模块、功能、任务和系统报告支持业务流程的执行,尽管只是在跨组织边界中,并且目不跨组织边界共享性能和实现知识	使用逻辑和物理应用程序组件以及应用程序模块、功能、任务和系统报告来成功实现和运用业务流程在组织边界行业边界,并允许基本的业务不可重复的,并允许基本的业务流程标准化	现在,跨组织边界清晰地定义和记录逻辑和物理应用程序组件以及应用程序模块、功能、任务和系统报告,知识共享跨越业务部门的战略、战术和运营管理层面	逻辑和物理应用程序组件以及应用程序模块、功能、任务和系统报告共享与所有业务流程的直接连接,并在所有面向流程的项目和开发计划中进行管理、控制和监视	逻辑和物理应用程序组件以及应用程序模块、功能、任务和系统报告的连接已成为跨组织边界的持续业务流程改进和优化的协作界面和优化的协作的持续循环的一部分

Process measurement	Process measurements are carried out largely sporadically and only in an ad hoc manner. Process measurement results and reporting is also entirely silo-based, which prevents knowledge sharing across business units.	Drawn from the successes of previous execution and implementation, process measurements are now being done on all executed business processes and allow for basic standardization across organizational boundaries.	Process measurements are explicitly defined and documented for how they should measure business executable business processes. Process measurement results and reporting happen across business units to enhance organizational learning.	Process measurements are being efficiently managed, controlled, and monitored in correspondence to relevant business measures, service measurements, as well as system measurements, and reporting is afterward delivered to all relevant process owners and stakeholders.	Process measurements are used as a central part in the collaborative feedback loop of continuous business process improvement across organizational boundaries. The process measurement results in combination with business measures, service measurements as well as system measurements allow for a much higher degree of knowledge-based business process development and optimization during process-oriented projects and initiatives.

Continued

（续表）

随时间变化的流程成熟度

	第1级 初始和不确定	第2级 重复和标准化	第3级 定义和觉醒	第4级 管理与启蒙	第5级 连续的提高
流程度量	流程度量主要是零星地进行的，而且只以一种特别的方式进行。流程度量结果和报告也是完全基于孤岛的，这阻止了跨业务单元的知识共享	从以前执行和实现的成功经验中得出，现在正在对所有执行的业务流程进行流程度量，并允许跨组织边界进行基本标准化	流程度量是显式定义的，并记录了它们应该如何度量可执行的业务流程。过程度量结果发和报告跨业务单元发生，以增强组织学习	根据相关的业务量、服务度量以及系统度量，对流程度量进行了有效的管理、控制和监视，然后将报告交付给所有相关的流程所有人和相关利益相关者	流程度量被用作跨组织边界的持续业务流程改进的协作循环流程的中心部分。流程度量结果与业务度量、服务度量以及系统度量相结合，从而在面向流程的项目和计划中实现更高程度的基于知识的业务流程开发和优化

Table 2 *The Business Process Management (BPM) Maturity Self-Assessment—Cont'd*

	Process Maturity Levels Over Time				
	Level 1 Initial and Uncertain	Level 2 Repeated and Standardize	Level 3 Defined and Awakening	Level 4 Managed and Enlightenment	Level 5 Continuous Improvement
Channel	Business, service, application, data, platform, and infrastructure channels are used only sporadically and largely ad hoc during business process execution. They are used only in silos, thus preventing knowledge sharing across organizational boundaries.	Business, service, application, data, platform, and infrastructure channels have become standardized and are now repeatable due to earlier successful implementations in process-oriented projects and development initiatives.	Business, service, application, data, platform, and infrastructure channels are defined and documented in detail across organizational boundaries on the strategic, tactical, and operational management levels and are central to knowledge sharing and learning across the enterprise.	Business, service, application, data, platform, and infrastructure channels are directly related to all business process operations and are managed, controlled, and monitored across organizational boundaries during process-oriented projects and initiatives.	All business process operations make use of business, service, application, data, platform, and infrastructure channels in the collaborative feedback loop during process-oriented projects and development initiatives. They also represent an important aspect of supporting the continuous improvement and optimization of existing business processes.
Data	Data components, entities, and tables are used actively throughout all business process operations, although only in silos (i.e., in single business units), preventing knowledge spread throughout the organization.	Data components, entities, and tables are being utilized across organizational boundaries during business process implementation. The data components, entities, and tables are fully understood, and are now repeatable for standardization projects around business process implementation.	Data components, entities, and tables have become clearly defined and documented across all strategic, tactical, and operational management levels to help support organizational learning around business process operations.	Data components, entities, and tables have a direct relationship to all relevant process objects, and their purpose is fully understood across the organizational business units.	The organization has become explicitly data-centric and data-driven in the collaborative feedback loop during business process implementation, development, and improvement projects and initiatives. Data aid in supporting business differentiation and supports enterprise-wide integration.

（续表）

随时间变化的流程成熟度

	第1级 初始和不确定	第2级 重复和标准化	第3级 定义和觉醒	第4级 管理与启蒙	第5级 连续的提高
渠道	业务、服务、应用程序、数据、平台和基础设施通道只在业务流程执行期间偶尔使用，而且主要是临时的使用。它们只在孤岛中使用，从而阻碍了知识的跨组织边界共享	业务、服务、应用程序、数据、平台和基础设施通道已经标准化，并且由于早期面向流程的项目和开发计划中的成功实现，它们现在是可重复的	业务、服务、应用程序、数据、平台和基础设施通道是跨组织边界的战略、战术和操作管理级别上详细定义和文档化的，是跨企业知识共享和学习的核心	业务、服务、应用程序、数据、平台和所有基础设施通道与所有业务流程操作都直接相关，并且在面向流程的项目和计划期间跨组织边界进行管理、控制和监视	在面向流程的项目和开发活动期间，所有业务流程操作都使用通道，服务、数据、平台通道和基础设施代表了支持现有业务流程的持续改进和优化的一个重要方面
数据	数据组件、实体和表在所有业务流程操作中都被主动使用，尽管只是在单个业务单元中（即在单个业务单元中），阻止知识在整个组织中传播	在业务流程实现期间，跨组织边界使用数据组件、实体和表。数据组件、实体和表已被完全理解，现在可以在围绕业务流程实现的标准化项目中重复使用	数据组件、实体和表已在所有战略、战术和运营管理级别中得到明确定义和记录，以帮助支持围绕组织围绕业务流程运营管理进行学习	数据组件、实体和表与所有相关的流程对象有直接关系，它们的目的在组织业务单元中得到充分理解	在业务流程实现、开发和改进项目和计划期间的协作反馈循环中，组织已经明确地成为以数据为中心的和数据驱动的。数据帮助支持业务分化和企业范围内的集成

Media	The use of business, application, data, platform, and infrastructure media during business process operations often occurs ad hoc and delivers performance on an initial basis. The usage also only happens in silos, so knowledge is never shared among business units.	Business process operations during implementation phases make use of business, application, data, platform, and infrastructure media under more standardized yet basic circumstances.	Business, application, data, platform, and infrastructure media have now been explicitly defined and documented in how they support business process operations and development, and the knowledge thereof is shared across organizational boundaries for increased strategic, tactical, and operational management agility.	Business, application, data, platform, and infrastructure media relate to all relevant process objects, and support business process execution across organizational boundaries. The use of media is fully understood by the organization. They are also managed, controlled, and monitored across the enterprise.	All business process operations make use of business, application, data, platform, and infrastructure media in the collaborative feedback loop during process-oriented projects and development initiatives and are of high importance in the support of continuous business process improvement and optimization.
Platform	Logical and physical platform components as well as platform devices and functions effectively support the development and execution of business process operations, although their use is largely ad hoc and occurs only in silos within the organization.	Logical and physical platform components as well as platform devices and functions support implementation of standardized business process developments and installments.	Logical and physical platform components as well as platform devices and functions are being clearly defined and documented across the organization to support organizational learning of business process operations and development.	Logical and physical platform components as well as platform devices and functions are efficiently managed and controlled across organizational boundaries. The platform objects are also directly related to all relevant process objects.	Logical and physical platform components as well as platform devices and functions serve as important aspects to support the continuous business process improvement and optimization during the collaborative feedback loop.

Continued

（续表）

随时间变化的流程成熟度

	第1级 初始和不确定	第2级 重复和标准化	第3级 定义和觉醒	第4级 管理与启蒙	第5级 连续的提高
媒介	在业务流程操作期间使用业务、应用程序、数据、平台、基础设施媒体常是临时的，并在初始基础上交付性能。这种用法也只出现在孤岛中，所以知识从来不会在业务单元之间共享	在实现阶段的业务流程操作在更加标准化的基本环境下使用业务、应用程序、数据、平台和基础设施媒体	业务、应用程序、数据、平台和基础设施又已经明确记录了它们如何支持业务流程操作和开发，并且跨组织边界共享这些知识，以提高成战术和操作管理的灵活性	业务、应用程序、数据、平台和基础设施与所有相关的流程对象相关，并在支持跨组织边界的业务流程执行。本组织完全理解媒体的使用。它们还在整个企业中进行管理、控制和监视	在面向流程的项目和开发活动中，所有业务流程操作都在协作循环中使用业务、应用程序、数据、平台和基础设施媒体，这对于支持的业务流程改进和优化非常重要
平台	逻辑和物理平台组件以及平台设备和功能有效地支持业务流程操作的开发和执行，尽管它们的开发和使用主要是临时的，并且主要在组织内的孤岛中发生	逻辑和物理平台组件以及平台设备和功能支持标准化业务流程开发和分期付款的实施	在整个组织中明确定义和记录逻辑和物理平台组件以及平台设备和功能，以支持组织学习业务流程操作和开发	逻辑和物理平台组件和功能以及平台设备可跨组织边界进行有效管理和控制。平台对象也与所有相关的流程对象直接相关	逻辑和物理平台设计、组件以及平台设备和功能是支持反馈循环期间持续业务流程改进和优化的重要方面

Table 2 *The Business Process Management (BPM) Maturity Self-Assessment—Cont'd*

	Level 1 Initial and Uncertain	Level 2 Repeated and Standardize	Level 3 Defined and Awakening	Level 4 Managed and Enlightenment	Level 5 Continuous Improvement
			Process Maturity Levels Over Time		
Infrastructure	Logical and physical infrastructure components as well as infrastructure devices, functions, and features effectively support the development and execution of business process operations, and help support the organization in doing so through networking capabilities.	Logical and physical infrastructure components as well as infrastructure devices, functions, and features aid the platform components in supporting the implementation of standardized business process developments and operations.	Logical and physical infrastructure components as well as infrastructure devices, functions, and features are being clearly defined and documented across the organization to support organizational learning of business process operations and development.	Logical and physical infrastructure components as well as infrastructure devices, functions and features are efficiently managed and controlled across organizational boundaries. The infrastructure objects are also directly related to all relevant process objects.	Logical and physical infrastructure components as well as infrastructure devices, functions, and features serve as important aspects to support the continuous business process improvement and optimization during the collaborative feedback loop.
Service	Business processes are directly supported by application, data, platform, and infrastructure services; however, they share no connection to the business services of the organization. Service delivery of executed business processes are neither measured nor controlled, and appear largely ad hoc and initial in their behavior.	Application, data, platform, and infrastructure services are used to support basic business process standardization across organizational boundaries due to the repetition of earlier successes in previous process-oriented projects and development initiatives.	Application, data, platform, and infrastructure services have been defined and documented in detail to support and enhance organizational learning of their function and purpose around business process operations and development.	Business, application, data, platform, and infrastructure services are now all directly related to all relevant process flows. Their function and purpose is fully managed, controlled, and monitored across organizational boundaries of the enterprise.	Business, application, data, platform, and infrastructure services are used as a central part in the collaborative feedback loop across organizational boundaries. The services also aid in supporting the organization during continuous business improvement and optimization.

（续表）

随时间变化的流程成熟度

	第1级 初始和不确定	第2级 重复和标准化	第3级 定义和觉醒	第4级 管理与启蒙	第5级 连续的提高
基础设施	逻辑和物理基础设施组件以及基础设施设备、功能和特性有效地支持业务流程操作的开发和执行,并通过网络功能帮助组织支持这样做	逻辑和物理基础设施组件以及基础设施设备、功能和特性帮助平台组件支持标准化业务流程开发和操作的实现	逻辑和物理基础设施组件以及基础设施设备、功能和特性在整个组织中得到了清晰的定义和文档化,以支持组织对业务流程操作和开发的学习	逻辑和物理基础设施组件以及基础设施设备、功能和特性可以跨组织边界有效地管理和控制。基础设施对象也与所有相关业务流程对象直接相关	逻辑和物理基础设施组件以及基础设施设备、功能和特性是支持协作反馈循环期间和持续业务流程改进和优化的重要方面
服务	业务流程由应用程序、数据、平台和基础设施服务直接支持;但是,它们与组织的业务服务没有共享连接。已执行业务流程的服务交付既不受量度量也不受控制,并且在其行为为中主要表现为中临时的和初始的	应用程序、数据、平台和基础设施服务用于支持跨组织边界的基本业务流程标准化,这是因为在以前运行流程的项目和重复了早期的成功	已经对应用程序、数据、平台和基础设施服务进行了详细的定义和文档化,以支持和增强组织对其功能和围绕业务流程操作和开发目的的学习	业务、应用程序、数据、平台和基础设施服务现在都与所有相关的流程直接相关。它们的功能和目的是跨企业的功能和组织边界进行全面管理、控制和监视	业务、应用程序、数据、基础设施服务被用作跨组织边界的协作反馈循环的中心部分。这些服务还有助于在持续的业务改进和优化期间支持组织

FROM MATURITY LEVEL ASSESSMENT TO MATURITY BENCHMARK

A BPM Maturity level assessment is essentially a way of describing the extent to which a process or function exists in context with the rest of the organization. This is important because the process activities and their context are the relationships that relate to the:

- Effective way of working
- Efficient operation
- Consistent performance
- Reliable value creation and realization.

Since the early 1990s, to develop new strategic direction and improve performance, organizations have analyzed their As Is situation to find out what they need to change to reach the desired To Be stage. Once the organization has finished the BPM maturity self-assessment and understand its BPM maturity of its As-Is situation, we have found that various organizations spend a tremendous amount of time, resources, and money to understand and benchmark the different aspects. We have, therefore, developed a standard BPM maturity benchmark that enables comparison of the different areas against one other. Given an immediate overview of the specific maturity level of the different areas and where the lowest maturity within an area is, this enables us to see the weakest maturity and how it impacts the other areas. In Figure 5, an example of such a BPM Maturity Benchmark is illustrated:

Organizations and people want to know the maturity and benchmark against the various functions related to processes for two main reasons. The first is to establish a baseline, that is, Where are we now? The second is to understand the potential for improvement and development. If the self-assessment generates a maturity value of less than 4 or 5, one could say that hope exists for improvement and development. Likewise in a benchmarking exercise if the various maturity assessments against each other reveal a huge gap, an organization can assume that hope exists for improvement and development. It is, however, vital to understand that such a benchmark cannot answer the following important questions:

- What should the maturity value be for this process in our organization now?
- What could a possible maturity development path look like?
- Which areas are impacted and improve when increasing the maturity in this specific area?

In addition to the above, one of the greatest challenges in such a process is the impact to the business in terms of the impact to the operating model, performance, and cost model as well as the service, value, and even revenue model. This is seen as very relevant because the various context areas assessed impact on the business

3.10.7　从成熟度级别评估到成熟度基准

BPM成熟度级别评估本质上是描述流程或功能在组织其他部分的背景中存在的程度的一种方法。这很重要,因为流程活动及其语义是与以下内容相关的关系:

- 有效的工作方式;
- 高效的操作;
- 一致的性能;
- 可靠的价值创造和实现。

自20世纪90年代初以来,为了制定新的战略方向和提高绩效,组织已经分析了他们的现状,找出他们需要改变什么以达到理想的待定阶段。一旦组织完成了BPM成熟度自我评估并了解了其BPM成熟度的现状,我们发现各个组织花费了大量的时间、资源和资金来理解和评估不同方面。因此,我们开发了一个标准的BPM成熟度基准,可以将不同领域相互比较。如果能够立即了解不同区域的特定成熟度水平以及区域内最低成熟度,这使我们能够看到最弱的成熟度以及它对其他区域的影响。在图5中,示出了这种BPM成熟度基准的示例。

组织和人员希望了解与流程相关的各种功能的成熟度和基准,主要有两个原因。第一是建立基线,即我们现在在哪里。第二是了解改进和发展的潜力。如果自我评估产生的成熟度值小于4或5,则可以说存在改进和发展的希望。同样,在基准测试中,如果相互之间的各种成熟度评估显示出巨大差距,那么组织可以假设存在改进和发展的希望。但是,了解这样的基准不能回答以下重要问题至关重要。

- 现在我们组织中的这个流程的成熟度值应该是多少?
- 可能的成熟度发展路径是什么样的?
- 在增加特定区域的成熟度时,哪些区域受到影响并得到改善?

除此之外,此类流程中最大的挑战之一是对运营模式、绩效和成本模型以及服务、价值甚至收入模型对业务的影响。这被视为"非常相关",因为各种背景领域以不同方式评估对业务的影响。图6是BPM成熟度开发路径的一个示例,

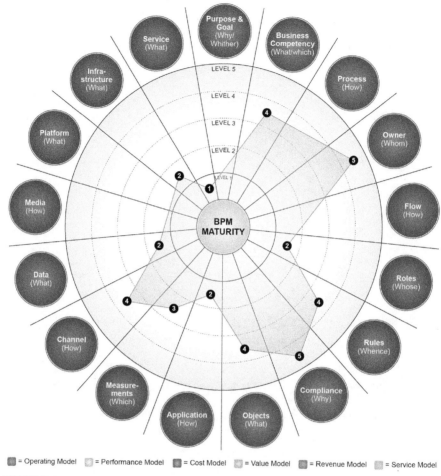

FIGURE 5

Example of a business process management (BPM) maturity benchmark.[40] (For interpretation of the references to color in this figure legend, the reader is referred to the online version of this book.)

in different ways. In Figure 6 is an example of a BPM maturity development path that specifies the existing maturity, which in this example is level 1, the identified impacted business aspects. In this example, the impacted business aspects of low maturity affect the revenue model, the value model, as well as the daily performance model. In addition, the time frame for development through the maturity levels is specified.

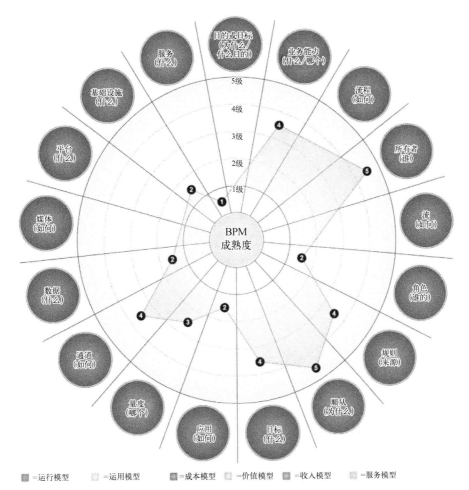

图5　BPM成熟度基准的示例[40]（有关本图例中颜色参考的解释，读者可参考本书的在线版本）

该路径指定现有成熟度，在此示例中为1级，即已识别的受影响业务方面。在此示例中，低成熟度的受影响业务方面会影响收入模型、价值模型以及日常绩效模型。此外，示例中还规定了通过成熟度级别进行开发的时间跨度。

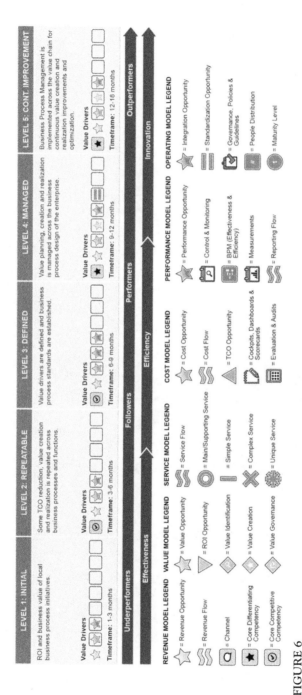

FIGURE 6

Example of a business process management (BPM) maturity development path complete with value drivers and time line for each maturity level.[41] (For interpretation of the references to color in this figure legend, the reader is referred to the online version of this book.)

图6　BPM成熟度开发路径示例，其中包含每个成熟度级别的价值驱动因素和时间线[14]（有关本图例中颜色参考的解释，读者可参考本书的在线版本）

Such a specific development path is seen as vital for any organization, especially because this is one of the weak points of general maturity models, for example, CMMI. General maturity models encourage the achievement of a higher maturity level in all aspects. We see this as wrong, and actually more hurtful of the development of the organization than helping it. An organization will and needs to have different levels of maturity in its various areas. Although, for example, core-differentiating aspects and the value-creating aspects of an organization need to be at maturity levels 4 and 5. However, the nondifferentiating, noncompeting aspects of the organization should not be at maturity level 4 or 5, as the cost to achieve a higher maturity level than 3 would be far greater than the possible gain. The best maturity could be level 2 at which it needs to be repeated and standardized; anything higher might not have the cost/value trade-off. Such a cost/value trade-off obviously needs to be closely analyzed by the organization, and this is exactly what the BPM maturity development path is about.

CONCLUSION

In this section, we have focused on maturity models, what they are, their historic development, how they could be used, and when BPM can use maturity concepts. We illustrated a detailed BPM maturity self-assessment, a benchmark among the various aspects that are related to the BPM maturity context, as well as a BPM maturity development path, all to enable hands-on practical guidance, assess one's maturity, and develop it. Without such a BPM Maturity assessment and a connected benchmark, the journey to BPM maturity will be difficult and frustrating. What we have provided here is a starting point for organizations to map out their development journey ahead of time and determine the proper number of rest stops along the way to the ultimate destination, which may or not be level 5. We believe this is the start of a great journey, and we wish you luck with your maturity development journey.

End Notes

1. CMMI for Software Development. CMMI-DEV. Carnegie Mellon University Software Engineering Institute.

2. Organizational Project Management Maturity Model (OPM3®) – Third Edition, 2013, Project Management Institute: http://www.pmi.org/PMBOK-Guide-and-Standards/Standards-Library-of-PMI-Global-Standards.aspx.

3. Curtis B., Hefley W. E., and Miller S., *People Capability Maturity Model*. CMU/SEI-95-MM-02 (Pittsburgh: Carnegie Mellon University, Software Engineering Institute, 1995). Available at: http://www.sei.cmu.edu/cmmi/tools/peoplecmm/.

4. http://www.p3m3-officialsite.com/nmsruntime/saveasdialog.aspx?lID=456&sID=166.

5. http://www.utdc.vuw.ac.nz/research/emm/.

6. Humphrey W. S., *Characterizing the Software Process: A Maturity Framework*. CMU/SEI-87-TR-11 (Pittsburgh: Carnegie Mellon University, Software Engineering Institute, 1987).

这种特定的发展路径被认为对任何组织都至关重要,特别是因为这是一般成熟度模型的弱点之一,如CMMI。一般成熟度模型鼓励在所有方面实现更高的成熟度水平。我们认为这是错误的,实际上更有害于组织的发展而不是帮助它。组织将在各个领域具有不同的成熟度。例如,虽然组织的核心差异化方面和价值创造方面需要处于成熟度第4级和第5级。但是,组织的非差异性、非竞争性方面不应该处于成熟度第4级或第5级,因为达到比第3级更高的成熟度水平的成本远远大于可能的收益。最佳成熟度可能是第2级,需要重复和标准化,任何更高的东西都可能没有成本/价值权衡。这样的成本/价值权衡显然需要由组织进行密切分析,这正是BPM成熟度发展路径的关键所在。

3.10.8 结论

在本节中,我们重点关注成熟度模型,它们是什么,它们的历史发展,如何使用它们,以及何时BPM可以使用成熟度概念。我们举例说明了详细的BPM成熟度自我评估,这是与BPM成熟度背景相关的各个方面的基准,以及BPM成熟度发展路径,所有这些都是为了实现实践指导,评估成熟度并开发它。如果没有这样的BPM成熟度评估和连接基准,BPM成熟的过程将是困难和令人沮丧的。我们在这里提供的是组织提前绘制他们的开发历程并确定在到达最终目的地的途中休息站的正确数量的起点,这可能是或不是第5级。我们相信这是一个开始,一个伟大的旅程,我们祝您在成熟的发展历程中好运。

参考文献

[1] CMMI for Software Development. CMMI-DEV. Carnegie Mellon University Software Engineering Institute.

[2] Organizational Project Management Maturity Model (OPM3®) — Third Edition, 2013, Project Management Institute: http://www.pmi.org/PMBOK-Guide-and-Standards/Standards-Library-of-PMI-Global-Standards.aspx.

[3] Curtis B., Hefley W. E., and Miller S., People Capability Maturity Model. CMU/SEI-95-MM-02 (Pittsburgh: Carnegie Mellon University, Software Engineering Institute, 1995). Available at: http://www.sei.cmu.edu/cmmi/tools/peoplecmm/.

[4] http://www.p3m3-officialsite.com/nmsruntime/saveasdialog.aspx?lID=456&sID=166.

[5] http://www.utdc.vuw.ac.nz/research/emm/.

[6] Humphrey W. S., Characterizing the Software Process: A Maturity Framework. CMU/SEI-87-TR-11 (Pittsburgh: Carnegie Mellon University, Software Engineering Institute, 1987).

7. Crosby P. B., *Quality is Free*, (New York: McGraw-Hill, 1979).

8. Hays D. W., "Quality improvement and its origin in scientific management," *Quality Progress* 27, no. 6 (May 1994): 89–90.

9. Deming W. E., *Out of the Crisis* (Cambridge, MA: MIT Center for Advanced Engineering Study, 1986).

10. Juran J. M., *Juran on Planning for Quality* (New York: Macmillan, 1988).

11. Emam K. and Goldenson D. R., (1999), An empirical review of software process assessments. NRC/ERB-1065 (NRC 43610). National Research Council Canada, Institute for Information Tech.

12. Humphrey W. S., *Managing the Software Process* (Reading, MA: Addison-Wesley, 1989).

13. Paulk M. C., Humphrey W. S., and Pandelios G. J., "Software process assessments: issues and lessons learned," in Proceedings of ISQE92, Juran Institute, March 1992, 4B/41–58.

14. LEADing Practice Maturity Reference Content [#LEAD-ES60003AL].

15. Kasse, Konrad M. D., Perdue J. R., Weber C. V., and Withey J. V., *Capability Maturity Model for Software*. CMU/SEI-91-TR-24 (Pittsburgh: Carnegie Mellon University, Software Engineering Institute, 1991).

16. Paulk M. C., Curtis B., Chrissis M. B., and Weber C. V., *Capability Maturity Model for Software, Version 1.1*. CMU/SEI-93-TR-24 (Pittsburgh: Carnegie Mellon University, Software Engineering Institute, 1993a).

17. Paulk M. C., Weber C. V., Curtis B., and Chrissis M. B., *The Capability Maturity Model: Guidelines for Improving the Software Process* (1995a).

18. Capability Maturity Model Version 1.0. CMU/SEI-94-HB-04 (Pittsburgh: Carnegie Mellon University, Software Engineering Institute).

19. Paulk M. C., Humphrey W. S., and Pandelios G. J., Software Process (1992).

20. Paulk M. C., Weber C. V., Garcia S. M., Chrissis M. B., and Bush M. W., *Key Practices of the Capability Maturity Model, Version 1.1*. CMU/SEI-93-TR-25 (Pittsburgh: Carnegie Mellon University, Software Engineering Institute, 1993b).

21. Paulk M. C., Curtis B., Chrissis M. B., Averill E. L., Bamberger J., T. C. Kasse, M. D. Konrad, J. R. Perdue, C. V. Weber, and J. V. Withey. 1991. *Capability Maturity Model for Software*. CMU/SEI-91-TR-24. Pittsburgh: Carnegie Mellon University, Software Engineering Institute.

22. SEI, *Process Maturity Profile: Software CMM 2005 End-Year Update* (Pittsburgh: Software Engineering Institute, Carnegie Mellon University, 2006).

23. Chrissis M. B., Konrad M. D., and Shrum S., *CMMI: Guidelines for Process Integration and Product Improvement*, second ed. (Boston: Addison-Wesley, 2006).

24. Humphrey W. S. and Sweet W. L., (1987b), A Method for Assessing the Software Engineering Capability of Contractors, Carnegie Mellon University, Software Engineering Institute, CMU/SEI-87-TR-23, September.

25. DOD, "Excerpts from Fall 1987 Report of the defense science board task force on military software," *ACM Ada Letters*, (July/August, 1988): 35–46.

26. SEI, *CMMI for Acquisition, Version 1.2*. CMU/SEI-2007-TR-017 (Pittsburgh: Carnegie Mellon University, Software Engineering Institute, 2007).

27. SEI, *CMMI for Services, Version 1.2*. CMU/SEI-2009-TR-001 (Pittsburgh: Carnegie Mellon University, Software Engineering Institute, 2009).

28. Gallagher B. P., Phillips M., Richter K. J., and Shrum S., CMMIACQ: Guidelines for Improving the Acquisition of Products and Services (Boston: Addison-Wesley Professional, 2009).

［7］ Crosby P. B., Quality is Free, (New York: McGraw-Hill, 1979).

［8］ Hays D. W., "Quality improvement and its origin in scientific management," Quality Progress 27, no. 6 (May 1994): 89−90.

［9］ Deming W. E., Out of the Crisis (Cambridge, MA: MIT Center for Advanced Engineering Study, 1986).

［10］ Juran J. M., Juran on Planning for Quality (New York: Macmillan, 1988).

［11］ Emam K. and Goldenson D. R., (1999), An empirical review of software process assessments. NRC/ERB-1065 (NRC 43610). National Research Council Canada, Institute for Information Tech.

［12］ Humphrey W. S., Managing the Software Process (Reading, MA: Addison-Wesley, 1989).

［13］ Paulk M. C., Humphrey W. S., and Pandelios G. J., "Software process assessments: issues and lessons learned," in Proceedings of ISQE92, Juran Institute, March 1992, 4B/41−58.

［14］ LEADing Practice Maturity Reference Content［#LEAD-ES60003AL］.

［15］ Kasse, Konrad M. D., Perdue J. R., Weber C. V., and Withey J. V., Capability Maturity Model for Software. CMU/SEI-91-TR-24 (Pittsburgh: Carnegie Mellon University, Software Engineering Institute, 1991).

［16］ Paulk M. C., Curtis B., Chrissis M. B., and Weber C. V., Capability Maturity Model for Software, Version 1.1. CMU/SEI-93-TR-24 (Pittsburgh: Carnegie Mellon University, Software Engineering Institute, 1993a).

［17］ Paulk M. C., Weber C. V., Curtis B., and Chrissis M. B., The Capability Maturity Model: Guidelines for Improving the Software Process (1995a).

［18］ Capability Maturity Model Version 1.0. CMU/SEI-94-HB-04 (Pittsburgh: Carnegie Mellon University, Software Engineering Institute).

［19］ Paulk M. C., Humphrey W. S., and Pandelios G. J., Software Process (1992).

［20］ Paulk M. C., Weber C. V., Garcia S. M., Chrissis M. B., and Bush M. W., Key Practices of the Capability Maturity Model, Version 1.1. CMU/SEI-93-TR-25 (Pittsburgh: Carnegie Mellon University, Software Engineering Institute, 1993b).

［21］ Paulk M. C., Curtis B., Chrissis M. B., Averill E. L., Bamberger J., T. C. Kasse, M. D. Konrad, J. R. Perdue, C. V. Weber, and J. V. Withey. 1991. Capability Maturity Model for Software. CMU/SEI-91-TR-24. Pittsburgh: Carnegie Mellon University, Software Engineering Institute.

［22］ SEI, Process Maturity Profile: Software CMM 2005 End-Year Update (Pittsburgh: Software Engineering Institute, Carnegie Mellon University, 2006).

［23］ Chrissis M. B., Konrad M. D., and Shrum S., CMMI: Guidelines for Process Integration and Product Improvement, second ed. (Boston: Addison-Wesley, 2006).

［24］ Humphrey W. S. and Sweet W. L., (1987b), A Method for Assessing the Software Engineering Capability of Contractors, Carnegie Mellon University, Software Engineering Institute, CMU/SEI-87-TR-23, September.

［25］ DOD, "Excerpts from Fall 1987 Report of the defense science board task force on military software," ACM Ada Letters, (July/August, 1988): 35−46.

［26］ SEI, CMMI for Acquisition, Version 1.2. CMU/SEI-2007-TR-017 (Pittsburgh: Carnegie Mellon University, Software Engineering Institute, 2007).

［27］ SEI, CMMI for Services, Version 1.2. CMU/SEI-2009-TR-001 (Pittsburgh: Carnegie Mellon University, Software Engineering Institute, 2009).

［28］ Gallagher B. P., Phillips M., Richter K. J., and Shrum S., CMMIACQ: Guidelines for Improving the Acquisition of Products and Services (Boston: Addison-Wesley Professional, 2009).

29. Paulk M. C., (2008), A taxonomy for improvement frameworks. World Congress for Software Quality, Bethesda, MD, 15–18 September.

30. Humphrey W. S., "Three process perspectives: organizations, teams, and people," *Annals of Software Engineering* 4, (2002): 39–72.

31. Britz G., Emerling D., Hare L., Hoerl R., and Shade J., *Statistical Thinking*. A Special Publication of the ASQC Statistics Division (Spring, 1996).

32. Pfeffer J. and Sutton R. I., *Hard Facts, Dangerous Half-Truths, & Total Nonsense: Profiting from Evidence-Based Management*. (Boston: Harvard Business School Press, 2006).

33. See note 14 above.

34. Besselman J. J., "A collection of software capability evaluation (SCE) findings: many lessons learned," in Proceedings of the Eighth Annual National Joint Conference on Software Quality and Productivity, Arlington, VA, March 1992, 196–215.

35. Krasner H., (2001), "Accumulating the body of evidence for the payoff of software process improvement – 1997," in *Software Process Improvement*, eds. Hunter R. B., and Thayer R. H., (New York: IEEE Computer Society Press, 2001), 519–539.

36. Austin, R. D., *Measuring and Managing Performance in Organizations* (New York: Dorset House Publishing, 1996).

37. "BPM Maturity Model is Important for Long Lasting BPM Success," Michael Melenovsky and Jim Sinur from http://www.brcommunity.com/b325.php.

38. See note 14 above.

39. See note 14 above.

40. Ibid.

41. Ibid.

[29] Paulk M. C., (2008), A taxonomy for improvement frameworks. World Congress for Software Quality, Bethesda, MD, 15−18 September.

[30] Humphrey W. S., "Three process perspectives: organizations, teams, and people," Annals of Software Engineering 4, (2002): 39−72.

[31] Britz G., Emerling D., Hare L., Hoerl R., and Shade J., Statistical Thinking. A Special Publication of the ASQC Statistics Division (Spring, 1996).

[32] Pfeffer J. and Sutton R. I., Hard Facts, Dangerous Half-Truths, & Total Nonsense: Profiting from Evidence-Based Management. (Boston: Harvard Business School Press, 2006).

[33] See note 14 above.

[34] Besselman J. J., "A collection of software capability evaluation (SCE) findings: many lessons learned," in Proceedings of the Eighth Annual National Joint Conference on Software Quality and Productivity, Arlington, VA, March 1992, 196−215.

[35] Krasner H., (2001), "Accumulating the body of evidence for the payoff of software process improvement — 1997," in Software Process Improvement, eds. Hunter R. B., and Thayer R. H., (New York: IEEE Computer Society Press, 2001), 519−539.

[36] Austin, R. D., Measuring and Managing Performance in Organizations (New York: Dorset House Publishing, 1996).

[37] "BPM Maturity Model is Important for Long Lasting BPM Success," Michael Melenovsky and Jim Sinur from http://www.brcommunity.com/b325.php.

[38] See note 14 above.

[39] See note 14 above.

[40] Ibid.

[41] Ibid.